T0093570

Introduction To Lagrangian Dynamics

Aron Wolf Pila

Introduction To Lagrangian Dynamics

Aron Wolf Pila
10 Hatechiyah Street, apt. 2
Kfar Saba, Israel

ISBN 978-3-030-22377-9 ISBN 978-3-030-22378-6 (eBook)
https://doi.org/10.1007/978-3-030-22378-6

This Springer imprint is published by the registered company Springer Nature Switzerland AG.
The registered company address is: Gewerbestrasse 11, 6330 Cham, Switzerland

Preface

This volume was intended as a short summary of the essentials of Lagrangian dynamics for undergraduate students of physics and engineering. A number of topics have been included in order to make the presentation compact, succinct, as well as comprehensive. The topics include:

(a) A Review of Classical Mechanics
(b) Holonomic and Non-holonomic Systems
(c) Virtual Work
(d) The Principle of D'Alembert for Dynamical Systems
(e) The Mathematics of Conservative Forces
(f) The Extended Hamilton's Principle
(g) Lagrange's Equations and Lagrangian Dynamics
(h) A Systematic Procedure for Generalized Forces
(i) Quasi-coordinates and Quasi-velocities
(j) Lagrangian Dynamics with Quasi-coordinates
(k) Lagrangian Dynamics with Quasi-coordinates-Prof. Ranjan Vepa's Approach

An ample number of examples have been included which demonstrate the techniques involved.

Dedication

This volume is dedicated to my wife, Leah, for the patience, understanding, and love she has shown me from the day we first met. She's been an inspiration and guiding light throughout my life and, my working career and into our retirement years.

Contents

1 Introduction 1
1.1 Introductory Remarks 1
1.2 Direction Cosines and Euler Angles of Rotation 3

2 Lagrangian Dynamics: Preliminaries 13
2.1 Angular Velocity of a Body and Linear Velocity of a Typical Particle Within That Body 13
2.2 Angular Velocity of a Body and Linear Velocity of a Typical Particle Within the Body: Examples 15
2.3 Most General Form of Kinetic Energy 21
2.4 Summary: Important Points Regarding Kinetic Energy 22
2.5 Examples: Kinetic Energy and Equations of Motion 23
2.6 Notation System Used in This Book 31
2.7 Angular Momentum of a Mass Particle 33
2.8 Rigid Body Angular Momentum 37
2.9 Linear and Angular Momenta and Their Derivatives 39
2.10 Work and Calculation of Kinetic and Potential Energies 40
2.11 Systems of Particles 45
2.12 Principle of Work and Energy for a Rigid Body 51
2.13 Angular Momentum of a Rigid Body in Three Dimensions 56

3 Lagrangian Dynamics 63
3.1 Definitions Required for the Study of Lagrangian Dynamics 64
3.2 Summary: Holonomic and Non-holonomic Systems 71
3.3 Virtual Work for Static Systems Only 75
3.4 The Principle of d'Alembert for Dynamical Systems 77
3.5 The Mathematics of Conservative Forces 78
3.6 The Extended Hamilton's Principle 80
3.7 Lagrange's Equations and Lagrangian Dynamics 82
3.8 Recap: Writing d'Alembert–Lagrangian Dynamics 103
3.9 Lagrange Multipliers for Constrained Systems 118
3.10 A Systematic Procedure for Generalized Forces 123

3.11 Practice Finding Equations of Motion—D'Alembert–Lagrange 139
3.12 A Note on Equivalent Forces and Torques 144
3.13 Lagrangians, Hamiltonians, and the Legendre Transformation 151

4 Quasi-Coordinates and Quasi-Velocities 163
4.1 Definitions and Recapitulation 164
4.2 Quasi-Coordinates and Quasi-Velocities 167
4.3 Lagrangian Dynamics with Quasi-Coordinates 177
4.4 Lagrangian Dynamics with Quasi-Coordinates: Prof. Ranjan Vepa's Approach 199
4.5 Lagrangian Dynamics in Quasi-Coordinates—Vepa's Approach—Origin Not at Mass Center 215

5 Conclusions 247

Bibliography 249

Index 253

List of Figures

Fig. 1.1 Definition of direction cosines l, m, n 3
Fig. 1.2 Body translating and rotating, while X, Y, Z frame rotates about O relative to the body 4
Fig. 1.3 Sketch of body, fixed at O, but free to rotate in any manner about this point 4

Fig. 2.1 Body translating and rotating, while X, Y, Z frame rotates about O relative to the body 14
Fig. 2.2 Illustration of the treatment of the angular velocity of a body and linear velocity of a typical mass particle. (**a**) Side view. (**b**) Top view 16
Fig. 2.3 Another example of the treatment of the angular velocity of a body and linear velocity of a typical mass particle 19
Fig. 2.4 Definition of velocities v_{ox}, v_{oy} 21
Fig. 2.5 Physical pendulum consisting of a lamina pivoted at p with the origin of the X, Y coordinates defined at three different locations 24
Fig. 2.6 Lamina with two applied forces—Lagrange's equations 25
Fig. 2.7 Double "pendulum" 27
Fig. 2.8 Definition of potential V and generalized forces F_θ, F_ϕ 29
Fig. 2.9 Rotating slender rod 30
Fig. 2.10 Illustration of notation systems 32
Fig. 2.11 Illustration of linear momentum $P_{B/O}$ 34
Fig. 2.12 Two link robot 35
Fig. 2.13 Carnival ride problem 38
Fig. 2.14 Motion of a particle with mass m subject to a force F and traveling along the curved path from point A_1 to point A_2 42
Fig. 2.15 Motion of a pendulum bob 44
Fig. 2.16 Internal forces cancel each other out 46

Fig. 2.17 Small displacements dr and dr' of the two particles differ but the components of these displacements along $A - B$ are equal—no net internal forces 52
Fig. 2.18 Work done by couple $M = dU = Fds_2 = Frd\theta$ 53
Fig. 2.19 Rigid body in plane motion—kinetic energy 54
Fig. 2.20 Rigid body in plane motion—noncentroidal rotation 55
Fig. 2.21 Rigid body angular momentum in three dimensions 56
Fig. 2.22 Rigid body angular momentum in three dimensions about a point O 59
Fig. 2.23 Kinetic energy of a rigid body in three dimensions 60

Fig. 3.1 A spherical pendulum whose length L may or may not vary 65
Fig. 3.2 A particle moving on a smooth surface 66
Fig. 3.3 Generic vehicle 68
Fig. 3.4 Double pendulum system 70
Fig. 3.5 Non-holonomic system—# of degrees of freedom $\leq$ # coord. needed to fully determine position of the ball 70
Fig. 3.6 Wheel rolling without slipping on a curved path—classic example of a non-holonomic system 73
Fig. 3.7 Mass–spring system 86
Fig. 3.8 Two degrees of freedom system composed of springs and masses 88
Fig. 3.9 Single and double pendulums 89
Fig. 3.10 Schematic diagram of a quadcopter 94
Fig. 3.11 Mass–spring dashpot (damper) single degree of freedom system 106
Fig. 3.12 Pendulum with a moveable mass and spring 108
Fig. 3.13 Pendulum with a moveable mass and spring—geometrical definitions I 109
Fig. 3.14 Pendulum with a moveable mass and spring-geometrical definitions II 109
Fig. 3.15 Cart with spring and pendulum 114
Fig. 3.16 Cart with spring, damper, and pendulum 117
Fig. 3.17 Cart with spring, damper, and pendulum-geometrical considerations 117
Fig. 3.18 System used to demonstrate the virtual work principle 122
Fig. 3.19 3D rigid body with n generalized coordinates acted upon by N non-conservative forces 124
Fig. 3.20 Generalized forces acting on a double pendulum 125
Fig. 3.21 Another example for the calculation of torques and kinetic energy 127
Fig. 3.22 Spring and mass on inclined face of moving cart 130
Fig. 3.23 Two carts connected by a spring 135
Fig. 3.24 Pendulum with a mass and spring—continued 137
Fig. 3.25 Puck sliding on a horizontal frictionless plane 140

Fig. 3.26 Generalized forces for sliding puck problem—I 142
Fig. 3.27 Geometry for equivalent forces and torques 144
Fig. 3.28 Generalized forces for hockey puck problem—II 145
Fig. 3.29 Pendulum with a plane of symmetry 147
Fig. 3.30 Atwood's machine 148
Fig. 3.31 Falling stick problem 150
Fig. 3.32 Strictly convex function used in the derivation of a simple legendre transformation 153

Fig. 4.1 Tricycle geometry and notation 187
Fig. 4.2 Geometry for velocities perpendicular to both front and rear wheels 188
Fig. 4.3 Time derivative of a rotation matrix—small angle rotations 203
Fig. 4.4 Velocity of point P with respect to center of mass point C 226

List of Symbols

l, m, n	Direction cosines
$\alpha_{ij}, i = 1, 2, 3, j = 1, 2, 3$	Direction cosines
i_1, i_2, i_3	Orthogonal unit vectors for stationary X_1, Y_1, Z_1 axes
$[x_{ea}, y_{ea}, z_{ea}]^T$	Position vector in inertial coordinates
$[x_b, y_b, z_b]^T$	Position vector in rotating body coordinates
X_1, Y_1, Z_1	x, y, z axes of inertial coordinate system
X, Y, Z	X, Y, Z axes of translating and rotating coordinates
X', Y', Z'	x, y, z axes of nonrotating coordinate system
v_x, v_y, v_z	Instantaneous velocities along X, Y, Z directions
x, y, z	Coordinates of m' along X, Y, Z directions
m'	Mass of an elemental particle
dm	Mass of an infinitesimal particle
M	Total system mass
ω	Angular rotation rate vector wrt inertial frame
$\omega_x, \omega_y, \omega_z$	Angular rates about X, Y, Z axes
T	Kinetic energy
V	Potential energy
I_x, I_y, I_z	Principal moments of inertia about X, Y, Z axes
I_{xy}	Product of inertia relative to the $X - Y$ axis
I_{xz}	Product of inertia relative to the $X - Z$ axis
I_{yz}	Product of inertia relative to the $Y - Z$ axis
$\overline{x}$	X axis distance from any origin point to c.g.
$\overline{y}$	Y axis distance from any origin point to c.g.
$\overline{z}$	Z axis distance from any origin point to c.g.
v_{ox}, v_{oy}, v_{oz}	Velocities of origin in X, Y, Z direction
$F_\theta = -\partial V/\partial\theta$	Generalized force derived from $V(\theta, \phi)$
$F_\phi = -\partial V/\partial\phi$	Generalized force derived from $V(\theta, \phi)$
$\mathcal{L} = T - V$	Lagrangian
ρ	Mass distribution per unit length
$R_{B/O}$	Position vector of point B wrt O

$R_{B/A}$	Position vector of point B wrt A
$\omega_{/O}, \omega_O$	Rotation of rigid body wrt O
O	Origin of inertial coordinate system
$P_{B/O}$	Particle linear momentum at point B wrt O
$h_{B/O}$	Particle angular momentum at point B wrt O
$h_{B/A}$	Particle angular momentum at point B wrt A
$\tau_{B/A}$	Torque at point B wrt A
$\tau_{/G}$	Torque wrt c.g.
$\dot{H}_G$	Rigid body angular momentum rate wrt c.g.
$\tau_{/A}$	Torque wrt point A
$v_{B/O}$	Velocity of point B wrt O
$v_{A/O}$	Velocity of point A wrt O
F_B	Sum of all the forces at point B
$\dot{H}_A$	Rigid body angular momentum rate wrt point A
$P_{G/O}$	Rigid body angular momentum of c.g. wrt point O
$\hat{e}_\theta$	Polar coordinates, unit vector in θ direction
$\hat{e}_r$	Polar coordinates, unit vector in r direction
$L = mv$	Linear momentum
$\dot{L} = ma$	Force
$H_O = r \times mv$	Angular momentum
$\dot{H}_O = r \times ma$	Moment or torque
U	Work
$U_{1\to 2}$	Work performed over path $A_1 \to A_2$
dr	Infinitesimal distance vector
k	Spring constant—Nt/m
F_t	Force tangential to curved trajectory
a_t	Acceleration tangential to curved trajectory
T	Kinetic energy
V	Potential energy
g	Acceleration due to gravity
$E = T + V$	Total mechanical energy = kinetic + potential energy
r_i'	Position of m_i at point P_i wrt axes $Gx'y'z'$
v_i'	Velocity of m_i at point P_i wrt axes $Gx'y'z'$
a_i'	Acceleration of m_i at point P_i wrt axes $Gx'y'z'$
a_i	Acceleration of m_i at point P_i wrt inertial axes $Oxyz$
$\overline{a}$	Acceleration of c.g wrt inertial axes $Oxyz$
H_G'	Angular momentum of system of particles about mass center G
H_G	Angular momentum of system of particles about origin of inertial coordinates $Oxyz$
M_{G_i}	Moment of mass m_i about axes $Gx'y'z'$
$\overline{I}$	Moment of inertia of the rigid body about the axis of rotation
$Gxyz$	Nonrotating centroidal axes ‖ inertial axes
H_G	Angular moment of inertia wrt $Gxyz$ frame

H_x	x axis component of H_G wrt $Gxyz$ frame
H_y	y axis component of H_G wrt $Gxyz$ frame
H_z	z axis component of H_G wrt $Gxyz$ frame
$f(x, y, z, t) = 0$	Shape function or constraint equation
n	Number of degrees of freedom
m	Number of holonomic constraints
$p = n - m$	No. dof—no. of holonomic constraints
dof	Degrees of freedom
a_{je}	Constraint coefficient
N	Number of rigid bodies in the plane
$3N$	No. of degrees of freedom of N rigid bodies in the plane without constraints
k	No. of holonomic constraints of N rigid bodies in the plane
$d = 3N - k$	Resulting no. of degrees of freedom of N rigid bodies in the plane
$f(q_1, q_2, \ldots) = 0$	Constraint equation for holonomic system
$f(q_1, q_2, \ldots, q_n, \dot{q}_1, \dot{q}_2, \ldots, \dot{q}_n,) = 0$	Non-holonomic constraint equation
$\sum_{e=1}^{n} a_{je} dq_e + a_{jt} dt = 0$	Non-holonomic constraint equation in Pfaffian form
$\overline{\delta W}$	Virtual work
R_i	ith Resultant force
F_i	ith Applied force
f_i	ith Constraint force
$\overline{\delta W}^{NC}$	Virtual work due to non-conservative forces
$\overline{\delta W}_C$	Virtual work due to conservative forces
δT	Variation of kinetic energy
$\delta \mathcal{L}$	Variation of Lagrangian
δT	Variation in kinetic energy
δV	Variation in potential energy
Q_k	Generalized non-conservative force
δ_k	Virtual displacement of generalized coordinate
τ_{m_i}	Torque due to ith motor and rotor
P	Electric motor power
v_h	Quadrotor hover velocity
A	Rotor disk area
D	Drag force
ω	Rotor speed
τ_{drag_i}	Torque due to blade profile drag on ith motor and rotor
$\mathcal{J}$	Inertia matrix of quadcopter's rotational kinetic energy
$\mathcal{U}$	Quadcopter's potential energy
$\dot{\xi}$	Quadcopter's inertial linear velocities
m_{quad}	Quadcopter's total mass

τ_{ψ}	Quadcopter's torque in ψ direction
τ_{ϕ}	Quadcopter's torque in ϕ direction
τ_{θ}	Quadcopter's torque in θ direction
η	Vector of generalized Euler angle coordinates
$\hat{V}(\eta, \dot{\eta})$	"Coriolis-centripetal" vector
$\tilde{\tau}$	Vector of torques in inertial coordinates
$\tilde{\tau}_{\phi}$	Torques in inertial coordinates: ϕ direction
$\tilde{\tau}_{\psi}$	Torques in inertial coordinates: ψ direction
$\tilde{\tau}_{\theta}$	Torques in inertial coordinates: θ direction
δ_j	Virtual displacement in j direction
L_0	Un-stretched length of spring
L_2	$\frac{1}{2}$ length of metal sleeve
x_1	Length from pivot point A to c.g of metal sleeve
Δh	Height differences for potential energy calculations
$\frac{L_1}{2}$	$\frac{1}{2}$ length of rod
$T_{rot-rod}$	Rotational kinetic energy of the rod
$T_{rot-sleeve}$	Rotational kinetic energy of the sleeve
$f(q_1, q_2, \ldots, q_n) = 0$	Holonomic constraint equation
δq_i	Variation of q_i
$\lambda(t)$	Lagrange multiplier
$\lambda_j(t)$	jth Lagrange multiplier
Q'	Generalized constraint forces
Q'_k	kth generalized constraint force
δW^{NC}	Virtual work of all non-conservative forces
Q_j	jth generalized force
Q_{θ}	Generalized force in generalized θ dir'n
Q_x, Q_y	Generalized force in generalized x, y dir'n
$\oint$	Integral over a closed path
F_c	Conservative force
F^{NC}	Non-conservative force
$V(r)$	Potential energy as a function of position r
$\mathcal{H}$	Hamiltonian function
p_i	Generalized momentum
$\sum_{e=1}^{n} a_{le}\delta q_e = 0$	Non-holonomic constraint equation
λ_l	Lagrange multiplier
$Q_e^{'} = \sum_{l=1}^{m} \lambda_l a_{le}$	Generalized constraint forces
$Q_e = Q_e + Q_e^{'}$	Generalized forces and constraint forces
γ_j	Quasi-coordinate
$\dot{\gamma}_j$	Quasi-velocity
$\dot{q}_j$	Generalized velocity
Θ_{ij}	Coefficients of the generalized coordinates q_k
$[\vartheta]$	Matrix of Θ_{ij} elements
$[\dot{\vartheta}]$	Matrix of $\dot{\Theta}_{ij}$ elements
$\{\dot{\Gamma}\}$	Vector of quasi-velocities $\dot{\gamma}_j$'s

$[\beta]^T = [\vartheta]^{-1}$	Inverse of $[\vartheta]$
$\{\dot{q}\}$	Vector of generalized velocities
$T(q; \dot{q})$	Standard kinetic energy
$\bar{T}(q; \dot{\Gamma})$	Modified kinetic energy
$\left\{\frac{\partial T}{\partial \dot{q}}\right\}$	Vector: coefficients of k.e. wrt generalized velocities
$\left\{\frac{\partial T}{\partial q}\right\}$	Vector: coefficients of k.e. wrt generalized coordinates
$\frac{\partial T}{\partial \dot{q}_k}$	Coefficient of k.e. wrt $\dot{q}_k$
$\frac{\partial T}{\partial q_k}$	Coefficient of k.e. wrt q_k
$\left\{\frac{\partial \bar{T}}{\partial \dot{\Gamma}}\right\}$	Vector: coefficients of modified k.e. wrt generalized velocities
$\left[\frac{\partial \vartheta}{\partial q_k}\right]$	Matrix of partial derivatives of $\Theta_{ij}, i, j, = 1, 2, \ldots, n$ wrt q_k
w_x, w_y, w_z	Angular rates in body coordinates
v, v_{Yi}, v_R	Tricycle's forward velocity, rear and front wheel lateral velocities
$\dot{T}_{a/b}(t)$	Time derivative of a rotation matrix
$\dot{T}_{I/B}(t)$	Body to inertial rotation matrix rate
$\Delta\theta, \Delta\psi, \Delta\phi$	Infinitesimal angles θ ψ, ϕ
Φ	Vector with $[\phi, \theta, \psi]^T$
I_{rotor}	Quad copter motor and rotor blade moment of inertia
C_D	Aerodynamic blade profile drag coefficient

Chapter 1
Introduction

Joseph Louis Lagrange was one of the greatest mathematicians of the eighteenth and early nineteenth centuries and he has left a remarkable legacy in both the fields of physics and mathematics. This volume begins by recounting the biographical highlights of his life and his contributions.

In the present chapter one of the cornerstones of the book in the form of the direction cosines and their relationship to the Euler angles is presented and elaborated upon. The direction cosines play an important role in the approach by Prof. Ranjan Vepa and are used extensively in Chap. 4.

1.1 Introductory Remarks

Joseph Louis Lagrange, originally Giuseppe Lodovico Lagrangia, was of French and Italian descent and was born in Turin in 1736 (see Rouse Ball [30, pp. 330–339]). Lagrange had originally intended to study law but while at college in Turin, he came across a tract by Halley which roused his enthusiasm for the analytical method. He thereupon applied himself to mathematics, and in his 17th year he became professor of mathematics in the royal military academy at Turin. Without assistance or guidance he entered upon a course of study which in 2 years placed him on a level with the greatest of his contemporaries. With the aid of his pupils he established a society which subsequently developed into the Turin Academy. Most of his earlier papers appear in the first five volumes of its transactions. At the age of 19 he communicated a general method of dealing with "isoperimetrical problems," known now as the calculus of variations to Euler. This commanded Euler's admiration, and the latter, for a time, courteously withheld some researches of his own on this subject from publication, so that the youthful Lagrange might complete his investigations and lay claim to being the first to posit the calculus of variations. Lagrange did quite as much as Euler towards the creation

A. W. Pila, *Introduction To Lagrangian Dynamics*,
https://doi.org/10.1007/978-3-030-22378-6_1

of the calculus of variations. The subject, as developed by Euler lacked an analytic foundation, and this Lagrange supplied. He separated the principles of this calculus from geometric considerations by which his predecessor had derived them. Euler had assumed as fixed, the limits of the integral, i.e. the extremities of the curve to be determined, but Lagrange removed this restriction and allowed all co-ordinates of the curve to vary at the same time. In 1766 Euler introduced the name "calculus of variations," and did much to improve this science along the lines marked out by Lagrange.

In the year 1766, Euler left Berlin for St. Petersburg, and he pointed to Lagrange as being the only man capable of filling his place. D'Alembert recommended him at the same time. Frederick the Great thereupon sent a message to Turin, expressing the wish of "the greatest king of Europe" to have "the greatest mathematician" at his court. Lagrange went to Berlin, and remained there for 20 years. Frederick the Great held him in high esteem, and frequently conversed with him on the advantages of perfect regularity of life. This led Lagrange to cultivate regular habits. He worked no longer each day than experience taught him he could, without breaking down. His papers were carefully thought out before he began writing, and when he wrote he did so without a single correction. During the 20 years in Berlin he crowded the transactions of the Berlin Academy with memoirs, and also wrote the epoch-making work called the Mécanique Analytique. The approach used by Lagrange will be the subject matter of this volume and will be presented in the subsequent chapters.

Newton's laws were formulated for a single particle and can be extended to systems of particles and rigid bodies. The equations of motion are expressed in terms of physical coordinates and forces, both quantities conveniently represented by vectors. For this reason, *Newtonian mechanics* is often referred to as *vectorial mechanics*. The main drawback of Newtonian mechanics is that it requires one free-body diagram for each of the masses in the system, thus necessitating the inclusion of reaction forces, the latter resulting from kinematical constraints ensuring that the individual bodies act together as a system. These reaction and constraint forces play the role of unknowns, which makes it necessary to work with a surplus of equations of motion, one additional equation for every unknown force. J.L. Lagrange reformulated Newton's Laws in a way that eliminates the need to calculate forces on isolated parts of a mechanical system. A different approach to mechanics, referred to as *analytical mechanics*, or *analytical dynamics*, considers the system as a whole, rather than the individual components separately, a process that excludes the reaction and constraint forces automatically. This approach, due to Lagrange, permits the formulation of problems of dynamics in terms of two scalar functions, the kinetic energy and the potential energy, and an infinitesimal expression, the virtual work performed by the non-conservative forces. Analytical mechanics represents a broader and more abstract approach, as the equations of motion are formulated in terms of generalized coordinates and generalized forces, which are not necessarily physical coordinates and forces, although in certain cases they can be chosen as such. Any convenient set of variables obeying *the constraints on a system* can be used to describe the motion. In this manner, the mathematical formulation is rendered independent of any special system of coordinates. There are only as many equations to solve as there are physically significant variables (see Meirovitch [24, pp. 262–263]).

1.2 Direction Cosines and Euler Angles of Rotation

The relationship between direction cosines and Euler angles is presented as background material to be used in the subsequent portions of this text. This chapter has been adopted from Wells [51, pp. 139–141 and Appendix A, pp. 343–344]. The direction cosines l, m, n of line Ob, relative to axes X, Y, Z are just $l = x/r$, $m = y/r$, $n = z/r$, where x, y, z are the X, Y, Z coordinates of the tip of r, where $r = \sqrt{(x^2 + y^2 + z^2)}$ (see Fig. 1.1). It then follows that $(x^2 + y^2 + z^2)/r^2 = (x^2 + y^2 + z^2)/(x^2 + y^2 + z^2) = 1$.

Assuming that coordinates X_1, Y_1, Z_1 form an inertial coordinate frame, while coordinates X, Y, Z are attached to a translating and rotating body, the angles between the X coordinate and coordinates X_1, Y_1, Z_1 are $\theta_{11}, \theta_{12}, \theta_{13}$, respectively. Hence the direction cosines between coordinate X and coordinates X_1, Y_1, Z_1 are $\alpha_{11} = \cos\theta_{11}$, $\alpha_{12} = \cos\theta_{12}$, $\alpha_{13} = \cos\theta_{13}$, respectively (see Fig. 1.2). The same relationships between the X coordinate and coordinates X_1, Y_1, Z_1 exist as for line Ob, that is:

$$\alpha_{11}^2 + \alpha_{12}^2 + \alpha_{13}^2 = 1 \tag{1.1}$$

We can similarly show that the direction cosines between coordinate Y and X_1, Y_1, Z_1, that is $\alpha_{21}, \alpha_{22}, \alpha_{23}$ and between coordinate Z and X_1, Y_1, Z_1, that is

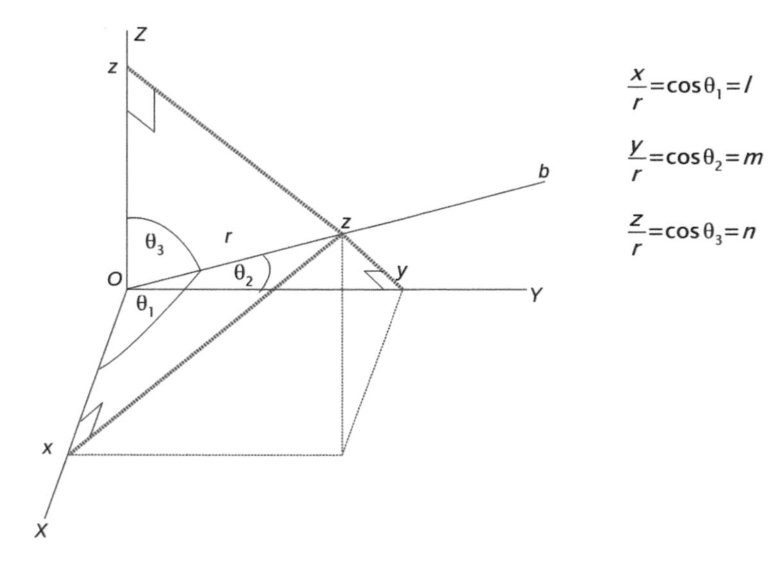

Fig. 1.1 Definition of direction cosines l, m, n

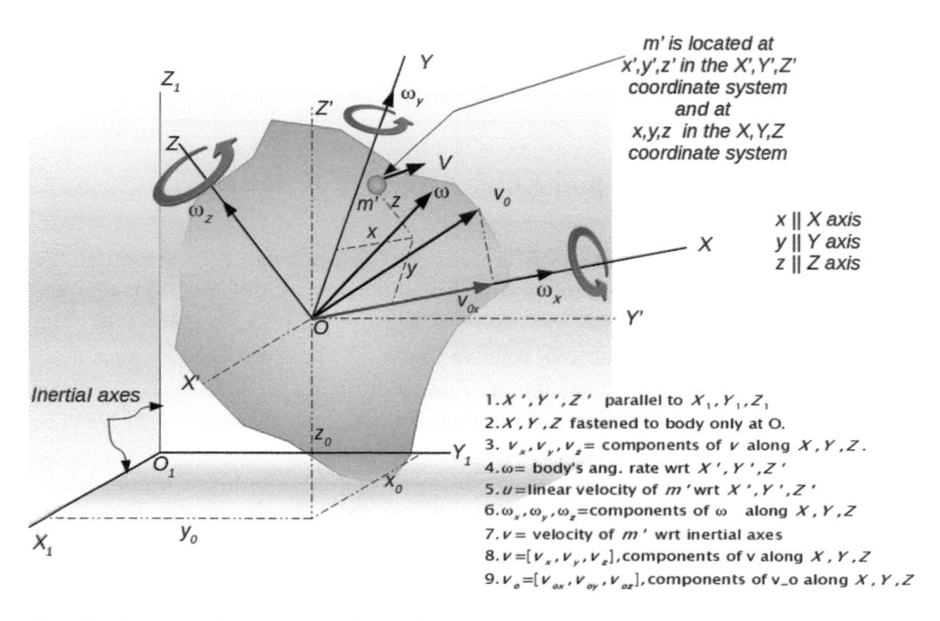
m' is located at
x',y',z' in the X',Y',Z'
coordinate system
and at
x,y,z in the X,Y,Z
coordinate system
x || X axis
y || Y axis
z || Z axis
Inertial axes
1. X', Y', Z' parallel to X1, Y1, Z1
2. X, Y, Z fastened to body only at O.
3. vx, vy, vz = components of v along X, Y, Z.
4. ω = body's ang. rate wrt X', Y', Z'
5. u = linear velocity of m' wrt X', Y', Z'
6. ωx, ωy, ωz = components of ω along X, Y, Z
7. v = velocity of m' wrt inertial axes
8. v = [vx, vy, vz], components of v along X, Y, Z
9. vo = [vox, voy, voz], components of v_o along X, Y, Z

Fig. 1.2 Body translating and rotating, while X, Y, Z frame rotates about O relative to the body

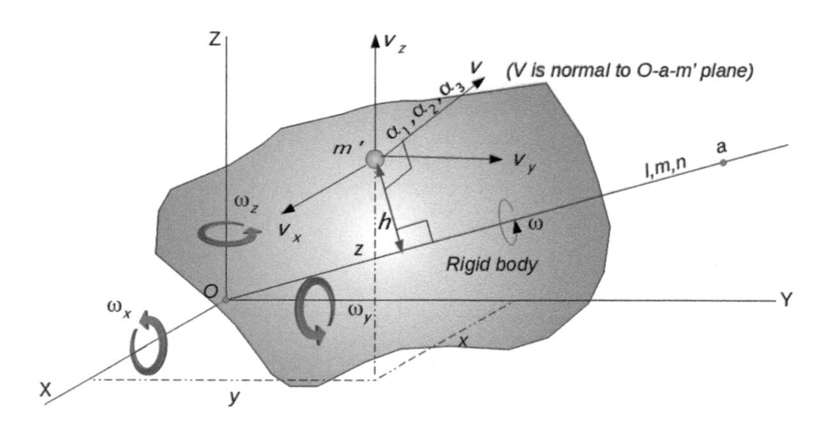
(V is normal to O-a-m' plane)
l,m,n
Rigid body

Fig. 1.3 Sketch of body, fixed at O, but free to rotate in any manner about this point

$\alpha_{31}, \alpha_{32}, \alpha_{33}$, respectively, obey the same relationship as in Eq. 1.1 or:

$$\alpha_{21}^2 + \alpha_{22}^2 + \alpha_{23}^2 = 1; \;\; \alpha_{31}^2 + \alpha_{32}^2 + \alpha_{33}^2 = 1 \tag{1.2}$$

Consider that the body in Fig. 1.3 is fixed at O, but is free to rotate in an arbitrary and random fashion about this point. All quantities under consideration will be measured relative to the **fixed inertial axis system** X, Y, Z. At a given instant of time, the body is undergoing rotation about some line Oa with an angular velocity of ω. As a consequence of this rotation, the mass particle m' possesses a linear velocity v

normal to the $Oa - m'$ plane, of magnitude $v = \omega h$, where h is the normal distance from m' to the rotating Oa axis. The axis of rotation Oa has direction cosines with respect to the fixed coordinate system X, Y, Z of l, m, n, respectively. Similarly, the velocity vector v has direction cosines $\alpha_1, \alpha_2, \alpha_3$ with respect to the X, Y, Z axes and its components along the X, Y, Z axes are, respectively, v_x, v_y, and v_z. From the discussion related to Fig. 1.1 above, it follows that: $\alpha_1 = v_x/v, \alpha_2 = v_y/v, \alpha_3 = v_z/v$. Similarly, ω is also composed of components along the X, Y, Z axes, that is: $\omega = [\omega_x, \omega_y, \omega_z]$. The direction cosines may then be shown to be; $l = \omega_x/\omega, m = \omega_y/\omega, n = \omega_z/\omega$. Recall that ω is directed along the line Oa. Now the velocity of the mass particle m' may be written in the form: $v = \omega \times r$, where $r = x\hat{i} + y\hat{j} + z\hat{k}$ and $\omega = \omega_x\hat{i} + \omega_y\hat{j} + \omega_z\hat{k}$. Performing the above vector multiplication results in:

$$v_x = \omega_y z - \omega_z y; \quad v_y = \omega_z x - \omega_x z; \quad v_z = \omega_x y - \omega_y x \tag{1.3}$$

However, $\alpha_1 = v_x/v = v_x/\omega h$. This implies that:

$$\alpha_1 = \frac{\omega_y z - \omega_z y}{\omega h} = \frac{\omega_y z}{\omega h} - \frac{\omega_z y}{\omega h} = \frac{mz - ny}{h} \tag{1.4}$$

due to the fact that $l = \omega_x/\omega$, $m = \omega_y/\omega$, and $n = \omega_z/\omega$. Hence the direction cosines α_1, α_2, and α_3 may be written as:

$$\begin{aligned}
\alpha_1 = v_x/v &= \frac{\omega_y z - \omega_z y}{\omega h} = \frac{\omega_y z}{\omega h} - \frac{\omega_z y}{\omega h} = \frac{mz - ny}{h} \\
\alpha_2 = v_y/v &= \frac{\omega_z x - \omega_x z}{\omega h} = \frac{\omega_z x}{\omega h} - \frac{\omega_x z}{\omega h} = \frac{nx - lz}{h} \\
\alpha_3 = v_z/v &= \frac{\omega_x y - \omega_y x}{\omega h} = \frac{\omega_x y}{\omega h} - \frac{\omega_y x}{\omega h} = \frac{ly - mx}{h}
\end{aligned} \tag{1.5}$$

The body in Fig. 1.2 is assumed to be rotating and translating with respect to the inertial coordinate frame X_1, Y_1, Z_1. The X, Y, Z coordinate system, with its origin attached to the rigid body, at O, rotates in a random fashion relative to the body. The X', Y', Z' axes whose origin is also located at O remain parallel to the inertial axes X_1, Y_1, Z_1. The coordinates of m' with respect to the X, Y, Z and X', Y', Z' axes, respectively, are: x, y, z and x', y', z'.

Letting ω represent the angular velocity of the body while u stands for the linear velocity of m', each measured relative to X', Y', Z', the components of the vectors ω and u, along the X', Y', Z' axes are designated as $\omega'_x, \omega'_y, \omega'_z$ and u'_x, u'_y, u'_z, respectively. Then akin to the fact established earlier that $v_x = \omega_y z - \omega_z y;\ v_y = \omega_z x - \omega_x z;\ v_z = \omega_x y - \omega_y x$, we have: $u'_x = \omega'_y z' - \omega'_z y';\ u'_y = \omega'_z x' - \omega'_x z';\ u'_z = \omega'_x y' - \omega'_y x'$. Allowing u_x, u_y, u_z to be the components of u along the momentary positions of the X, Y, Z axis frame, we can write:

$$
\begin{aligned}
u_x &= u'_x\alpha_{11} + u'_y\alpha_{12} + u'_z\alpha_{13} \\
u_y &= u'_x\alpha_{21} + u'_y\alpha_{22} + u'_z\alpha_{23} \\
u_z &= u'_x\alpha_{31} + u'_y\alpha_{32} + u'_z\alpha_{33}
\end{aligned}
\tag{1.6}
$$

where $\alpha_{11}, \alpha_{12}, \alpha_{13}$ are the direction cosines of X relative to X', Y', Z', $\alpha_{21}, \alpha_{22}, \alpha_{23}$ are the direction cosines of Y relative to X', Y', Z', and $\alpha_{31}, \alpha_{32}, \alpha_{33}$ are the direction cosines of Z relative to X', Y', Z'. Equation 1.6 may be understood by taking the partial derivative of u_x with respect to u'_x, which results in: $\frac{\partial u_x}{\partial u'_x} = \alpha_{11}$. In other words, the cosine of the angle between u_x and u'_x is the same as the cosine of the angle between the X axis and the X' axis. This statement also holds for the cosine of the angle between the X and the Y' axes, etc. Another interpretation of the above equation is that u_x is the sum of the geometric projections onto the X axis of the velocities u'_x, u'_y and u'_z. A similar situation holds for u_y and u_z. Thus we have:

$$
\begin{aligned}
u_x &= (\omega'_y z' - \omega'_z y')\alpha_{11} + (\omega'_z x' - \omega'_x z')\alpha_{12} + (\omega'_x y' - \omega'_y x')\alpha_{13} \\
u_y &= (\omega'_y z' - \omega'_z y')\alpha_{21} + (\omega'_z x' - \omega'_x z')\alpha_{22} + (\omega'_x y' - \omega'_y x')\alpha_{23} \\
u_z &= (\omega'_y z' - \omega'_z y')\alpha_{31} + (\omega'_z x' - \omega'_x z')\alpha_{32} + (\omega'_x y' - \omega'_y x')\alpha_{33}
\end{aligned}
\tag{1.7}
$$

The relationship between the X' coordinate relative to X, Y, Z coordinates may similarly be shown to be of the form: $x' = x\alpha_{11} + y\alpha_{21} + z\alpha_{31}$ where $\alpha_{11}, \alpha_{21}, \alpha_{31}$ are the direction cosines of X' relative to X, Y, Z. We may show that for all three coordinates X', Y', Z' relative to the X, Y, Z coordinates, the relationship is the following:

$$
\begin{aligned}
x' &= x\alpha_{11} + y\alpha_{21} + z\alpha_{31} \\
y' &= x\alpha_{12} + y\alpha_{22} + z\alpha_{32} \\
z' &= x\alpha_{13} + y\alpha_{23} + z\alpha_{33}
\end{aligned}
\tag{1.8}
$$

where $\alpha_{11}, \alpha_{21}, \alpha_{31}$ are the direction cosines of X' relative to X, Y, Z, $\alpha_{12}, \alpha_{22}, \alpha_{32}$ are the direction cosines of Y' relative to X, Y, Z, and $\alpha_{13}, \alpha_{23}, \alpha_{33}$ are the direction cosines of Z' relative to X, Y, Z. Similarly, for angular rates ω'_x, $\omega'_x\omega'_x$, we have:

$$
\begin{aligned}
\omega'_x &= \omega_x\alpha_{11} + \omega_y\alpha_{21} + \omega_z\alpha_{31} \\
\omega'_y &= \omega_x\alpha_{12} + \omega_y\alpha_{22} + \omega_z\alpha_{32} \\
\omega'_z &= \omega_x\alpha_{13} + \omega_y\alpha_{23} + \omega_z\alpha_{33}
\end{aligned}
\tag{1.9}
$$

Using the identities:

$$\begin{aligned} u_x &= (\omega'_y z' - \omega'_z y')\alpha_{11} + (\omega'_z x' - \omega'_x z')\alpha_{12} + (\omega'_x y' - \omega'_y x')\alpha_{13} \\ u_y &= (\omega'_y z' - \omega'_z y')\alpha_{21} + (\omega'_z x' - \omega'_x z')\alpha_{22} + (\omega'_x y' - \omega'_y x')\alpha_{23} \\ u_z &= (\omega'_y z' - \omega'_z y')\alpha_{31} + (\omega'_z x' - \omega'_x z')\alpha_{32} + (\omega'_x y' - \omega'_y x')\alpha_{33} \end{aligned} \tag{1.10}$$

and the values for x', $y'z'$ and ω'_x, ω'_y, ω'_z in Eqs. 1.8 and 1.9, we have:

$$\begin{aligned} u_x &= (\omega'_y z' - \omega'_z y')\alpha_{11} + (\omega'_z x' - \omega'_x z')\alpha_{12} + (\omega'_x y' - \omega'_y x')\alpha_{13} \\ \Rightarrow u_x &= (\omega'_y[x\alpha_{13} + y\alpha_{23} + z\alpha_{33}] - \omega'_z[x\alpha_{12} + y\alpha_{22} + z\alpha_{32}])\alpha_{11} \\ &\quad + (\omega'_z[x\alpha_{11} + y\alpha_{21} + z\alpha_{31}] - \omega'_x[x\alpha_{13} + y\alpha_{23} + z\alpha_{33}])\alpha_{12} \\ &\quad + (\omega'_x[x\alpha_{12} + y\alpha_{22} + z\alpha_{32}] - \omega'_y[x\alpha_{11} + y\alpha_{21} + z\alpha_{31}])\alpha_{13} \end{aligned} \tag{1.11}$$

It turns out that the coefficient which multiplies x is zero, or $\frac{\partial u_x}{\partial x} = 0$. This may be seen from the following expression:

$$\begin{aligned} \frac{\partial u_x}{\partial x} &= \alpha_{13}(\alpha_{12}[\alpha_{11}\omega_x + \alpha_{21}\omega_y + \alpha_{31}\omega_z] - \alpha_{11}[\alpha_{12}\omega_x + \alpha_{22}\omega_y + \alpha_{32}\omega_z]) \\ &\quad - \alpha_{12}(\alpha_{13}[\alpha_{11}\omega_x + \alpha_{21}\omega_y + \alpha_{31}\omega_z] - \alpha_{11}[\alpha_{13}\omega_x + \alpha_{23}\omega_y + \alpha_{33}\omega_z]) \\ &\quad + \alpha_{11}(\alpha_{13}[\alpha_{12}\omega_x + \alpha_{22}\omega_y + \alpha_{32}\omega_z] - \alpha_{12}[\alpha_{13}\omega_x + \alpha_{23}\omega_y + \alpha_{33}\omega_z]) \\ &= (\alpha_{13}\alpha_{12} - \alpha_{12}\alpha_{13})[\alpha_{11}\omega_x + \alpha_{21}\omega_y + \alpha_{31}\omega_z] \\ &\quad + (\alpha_{11}\alpha_{13} - \alpha_{13}\alpha_{11})[\alpha_{12}\omega_x + \alpha_{22}\omega_y + \alpha_{32}\omega_z] \\ &\quad + (\alpha_{11}\alpha_{12} - \alpha_{12}\alpha_{11})[\alpha_{13}\omega_x + \alpha_{23}\omega_y + \alpha_{33}\omega_z] = 0 \end{aligned}$$

thus implying that u_x is of the form:

$$\begin{aligned} u_x &= \alpha_{11}\alpha_{23}\alpha_{32}\omega_z y - \alpha_{11}\alpha_{22}\alpha_{33}\omega_z y + \alpha_{12}\alpha_{21}\alpha_{33}\omega_z y - \alpha_{12}\alpha_{23}\alpha_{31}\omega_z y \\ &\quad - \alpha_{13}\alpha_{21}\alpha_{32}\omega_z y + \alpha_{13}\alpha_{22}\alpha_{31}\omega_z y + \alpha_{11}\alpha_{22}\alpha_{33}\omega_y z - \alpha_{11}\alpha_{23}\alpha_{32}\omega_y z \\ &\quad - \alpha_{12}\alpha_{21}\alpha_{33}\omega_y z + \alpha_{12}\alpha_{23}\alpha_{31}\omega_y z + \alpha_{13}\alpha_{21}\alpha_{32}\omega_y z - \alpha_{13}\alpha_{22}\alpha_{31}\omega_y z \\ &= (\omega_y z - \omega_z y)[\alpha_{11}\alpha_{22}\alpha_{33} - \alpha_{12}\alpha_{21}\alpha_{33}] \\ &\quad + (\omega_y z - \omega_z y)[\alpha_{13}\alpha_{21}\alpha_{32} - \alpha_{11}\alpha_{23}\alpha_{32}] \\ &\quad + (\omega_y z - \omega_z y)[\alpha_{12}\alpha_{23}\alpha_{31} - \alpha_{13}\alpha_{22}\alpha_{31}] \end{aligned}$$

which may be simplified as follows:

$$u_x = (\omega_y z - \omega_z y)\left(\underbrace{(\alpha_{11}\alpha_{22} - \alpha_{12}\alpha_{21})}_{=\alpha_{33}}\alpha_{33} + \underbrace{(\alpha_{13}\alpha_{21} - \alpha_{11}\alpha_{23})}_{=\alpha_{32}}\alpha_{32}\right.$$

$$\left. + \underbrace{(\alpha_{12}\alpha_{23} - \alpha_{13}\alpha_{22})}_{=\alpha_{31}}\alpha_{31}\right) = \underbrace{\left(\alpha_{31}^2 + \alpha_{32}^2 + \alpha_{33}^2\right)}_{=1}(\omega_y z - \omega_z y) \tag{1.12}$$

since $u_x = \omega_y z - \omega_z y$. Similarly u_y and u_z are:

$$u_y = (\omega_z x - \omega_x z)\left(\underbrace{(\alpha_{12}\alpha_{31} - \alpha_{11}\alpha_{32})}_{=\alpha_{23}}\alpha_{23} + \underbrace{(\alpha_{11}\alpha_{33} - \alpha_{13}\alpha_{31})}_{=\alpha_{22}}\alpha_{22}\right.$$

$$\left. + \underbrace{(\alpha_{13}\alpha_{32} - \alpha_{12}\alpha_{33})}_{=\alpha_{21}}\alpha_{21}\right) = \underbrace{\left(\alpha_{23}^2 + \alpha_{22}^2 + \alpha_{21}^2\right)}_{=1}(\omega_z x - \omega_x z)$$

$$u_z = (\omega_y x - \omega_x y)\left(\underbrace{(\alpha_{22}\alpha_{31} - \alpha_{21}\alpha_{32})}_{=\alpha_{13}}\alpha_{13} + \underbrace{(\alpha_{21}\alpha_{33} - \alpha_{23}\alpha_{31})}_{=\alpha_{12}}\alpha_{12}\right.$$

$$\left. + \underbrace{(\alpha_{23}\alpha_{32} - \alpha_{33}\alpha_{22})}_{=\alpha_{11}}\alpha_{11}\right) = \underbrace{\left(\alpha_{13}^2 + \alpha_{12}^2 + \alpha_{11}^2\right)}_{=1}(\omega_y z - \omega_z y)$$

The identities for α_{31}, α_{32}, and α_{33} appear in Wells' book [51, pp. 343] and will be developed in the sequel.

Let i_1, i_2, i_3 be the orthogonal unit vectors along the X_1, Y_1, Z_1 axes, respectively, and e_1, e_2, e_3 be the orthogonal unit vectors along the X, Y, Z axes. The direction cosines between the i_1 and e_1, e_2, and e_3 unit vectors are accordingly: α_{11}, α_{21} and α_{31}. The i_1, i_2, and i_3 vectors may then be written in terms of the e_1, e_2, and e_3 vectors and the corresponding direction cosines between the two systems of unit vectors as follows:

$$\begin{aligned} i_1 &= \alpha_{11}e_1 + \alpha_{21}e_2 + \alpha_{31}e_3 \\ i_2 &= \alpha_{12}e_1 + \alpha_{22}e_2 + \alpha_{32}e_3 \\ i_3 &= \alpha_{13}e_1 + \alpha_{23}e_2 + \alpha_{33}e_3 \end{aligned} \tag{1.13}$$

Similarly, the e_1, e_2, and e_3 unit vectors may be expressed in terms of the i_1, i_2, i_3 unit vectors and the corresponding direction cosines between the two systems of unit orthogonal vectors as follows:

$$\begin{aligned} e_1 &= \alpha_{11} i_1 + \alpha_{12} i_2 + \alpha_{13} i_3 \\ e_2 &= \alpha_{21} i_1 + \alpha_{22} i_2 + \alpha_{23} i_3 \\ e_3 &= \alpha_{31} i_1 + \alpha_{32} i_2 + \alpha_{33} i_3 \end{aligned} \tag{1.14}$$

Since the unit vectors are orthogonal we have: $e_1 \cdot e_2 = 0;\ e_1 \cdot e_3 = 0;\ e_2 \cdot e_3 = 0;\ e_1 \cdot e_1 = 1;\ e_2 \cdot e_2 = 1;\ e_3 \cdot e_3 = 1$. Similarly for the i_1, i_2, i_3 orthogonal unit vectors we have: $i_1 \cdot i_2 = 0;\ i_1 \cdot i_3 = 0;\ i_2 \cdot i_3 = 0;\ i_1 \cdot i_1 = 1;\ i_2 \cdot i_2 = 1;\ i_3 \cdot i_3 = 1$. The dot products of the vectors $i_1 \cdot i_1, i_2 \cdot i_2$ and $i_3 \cdot i_3$ will yield the following:

$$\begin{aligned} i_1 \cdot i_1 &= (\alpha_{11} e_1 + \alpha_{21} e_2 + \alpha_{31} e_3) \cdot (\alpha_{11} e_1 + \alpha_{21} e_2 + \alpha_{31} e_3) \\ &\Rightarrow 1 = \alpha_{11}^2 + \alpha_{21}^2 + \alpha_{31}^2 \\ i_2 \cdot i_2 &= (\alpha_{12} e_1 + \alpha_{22} e_2 + \alpha_{32} e_3) \cdot (\alpha_{12} e_1 + \alpha_{22} e_2 + \alpha_{32} e_3) \\ &\Rightarrow 1 = \alpha_{12}^2 + \alpha_{22}^2 + \alpha_{32}^2 \\ i_3 \cdot i_3 &= (\alpha_{13} e_1 + \alpha_{23} e_2 + \alpha_{33} e_3) \cdot (\alpha_{13} e_1 + \alpha_{23} e_2 + \alpha_{33} e_3) \\ &\Rightarrow 1 = \alpha_{13}^2 + \alpha_{23}^2 + \alpha_{33}^2 \end{aligned} \tag{1.15}$$

Similarly the dot products of the vectors $i_1 \cdot i_2, i_1 \cdot i_2$ and $i_2 \cdot i_3$ result in:

$$\begin{aligned} i_1 \cdot i_2 &= (\alpha_{11} e_1 + \alpha_{21} e_2 + \alpha_{31} e_3) \cdot (\alpha_{12} e_1 + \alpha_{22} e_2 + \alpha_{32} e_3) \\ &\Rightarrow 0 = \alpha_{11}\alpha_{12} + \alpha_{21}\alpha_{22} + \alpha_{31}\alpha_{31} \\ i_1 \cdot i_3 &= (\alpha_{11} e_1 + \alpha_{21} e_2 + \alpha_{31} e_3) \cdot (\alpha_{13} e_1 + \alpha_{23} e_2 + \alpha_{33} e_3) \\ &\Rightarrow 0 = \alpha_{11}\alpha_{13} + \alpha_{21}\alpha_{23} + \alpha_{31}\alpha_{33} \\ i_2 \cdot i_3 &= (\alpha_{12} e_1 + \alpha_{22} e_2 + \alpha_{32} e_3) \cdot (\alpha_{13} e_1 + \alpha_{23} e_2 + \alpha_{33} e_3) \\ &\Rightarrow 0 = \alpha_{12}\alpha_{13} + \alpha_{22}\alpha_{23} + \alpha_{32}\alpha_{33} \end{aligned} \tag{1.16}$$

The same procedure is employed on the e_1, e_2, and e_3 vectors, that is:

$$\begin{aligned} e_1 \cdot e_1 &= (\alpha_{11} i_1 + \alpha_{12} i_2 + \alpha_{13} i_3) \cdot (\alpha_{11} i_1 + \alpha_{12} i_2 + \alpha_{13} i_3) \\ &\Rightarrow 1 = \alpha_{11}^2 + \alpha_{12}^2 + \alpha_{13}^2 \\ e_2 \cdot e_2 &= (\alpha_{21} i_1 + \alpha_{22} i_2 + \alpha_{23} i_3) \cdot (\alpha_{21} i_1 + \alpha_{22} i_2 + \alpha_{23} i_3) \\ &\Rightarrow 1 = \alpha_{21}^2 + \alpha_{22}^2 + \alpha_{23}^2 \end{aligned}$$

$$
\begin{aligned}
e_3 \cdot e_3 &= (\alpha_{31} i_1 + \alpha_{32} i_2 + \alpha_{33} i_3) \cdot (\alpha_{31} i_1 + \alpha_{32} i_2 + \alpha_{33} i_3) \\
&\Rightarrow 1 = \alpha_{31}^2 + \alpha_{32}^2 + \alpha_{33}^2 \\
e_1 \cdot e_2 &= (\alpha_{11} i_1 + \alpha_{12} i_2 + \alpha_{13} i_3) \cdot (\alpha_{21} i_1 + \alpha_{22} i_2 + \alpha_{23} i_3) \\
&\Rightarrow 0 = \alpha_{11}\alpha_{21} + \alpha_{12}\alpha_{22} + \alpha_{13}\alpha_{23} \\
e_1 \cdot e_3 &= (\alpha_{11} i_1 + \alpha_{12} i_2 + \alpha_{13} i_3) \cdot (\alpha_{31} i_1 + \alpha_{32} i_2 + \alpha_{33} i_3) \\
&\Rightarrow 0 = \alpha_{11}\alpha_{31} + \alpha_{12}\alpha_{32} + \alpha_{13}\alpha_{33} \\
e_2 \cdot e_3 &= (\alpha_{21} i_1 + \alpha_{22} i_2 + \alpha_{23} i_3) \cdot (\alpha_{31} i_1 + \alpha_{32} i_2 + \alpha_{33} i_3) \\
&\Rightarrow 0 = \alpha_{21}\alpha_{31} + \alpha_{22}\alpha_{32} + \alpha_{23}\alpha_{33}
\end{aligned}
\tag{1.17}
$$

The following has been assumed:

(a) X_1, Y_1, Z_1 is a fixed and stationary coordinate system.
(b) X, Y, Z is a body fixed coordinate system where the body rotates around some fixed point O.
(c) The direction cosines between the X coordinate and coordinates X_1, Y_1, Z_1 are $\alpha_{11}, \alpha_{12}, \alpha_{13}$.
(d) Similarly, for the direction cosines between the Y coordinate and coordinates X_1, Y_1, Z_1, we have: $\alpha_{21}, \alpha_{22}, \alpha_{23}$, etc.

Hence, the transformation of a vector from the stationary X_1, Y_1, Z_1 coordinate system to the rotating X, Y, Z coordinate system can be written in matrix form as follows:

$$
\begin{bmatrix} x_b \\ y_b \\ z_b \end{bmatrix} = \begin{bmatrix} \alpha_{11} & \alpha_{12} & \alpha_{13} \\ \alpha_{21} & \alpha_{22} & \alpha_{23} \\ \alpha_{31} & \alpha_{32} & \alpha_{33} \end{bmatrix} \begin{bmatrix} x_{ea} \\ y_{ea} \\ z_{ea} \end{bmatrix} \tag{1.18}
$$

where the vector $\begin{bmatrix} x_{ea} \\ y_{ea} \\ z_{ea} \end{bmatrix}$ is in the X_1, Y_1, Z_1 coordinate frame and the vector $\begin{bmatrix} x_b \\ y_b \\ z_b \end{bmatrix}$ is in the X, Y, Z coordinate frame. The rotation of the vector X_1, Y_1, Z_1 coordinates into the vector in X, Y, Z coordinates can be described by Euler angular transformations in matrix form as follows:

$$
\begin{bmatrix} x_b \\ y_b \\ z_b \end{bmatrix} = \begin{bmatrix} \cos\theta\cos\psi & \cos\theta\sin\psi & -\sin\theta \\ \sin\phi\sin\theta\cos\psi - \cos\phi\sin\psi & \sin\phi\sin\theta\sin\psi + \cos\phi\cos\psi & \sin\phi\cos\theta \\ \cos\phi\sin\theta\cos\psi + \sin\phi\sin\psi & \cos\phi\sin\theta\sin\psi - \sin\phi\cos\psi & \cos\phi\cos\theta \end{bmatrix} \begin{bmatrix} x_{ea} \\ y_{ea} \\ z_{ea} \end{bmatrix} \tag{1.19}
$$

Equating the two transformations we note that:

$$\begin{aligned}
&\alpha_{11} = \cos\theta\cos\psi;\ \ \alpha_{12} = \cos\theta\sin\psi;\ \ \alpha_{13} = -\sin\theta \\
&\alpha_{21} = \sin\phi\sin\theta\cos\psi - \cos\phi\sin\psi;\ \ \alpha_{22} = \sin\phi\sin\theta\sin\psi + \cos\phi\cos\psi \\
&\alpha_{23} = \sin\phi\cos\theta;\ \ \alpha_{31} = \cos\phi\sin\theta\cos\psi + \sin\phi\sin\psi \\
&\alpha_{32} = \cos\phi\sin\theta\sin\psi - \sin\phi\cos\psi;\ \ \alpha_{33} = \cos\phi\cos\theta
\end{aligned} \tag{1.20}$$

Taking the product of $\alpha_{22} * \alpha_{33}$

$$\begin{aligned}
\alpha_{22} * \alpha_{33} &= (\sin\phi\sin\theta\sin\psi + \cos\phi\cos\psi) * (\cos\phi\cos\theta) \\
&= \cos\phi\cos\theta\sin\phi\sin\theta\sin\psi + \cos\theta\cos^2\phi\cos\psi
\end{aligned} \tag{1.21}$$

Similarly the product of $-\alpha_{23} * \alpha_{32}$

$$\begin{aligned}
-\alpha_{23} * \alpha_{32} &= -(\sin\phi\cos\theta) * (\cos\phi\sin\theta\sin\psi - \sin\phi\cos\psi) \\
&= -(\cos\phi\cos\theta\sin\phi\sin\theta\sin\psi) + \cos\theta\sin^2\phi\cos\psi
\end{aligned} \tag{1.22}$$

Calculating $\alpha_{22} * \alpha_{33} - \alpha_{23} * \alpha_{32}$, we have:

$$\begin{aligned}
\alpha_{22} * \alpha_{33} - \alpha_{23} * \alpha_{32} &= (\cos\phi\cos\theta\sin\phi\sin\theta\sin\psi) \\
&\quad - (\cos\phi\cos\theta\sin\phi\sin\theta\sin\psi) \\
&\quad + \cos\theta\cos^2\phi\cos\psi + \cos\theta\sin^2\phi\cos\psi \\
&= \cos\theta\cos\psi = \alpha_{11}
\end{aligned} \tag{1.23}$$

Similarly for α_{12}, we have: $\alpha_{12} = \alpha_{23} * \alpha_{31} - \alpha_{33} * \alpha_{21} = \cos\theta\sin\psi$, which is expanded in the following equation:

$$\begin{aligned}
\alpha_{23} * \alpha_{31} &= (\sin\phi\cos\theta) * (\cos\phi\sin\theta\cos\psi + \sin\phi\sin\psi) \\
&= \sin\phi\cos\theta\cos\phi\sin\theta\cos\psi + \sin^2\phi\cos\theta\sin\psi \\
&\quad - \alpha_{33} * \alpha_{21} = -(\cos\phi\cos\theta)(\sin\phi\sin\theta\cos\psi - \cos\phi\sin\psi) \\
&= \cos^2\phi\cos\theta\sin\psi - \cos\phi\cos\theta\sin\phi\sin\theta\cos\psi \\
&\Rightarrow \alpha_{23} * \alpha_{31} - \alpha_{33} * \alpha_{21} = \cos\theta\sin\psi
\end{aligned} \tag{1.24}$$

And finally for $\alpha_{21} = \alpha_{32} * \alpha_{13} - \alpha_{12} * \alpha_{33}$, the result is:

$$\begin{aligned}
\alpha_{32} * \alpha_{13} &= (\sin\phi\cos\psi - \cos\phi\sin\theta\sin\psi)(\sin\theta) \\
&= \sin\theta\sin\phi\cos\psi - \sin^2\theta\cos\phi\sin\psi \\
&\quad - \alpha_{12} * \alpha_{33} = -(\cos\theta\sin\psi)(\cos\phi\cos\theta) = -\cos^2\theta\sin\psi\cos\phi \\
&\quad \Rightarrow \alpha_{21} = \sin\theta\sin\phi\cos\psi - \sin^2\theta\cos\phi\sin\psi - \cos^2\theta\sin\psi\cos\phi \\
&= \sin\theta\sin\phi\cos\psi - \cos\phi\sin\psi
\end{aligned} \tag{1.25}$$

Following a similar procedure, all of the identities between the Euler angles and the direction cosines are as follows:

1. $\alpha_{11} = \alpha_{22} * \alpha_{33} - \alpha_{23} * \alpha_{32} = \cos\theta\cos\psi$
2. $\alpha_{12} = \alpha_{23} * \alpha_{31} - \alpha_{33} * \alpha_{21} = \cos\theta\sin\psi$
3. $\alpha_{13} = \alpha_{21} * \alpha_{32} - \alpha_{31} * \alpha_{22} = -\sin\theta$
4. $\alpha_{21} = \alpha_{32} * \alpha_{13} - \alpha_{12} * \alpha_{33} = \sin\phi\sin\theta\cos\psi - \cos\phi\sin\psi$
5. $\alpha_{22} = \alpha_{33} * \alpha_{11} - \alpha_{13} * \alpha_{31} = \sin\phi\sin\theta\sin\psi + \cos\phi\cos\psi$
6. $\alpha_{23} = \alpha_{31} * \alpha_{12} - \alpha_{11} * \alpha_{32} = \sin\phi\cos\theta$
7. $\alpha_{31} = \alpha_{12} * \alpha_{23} - \alpha_{22} * \alpha_{13} = \cos\phi\sin\theta\cos\psi + \sin\phi\sin\psi$
8. $\alpha_{32} = \alpha_{13} * \alpha_{21} - \alpha_{11} * \alpha_{23} = \cos\phi\sin\theta\sin\psi - \sin\phi\cos\psi$
9. $\alpha_{33} = \alpha_{11} * \alpha_{22} - \alpha_{12} * \alpha_{21} = \cos\phi\cos\theta$

Chapter 2
Lagrangian Dynamics: Preliminaries

The notions taken from classical mechanics, which are required for an understanding of Lagrangian Dynamics, are introduced in the present chapter. These notions include linear and angular velocities, linear and angular momenta and their derivatives with respect to time, as well as kinetic and potential energies and work. The chapter includes many examples of a didactic nature. These topics have been chosen to serve both as a review of classical mechanics, especially the concepts of work and kinetic and potential energy, and to set the framework for the study of Lagrangian Dynamics in Chap. 3.

2.1 Angular Velocity of a Body and Linear Velocity of a Typical Particle Within That Body (See Wells [51, pp. 140–142])

A body with axes centered at point O rotates with angular rate ω and translates with velocity v_o with respect to an inertial axis system X_1, Y_1, Z_1 (see Fig. 2.1—note that Figs. 1.2 and 2.1 are the same). An axis system X, Y, Z, attached to the body at point O, rotates in an arbitrary fashion with respect to the body. At every instant, the motion of the X, Y, Z coordinate frame is "frozen" with respect to the rotating body. The velocity v_o of the origin may be written in terms of the instantaneous components of the X, Y, Z axes as: $v_o = [v_{ox}, v_{oy}, v_{oz}]^T$. Similarly the rotation rate ω of the body in the instantaneous X, Y, Z components may be written as: $\omega = [\omega_x, \omega_y, \omega_z]^T$. In the rotating X, Y, Z axis frame, the instantaneous location of a point mass m' is $[x, y, z]^T$. The velocity of point mass m' with respect to inertial axes X_1, Y_1, Z_1, but expressed in X, Y, Z coordinates, is therefore:

$$v_x = v_{ox} + (\omega_y z - \omega_z y), \quad v_y = v_{oy} + (\omega_z x - \omega_x z), \quad v_z = v_{oz} + (\omega_x y - \omega_y x) \tag{2.1}$$

A. W. Pila, *Introduction To Lagrangian Dynamics*,
https://doi.org/10.1007/978-3-030-22378-6_2

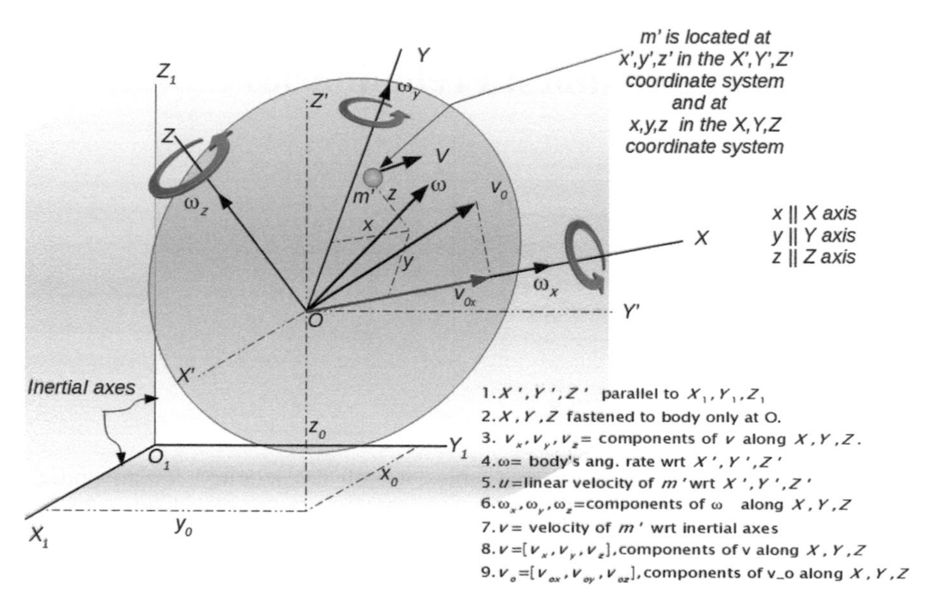
m' is located at
x',y',z' in the X',Y',Z'
coordinate system
and at
x,y,z in the X,Y,Z
coordinate system
x || X axis
y || Y axis
z || Z axis
Inertial axes
1. X',Y',Z' parallel to X1,Y1,Z1
2. X,Y,Z fastened to body only at O.
3. vx,vy,vz= components of v along X,Y,Z.
4. ω= body's ang. rate wrt X',Y',Z'
5. u=linear velocity of m' wrt X',Y',Z'
6. ωx,ωy,ωz=components of ω along X,Y,Z
7. v= velocity of m' wrt inertial axes
8. v=[vx,vy,vz],components of v along X,Y,Z
9. vo=[vox,voy,voz],components of v_o along X,Y,Z

Fig. 2.1 Body translating and rotating, while X, Y, Z frame rotates about O relative to the body

The full significance of Eq. 2.1 may be understood by considering the following points (not necessarily in order of importance) together with a study of several examples to follow.

1. The origin O is attached to any arbitrarily chosen point within the body, about which the body and the X, Y, Z coordinate system rotate.
2. Equation 2.1 is valid even though the X, Y, Z frame (origin fixed at O) may rotate relative to the body. It is also possible that this frame may be "body-fixed," that is, rigidly attached to the body. When this is not the case, the x, y, z vary, while in the body-fixed case, they are constant. Normally, body-fixed axes are utilized.
3. Components of v_O, the inertial velocity of O, are expressed along instantaneous directions of X, Y, Z, that is: $v_O = [v_{Ox}, v_{Oy}, v_{Oz}]^T$ (see Fig. 2.1 where v_{Ox} is illustrated).
4. Regardless of where O is located, v_{Ox}, v_{Oy}, v_{Oz} is the same for all of the mass particles which make up the rigid body. Hence v_o expresses the linear velocity of the object as a whole.
5. The total angular velocity of the body ω *is always measured with respect to an inertial frame*, i.e., relative to a non-rotating axis frame such as X', Y', Z' in Fig. 2.1.
6. ω is expressed in X, Y, Z components ω_x, ω_y, ω_z, that is, along the instantaneous directions of coordinates X, Y, Z.

7. Regardless of where the origin O is located in the body, ω always has the same magnitude and direction. Whatever the location of O, ω is directed along some imaginary line passing through O. This implies that this vector can be relocated to any other origin within the body without undergoing a change in magnitude or direction.
8. Subject to forces acting on the body, ω and v_o may change their magnitudes and directions. Under these circumstances, their directions, relative to the body, are not fixed.
9. Equation 2.1 is used to write a general expression for the kinetic energy of a rigid body as shown in the sequel (see Eq. 2.15).

2.2 Angular Velocity of a Body and Linear Velocity of a Typical Particle Within the Body: Examples

Example 1 [51, pp. 143–145]

The rigid body suspended within a supporting frame, in Fig. 2.2, rotates about a vertical shaft $\mathbf{A} - \mathbf{O_1}$ with angular velocity $d\psi/dt$. The body can simultaneously rotate with an angular velocity $d\phi/dt$ about a shaft, supported by bearings B_1, B_2, which is offset from the vertical by a constant angle θ. In the top view of Fig. 2.2b, ψ is read off relative to the inertial X_1 axis. The angle ϕ between the horizontal line $a - b$ and the rotating axis X is measured in the planar section of the body normal to line $O - Z$. The total angular velocity ω of the body is the vector sum of $d\phi/dt$ and $d\psi/dt$. The aim is to determine the components of ω and the linear velocity of an arbitrary mass particle within the body at various locations of the rotating X, Y, Z frame.

a. (a) body-fixed axes X, Y, Z as depicted in Fig. 2.2,
   (b) origin O at the intersection of the vertical $A - O_1$ line and the $B_1 - B_2$ axis
   (c) $\dot{\phi}$ is a vector pointing in the Z axis direction
   (d) $\dot{\psi}$ is directed along the vertical line $A - O$ or Z_1 axis
   (e) X, Y, Z components of ω: $\omega_{ax}, \omega_{ay}, \omega_{az}$, are calculated from components of $\dot{\psi}$ and $\dot{\phi}$ along the X, Y, Z axes

   This implies that

$$\omega_{ax} = \dot{\psi} \sin\theta \sin\phi; \quad \omega_{ay} = \dot{\psi} \sin\theta \cos\phi; \quad \omega_{az} = \dot{\phi} + \dot{\psi} \cos\theta \tag{2.2}$$

$\omega_{ax}, \omega_{ay}, \omega_{az}$ are the angular velocities along "instantaneous" coordinates X, Y, Z. With O located as in Fig. 2.2, v_o, the velocity of point O with respect to an inertial frame is zero, that is, $v_{ox} = v_{oy} = v_{oz} = 0$. Hence, the components

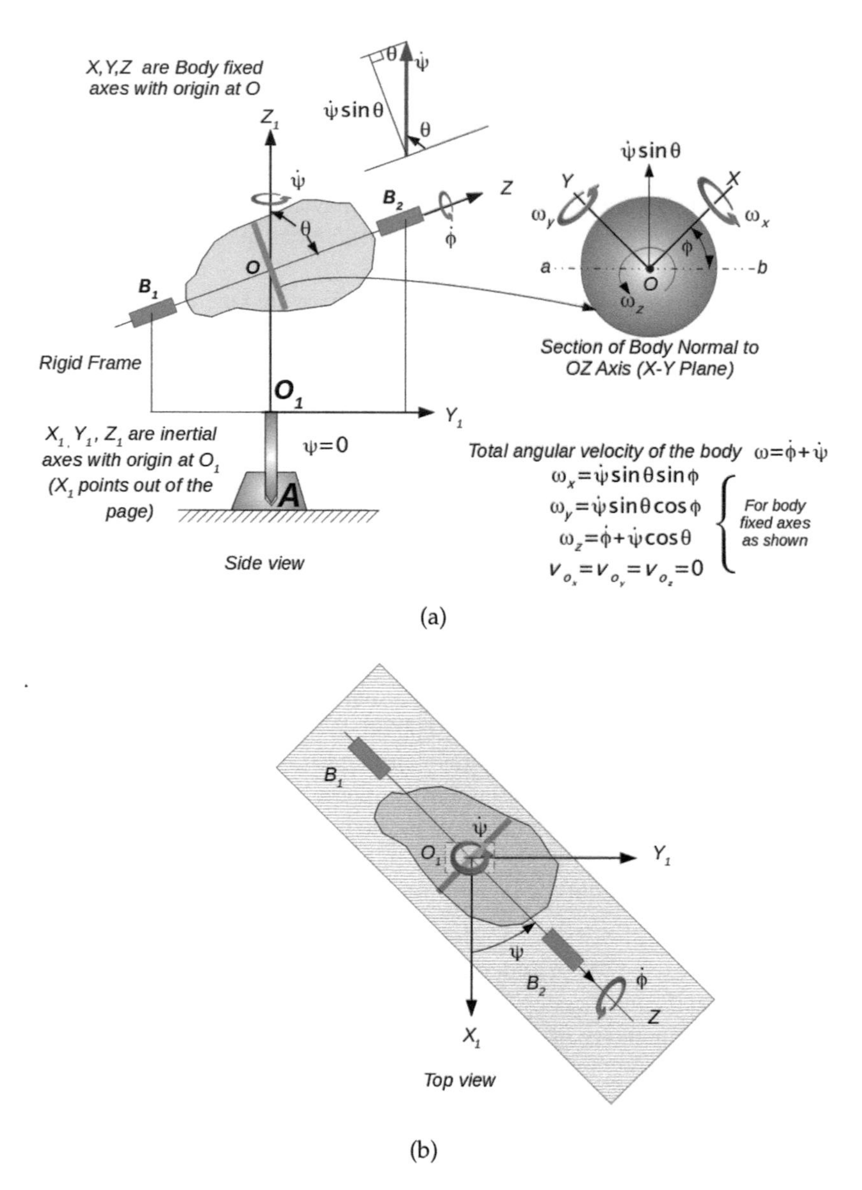

Fig. 2.2 Illustration of the treatment of the angular velocity of a body and linear velocity of a typical mass particle. (**a**) Side view. (**b**) Top view

of the velocity of m' relative to an inertial frame, expressed in terms of the *instantaneous directions of* X, Y, Z (see expression 2.1) are:

$$
\begin{aligned}
v_x &= \omega_y z - \omega_z y = \dot{\psi} z \sin\theta \cos\phi - (\dot{\phi} + \dot{\psi}\cos\theta) y \\
v_y &= \omega_z x - \omega_x z = (\dot{\phi} + \dot{\psi}\cos\theta) x - \dot{\psi} z \sin\theta \sin\phi \\
v_z &= \omega_x y - \omega_y x = \dot{\psi} y \sin\theta \sin\phi - \dot{\psi} x \sin\theta \cos\phi
\end{aligned}
\tag{2.3}
$$

where x, y, z are the coordinates of m' in the X, Y, Z axis system.

b. (a) The origin of the X, Y, Z coordinate frame is relocated to some arbitrary position along the $O - Z$ axis, point p_1, where the distance from O to p_1 is l.
   (b) All axes remain parallel to their original positions and the total angular velocity remains unchanged.
   (c) Shifting of $d\psi/dt$ to a vertical line through p_1 and taking components of $d\psi/dt$ and $d\phi/dt$ result in the same expressions as in Eq. 2.2, since the coordinate frame is parallel to its original position.
   (d) However, $v_o = d\psi/dt \times l = d\psi/dt \quad l \sin\theta$ (l = distance $O - p_1$) and it is directed along $O - b$ line in the planar section normal to line $O - Z$.

The components of the velocity of p_1 (the new origin) with respect to an inertial frame, and in terms of the instantaneous X, Y, Z axes are:

$$
v_{ox} = \dot{\psi} l \sin\theta \cos\phi; \quad v_{oy} = -\dot{\psi} l \sin\theta \sin\phi; \quad v_{oz} = 0 \tag{2.4}
$$

Note that 2.4 results from 2.3 directly by letting $z = l, x = y = 0$. Therefore, the components of the velocity of m' relative to the inertial coordinate system and expressed in the instantaneous coordinates X, Y, Z, with the origin at p_1, are:

$$
\begin{aligned}
v_x &= \dot{\psi} l \sin\theta \cos\phi + [\dot{\psi} z \sin\theta \cos\phi - (\dot{\phi} + \dot{\psi}\cos\theta) y] \\
v_y &= -\dot{\psi} l \sin\theta \sin\phi + [(\dot{\phi} + \dot{\psi}\cos\theta) x - \dot{\psi} z \sin\theta \sin\phi] \\
v_z &= \dot{\psi} y \sin\theta \sin\phi - \dot{\psi} x \sin\theta \cos\phi
\end{aligned}
\tag{2.5}
$$

where x, y, z are measured along the instantaneous X, Y, Z coordinates. The significance of v_x, v_y, v_z is the following: *a stationary observer located on the base A measures the velocity v of mass particle m' relative to inertial coordinates X_1, Y_1, Z_1. Then v_x, v_y, v_z, as given by 2.5, are components of v expressed in the body-fixed X, Y, Z axis system, at the instant that this axis system occupies this particular position and the observer takes the measurement.*

c. With the origin of the body-fixed X, Y, Z frame shifted to p_2 (any arbitrary point within the body) and with each axis parallel to its position in case a, the vectors $\dot{\psi}$ and $\dot{\phi}$ must be moved from the positions shown in Fig. 2.2 to parallel lines passing through p_2. Hence it is obvious that $\omega_{cx}, \omega_{cy}, \omega_{cz}$, the angular velocities with the origin at p_2 are equal to $\omega_{ax}, \omega_{ay}, \omega_{az}$, respectively, the angular velocities as in case a. A helpful way of determining v_{ox}, v_{oy}, v_{oz} for this situation is the following:

(a) let x_2, y_2, z_2 be coordinates of p_2 relative to X, Y, Z when the origin of the X, Y, Z coordinate system is at O (as in case a).
(b) then, applying 2.3, v_o is given by:

$$\begin{aligned} v_{ox} &= \dot{\psi} z_2 \sin\theta \cos\phi - (\dot{\phi} + \dot{\psi}\cos\theta) y_2 \\ v_{oy} &= (\dot{\phi} + \dot{\psi}\cos\theta) x_2 - \dot{\psi} z_2 \sin\theta \sin\phi \\ v_{oz} &= \dot{\psi} y_2 \sin\theta \sin\phi - \dot{\psi} x_2 \sin\theta \cos\phi \end{aligned} \tag{2.6}$$

Hence v_x, v_y, v_z, respectively, for case c are given by:

$$\begin{aligned} v_x &= \underbrace{\dot{\psi} z_2 \sin\theta \cos\phi - (\dot{\phi} + \dot{\psi}\cos\theta) y_2}_{v_{ox}} + [\dot{\psi} z \sin\theta \cos\phi - (\dot{\phi} + \dot{\psi}\cos\theta) y] \\ v_y &= \underbrace{(\dot{\phi} + \dot{\psi}\cos\theta) x_2 - \dot{\psi} z_2 \sin\theta \sin\phi}_{v_{oy}} + [\dot{\phi} + \dot{\psi}\cos\theta) x - \dot{\psi} z \sin\theta \sin\phi] \\ v_z &= \underbrace{\dot{\psi} y_2 \sin\theta \sin\phi - \dot{\psi} x_2 \sin\theta \cos\phi}_{v_{oz}} + [\dot{\psi} y \sin\theta \sin\phi - \dot{\psi} x \sin\theta \cos\phi] \end{aligned} \tag{2.7}$$

where x, y, z in 2.7, the location of the mass particle m', is measured relative to the origin at p_2, but expressed in X, Y, Z coordinates. The origin p_2, with coordinates x_2, y_2, z_2, is also measured with respect to O and expressed in X, Y, Z coordinates.

Example 2 [51, pp. 145–146]

The disk D, in Fig. 2.3, rotates freely about the shaft $b - c$ with angular velocity $d\phi/dt$, where the angle ϕ is read off relative to the shaft $b - c$—see pointer p_2. Simultaneously, the shaft $a - b$ rotates with angular velocity $d\psi/dt$ where ψ is the angle between the fixed $X_1 - Z_1$ plane and the rotating plane which is normal to the $a - b$ shaft—see pointer p_1. The complete angular velocity ω of D is the vector sum of $d\phi/dt$ and $d\psi/dt$. Relocating $d\psi/dt$ to O, the origin, and taking components along the body-fixed X, Y, Z axes, it is apparent that, just as in the example on page 15 (Eq. 2.2),

$$\omega_x = \dot{\psi} \sin\theta \sin\phi; \quad \omega_y = \dot{\psi} \sin\theta \cos\phi; \quad \omega_z = \dot{\phi} + \dot{\psi}\cos\theta \tag{2.8}$$

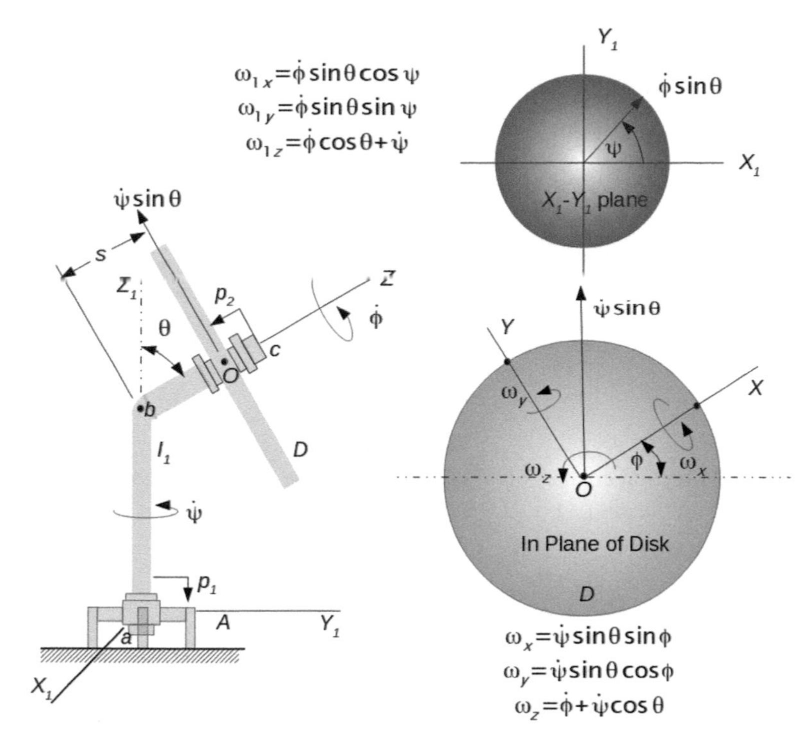

Fig. 2.3 Another example of the treatment of the angular velocity of a body and linear velocity of a typical mass particle

The components $\omega_{1x}, \omega_{1y}, \omega_{1z}$ along the fixed X_1, Y_1, Z_1 axes are:

$$\omega_{1x} = \dot{\phi} \sin\theta \cos\psi; \quad \omega_{1y} = \dot{\phi} \sin\theta \sin\psi; \quad \omega_{1z} = \dot{\phi} \cos\theta + \dot{\psi} \tag{2.9}$$

The magnitude of the two angular velocity vectors in Eqs. 2.8 and 2.9 are one and the same. The magnitude of the total angular velocity of disk D is therefore given by:

$$\begin{aligned}\omega &= \sqrt{\omega_x^2 + \omega_y^2 + \omega_z^2} = \sqrt{\omega_{1x}^2 + \omega_{1y}^2 + \omega_{1z}^2} \\ &= \sqrt{\dot{\phi}^2 \sin^2\theta \cos^2\psi + \dot{\phi}^2 \sin^2\theta \sin^2\psi + \dot{\phi}^2 \cos^2\theta + \dot{\psi}^2 + 2\dot{\phi}\dot{\psi}\cos\theta} \\ &= \sqrt{\dot{\phi}^2 \sin^2\theta + \dot{\phi}^2 \cos^2\theta + \dot{\psi}^2 + 2\dot{\phi}\dot{\psi}\cos\theta} = \sqrt{\dot{\phi}^2 + \dot{\psi}^2 + 2\dot{\phi}\dot{\psi}\cos\theta}\end{aligned} \tag{2.10}$$

The direction of ω relative to the moving X, Y, Z axes is determined by the direction cosines l, m, n which may be shown to be: $l = \omega_x/\omega \quad m = \omega_y/\omega \quad n = \omega_z/\omega$, where $\omega = \sqrt{\omega_x^2 + \omega_y^2 + \omega_z^2}$. Using the relations in Eqs. 2.8 and 2.10, the direction cosines l, m, n become:

$$
\begin{aligned}
l &= \frac{\omega_x}{\omega} = \frac{\dot{\psi}\sin\theta\sin\phi}{\sqrt{\dot{\phi}^2 + \dot{\psi}^2 + 2\dot{\phi}\dot{\psi}\cos\theta}} \\
m &= \frac{\omega_y}{\omega} = \frac{\dot{\psi}\sin\theta\cos\phi}{\sqrt{\dot{\phi}^2 + \dot{\psi}^2 + 2\dot{\phi}\dot{\psi}\cos\theta}} \\
n &= \frac{\omega_z}{\omega} = \frac{\dot{\phi} + \dot{\psi}\cos\theta}{\sqrt{\dot{\phi}^2 + \dot{\psi}^2 + 2\dot{\phi}\dot{\psi}\cos\theta}}
\end{aligned}
\tag{2.11}
$$

Similarly, the direction cosines of ω with respect to the X_1, Y_1, Z_1 coordinate frame are obtained with the aid of Eqs. 2.9 and 2.10 as follows:

$$
\begin{aligned}
l_1 &= \frac{\omega_{1x}}{\omega} = \frac{\dot{\phi}\sin\theta\cos\psi}{\sqrt{\dot{\phi}^2 + \dot{\psi}^2 + 2\dot{\phi}\dot{\psi}\cos\theta}} \\
m_1 &= \frac{\omega_{1y}}{\omega} = \frac{\dot{\phi}\sin\theta\sin\psi}{\sqrt{\dot{\phi}^2 + \dot{\psi}^2 + 2\dot{\phi}\dot{\psi}\cos\theta}} \\
n_1 &= \frac{\omega_{1z}}{\omega} = \frac{\dot{\phi}\cos\theta + \dot{\psi}}{\sqrt{\dot{\phi}^2 + \dot{\psi}^2 + 2\dot{\phi}\dot{\psi}\cos\theta}}
\end{aligned}
\tag{2.12}
$$

The linear velocity relative to the inertial frame, expressed in X, Y, Z components, of a typical particle in D, is found exactly as in case b on page 17 and the procedure will be illustrated below. The origin of the X, Y, Z axis system is at the center of disk D which is rotating with angular velocity of $\dot{\psi}\sin\theta$ (see Fig. 2.4). The velocities at the origin O are obtained from Eq. 2.3 with $x = 0$, $y = 0$, $z = s$ and are therefore:

$$
v_{ox} = s\dot{\psi}\sin\theta\cos\phi \quad v_{oy} = -s\dot{\psi}\sin\theta\sin\phi \quad v_{oz} = 0 \tag{2.13}
$$

For a mass particle m', located at x, y, z in the X, Y, Z coordinate system (on the disk D), the velocities v_x, v_y, v_z may be determined as in Eq. 2.1 as follows:

$$
v_x = v_{ox} + (\omega_y z - \omega_z y), \quad v_y = v_{oy} + (\omega_z x - \omega_x z), \quad v_z = v_{oz} + (\omega_x y - \omega_y x)
$$

$$
v_{ox} = s\dot{\psi}\sin\theta\cos\phi \quad v_{oy} = -s\dot{\psi}\sin\theta\sin\phi \quad v_{oz} = 0
$$

$$
\omega_x = \dot{\psi}\sin\theta\sin\phi, \quad \omega_y = \dot{\psi}\sin\theta\cos\phi, \quad \omega_z = \dot{\phi} + \dot{\psi}\cos\theta
$$

$$
x = 0, \quad y = 0, \quad z = s \Rightarrow
$$

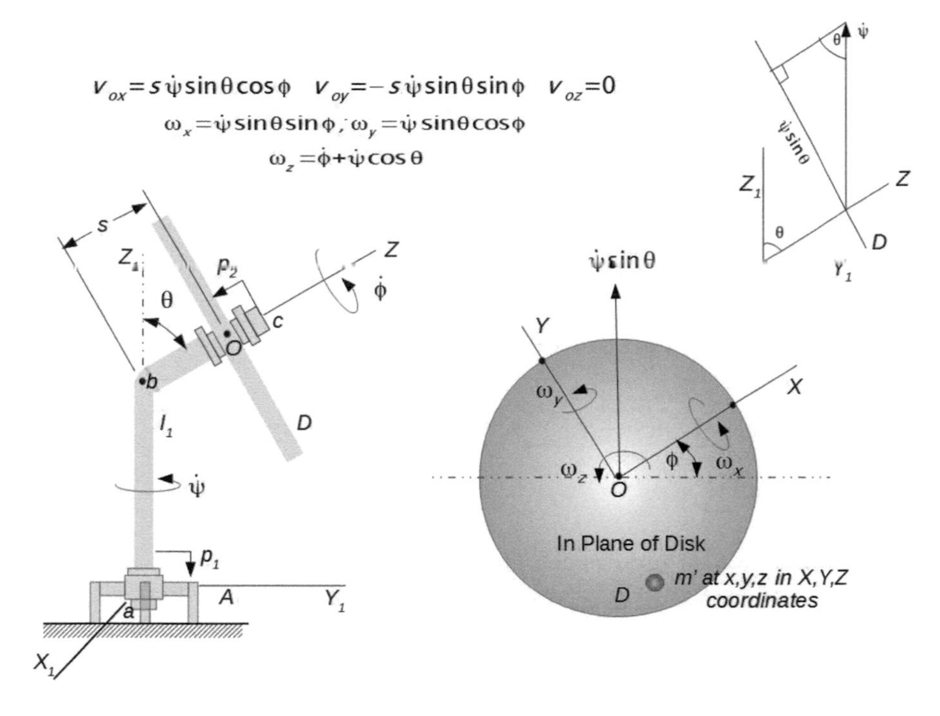

Fig. 2.4 Definition of velocities v_{ox}, v_{oy}

$$
\begin{aligned}
v_x &= s\dot{\psi}\sin\theta\cos\phi + z\dot{\psi}\sin\theta\cos\phi - y(\dot{\phi} + \dot{\psi}\cos\theta) \\
v_y &= -s\dot{\psi}\sin\theta\sin\phi + x(\dot{\phi} + \dot{\psi}\cos\theta) - z\dot{\psi}\sin\theta s \in \phi \\
v_z &= y\dot{\psi}\sin\theta\sin\phi - x\dot{\psi}\sin\theta\cos\phi
\end{aligned}
\tag{2.14}
$$

2.3 Most General Form of Kinetic Energy

When interpreted as in Eq. 2.1 above (repeated here for convenience), the preceding equations for v_x, v_y, v_z express the velocity of any particle in a rigid body (see Fig. 2.1), relative to inertial coordinates, but expressed in the components of the rotating and translating coordinates X, Y, Z.

$$v_x = v_{ox} + (\omega_y z - \omega_z y), \quad v_y = v_{oy} + (\omega_z x - \omega_x z), \quad v_z = v_{oz} + (\omega_x y - \omega_y x)$$

It thus turns out that a general expression for the kinetic energy T is obtained by inserting these relations into $T = 1/2\sum m'(v_x^2 + v_y^2 + v_z^2)$, where m' is an elemental mass, , and the summation takes place over all of the elemental mass particles which make up the rigid body.

On collecting terms, we have:

$$
\begin{aligned}
T = &\frac{1}{2}M \underbrace{(v_{ox}^2 + v_{oy}^2 + v_{oz}^2)}_{v_o^2} + \frac{1}{2}\omega_x^2 \sum m'(y^2 + z^2) \\
&+ \frac{1}{2}\omega_y^2 \sum m'(x^2 + z^2) \\
&+ \frac{1}{2}\omega_z^2 \sum m'(x^2 + y^2) \\
&+ v_{ox}\left(\omega_y \sum m'z - \omega_z \sum m'y\right) + v_{oy}\left(\omega_z \sum m'x - \omega_x \sum m'z\right) \\
&+ v_{oz}\left(\omega_x \sum m'y - \omega_y \sum m'x\right) - \omega_x\omega_y \sum m'xy - \omega_x\omega_z \sum m'xz \\
&- \omega_y\omega_z \sum m'yz
\end{aligned} \tag{2.15}
$$

where M is the total mass of the body, m' is an elemental mass particle and the summations are carried out over all of the elemental masses. Replacing the summations with integrals and the elemental mass m' with an infinitesimal mass dm, the kinetic energy T may be shown to be equivalent to:

$$
\begin{aligned}
T = &\frac{1}{2}Mv_o^2 + \frac{1}{2}\left[\omega_x^2 I_x + \omega_y^2 I_y + \omega_z^2 I_z - 2\omega_x\omega_y I_{xy} - 2\omega_x\omega_z I_{xz} - 2\omega_y\omega_z I_{yz}\right] \\
&+M\left[v_{ox}(\omega_y\overline{z} - \omega_z\overline{y}) + v_{oy}(\omega_z\overline{x} - \omega_x\overline{z}) + v_{oz}(\omega_x\overline{y} - \omega_y\overline{x})\right]
\end{aligned} \tag{2.16}
$$

where $\overline{x}, \overline{y}, \overline{z}$, the distance from the origin to the centers of gravity of the rigid body are: $\overline{x} = \int xdm/M, \overline{y} = \int ydm/M, \overline{z} = \int zdm/M$, the moments of inertia I_x, I_y, I_z are, respectively: $I_x = \int(y^2+z^2)dm,\ \ I_y = \int(x^2+z^2)dm,\ \ I_z = \int(x^2+y^2)dm$ and where the products of inertia I_{xy}, I_{xz}, I_{yz} are: $I_{xy} = \int xydm,\ \ I_{xz} = \int xzdm,\ \ I_{yz} = \int yzdm$. It may be shown that:

$$
\left[\omega_x^2 I_x + \omega_y^2 I_y + \omega_z^2 I_z - 2\omega_x\omega_y I_{xy} - 2\omega_x\omega_z I_{xz} - 2\omega_y\omega_z I_{yz}\right] = \omega^T I\omega
$$

2.4 Summary: Important Points Regarding Kinetic Energy (See Dare Wells [51, pp. 148–149])

a. It was previously noted that ω is the angular velocity of the body *relative to inertial coordinates* and v_o is the linear velocity of the origin O of the X, Y, Z coordinate system with respect to an inertial coordinate system. The angular velocity terms $\omega_x, \omega_y, \omega_z$, and the velocity components v_{ox}, v_{oy}, v_{oz}, of the origin O are *expressed along the instantaneous directions* of X, Y, Z.

b. The moment and product of inertia terms I_x, I_{xy}, etc., and the center of gravity components $\overline{x}, \overline{y}, \overline{z}$, must be calculated with respect to the instantaneous rotating and translating X, Y, Z axes (see Fig. 2.1, page 14).
c. With the angular and linear velocity vectors ω, v_o, respectively, and the moments of inertia, resolved as stated above, the kinetic energy expression 2.16 is valid for the X, Y, Z axis frame, *whether it is body-fixed or performing any arbitrary angular motion about O relative to the rigid body*. This also includes the case where X, Y, Z may be rigidly attached to the body or fixed in direction. The underlying assumption is that in all cases, the origin O is attached to the body.
d. When the X, Y, Z frame rotates relative to the body, x, y, z in 2.16 vary as do I_x, I_{xy}, etc., as well as $\overline{x}, \overline{y}, \overline{z}$. These inertial variables vary with the motion.
e. The X, Y, Z frame may be body-fixed, which implies that the inertial terms I_x, I_{xy}, etc., and $\overline{x}, \overline{y}, \overline{z}$ are constant. Hence body-fixed axes are desirable and are almost always used. In any event, statements made under paragraph (a) must be borne in mind.
f. Under certain conditions, for example, with O at the center of mass, Eq. 2.16 can be significantly simplified. This condition implies that $\overline{x} = \overline{y} = \overline{z} = 0$ and the terms of T which include $\overline{x}, \overline{y}, \overline{z}$ become zero. The coordinate frame need not be rigidly fastened to the body (except at O).
g. If *any point* in the body is fixed in relation to an inertial frame and O is located at this point, $v_{ox} = v_{oy}, = v_{oz} = 0$ and all the terms with v_{ox}, v_{oy}, v_{oz} are zero.
h. If O is at the body's center of mass and body-fixed X, Y, Z axes are defined to be along the body's principal axes of inertia, where $I_{xy} = I_{xz} = I_{yz} = 0$, then the kinetic energy T becomes: $T = \frac{1}{2}Mv_{\text{c.m.}}^2 + \frac{1}{2}(I_x\omega_x^2 + I_y\omega_y^2 + I_z\omega_z^2)$ where I_x, I_y, and I_z are constants.

2.5 Examples: Kinetic Energy and Equations of Motion

In the following group of examples *body-fixed axes have been employed throughout*. This is in general the most convenient procedure.

Example 3 [51, p. 150]

Three examples of a physical pendulum, consisting of a thin plate of sheet metal (lamina), pivoted at p and free to swing vertically through angle θ, are shown in Fig. 2.5. Expressions for the kinetic energy T, with the origin at three different points on the lamina, will be used to better understand Eq. 2.16.

1. Axes X, Y, Z are located as in Fig. 2.5a, with the origin at the pivot point p. The Z axis is normal to the plane of the paper. From this setup, it is apparent that $\omega_x = \omega_y = 0$, $\omega_z = d\theta/dt$. Since the origin is stationary, $v_{ox} = v_{oy} = v_{oz} = 0$. Hence Eq. 2.16 becomes: $T = \frac{1}{2}I_z\dot{\theta}^2$ as anticipated. I_z is taken around the Z axis at the pivot point.

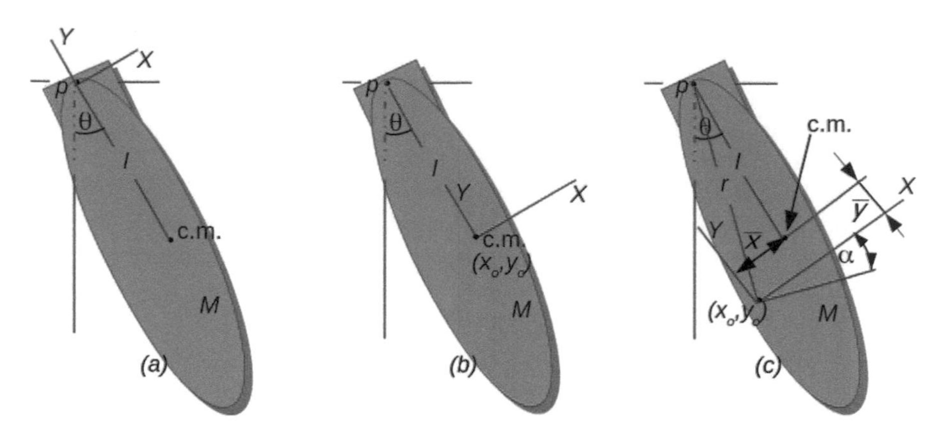

Fig. 2.5 Physical pendulum consisting of a lamina pivoted at p with the origin of the X, Y coordinates defined at three different locations

2. In Fig. 2.5b the origin is at the center of mass point, designated as c.m. Once more, $\omega_x = \omega_y = 0, \omega_z = \dot{\theta}$. Thus $v_{ox} = l\dot{\theta}, v_{oy} = v_{oz} = 0, \overline{x} = \overline{y} = \overline{z} = 0$. The velocity v_{ox} is derived from the expression $v = \omega \times R$, where $\omega = d\theta/dt$ and $R = l$. Hence from Eq. 2.16, $T = \frac{1}{2}Ml^2\dot{\theta}^2 + \frac{1}{2}I_z\dot{\theta}^2$. I_z is taken around the Z axis at the c.m. point in this case. Notice that the term $\frac{1}{2}Ml^2$ has been added to the moment of inertia I_z and this is just a restatement of the parallel axis theorem.
3. In Fig. 2.5c the axes are more generally oriented. As in the preceding two cases, $\omega_x = \omega_y = 0, \omega_z = \dot{\theta}$. However $v_{ox} = r\dot{\theta}\cos\alpha, v_{oy} = -r\dot{\theta}\sin\alpha, v_{oz} = 0$ and $\overline{x}, \overline{y}$ are the distances along the X and Y axes, respectively, from the new origin O to the c.m. point. Thus, from Eq. 2.16, the kinetic energy T is:

$$T = \frac{1}{2}Mr^2\dot{\theta}^2 + \frac{1}{2}I_z\dot{\theta}^2 - Mr\dot{\theta}^2(\overline{x}\sin\alpha + \overline{y}\cos\alpha)$$

where I_z is now about the Z axis at the instant considered. Note that the I_z appearing in situations 1, 2, and 3 are different in each case.

As an aside, the equations of motion for case (3), using the d'Alembert–Lagrange formulation will now be calculated. A detailed derivation of the d'Alembert–Lagrange equations will be carried out in the sequel. In each of the above cases $F_\theta = -Mgl\sin\theta$. The equation for potential energy is:

$$F_\theta = -\frac{\partial V}{\partial \theta} = -Mgl\sin\theta \Rightarrow V = \int Mgl\sin\theta d\theta = -Mgl\cos\theta \tag{2.17}$$

For case (3) The Lagrangian becomes:

$$\mathcal{L} = T - V = \frac{1}{2}Mr^2\dot{\theta}^2 + \frac{1}{2}I_z\dot{\theta}^2 - Mr\dot{\theta}^2(\overline{x}\sin\alpha + \overline{y}\cos\alpha) + Mgl\cos\theta = 0 \tag{2.18}$$

Hence the equations of motion for case (3) are:

$$
\begin{aligned}
\frac{d}{dt}\left(\frac{\partial \mathcal{L}}{\partial \dot{\theta}}\right) - \frac{\partial \mathcal{L}}{\partial \theta} &= 0; \quad -\frac{\partial \mathcal{L}}{\partial \theta} = Mgl\sin\theta \\
\frac{\partial \mathcal{L}}{\partial \dot{\theta}} &= Mr^2\dot{\theta} + I_z\dot{\theta} - 2Mr\dot{\theta}(\overline{x}\sin\alpha + \overline{y}\cos\alpha) \\
\frac{d}{dt}\left(\frac{\partial \mathcal{L}}{\partial \dot{\theta}}\right) &= Mr^2\ddot{\theta} + I_z\ddot{\theta} - 2Mr\ddot{\theta}(\overline{x}\sin\alpha + \overline{y}\cos\alpha) \\
\frac{d}{dt}\left(\frac{\partial \mathcal{L}}{\partial \dot{\theta}}\right) - \frac{\partial \mathcal{L}}{\partial \theta} &= Mr^2\ddot{\theta} + I_z\ddot{\theta} - 2Mr\ddot{\theta}(\overline{x}\sin\alpha + \overline{y}\cos\alpha) + Mgl\sin\theta = 0 \\
&\Rightarrow \ddot{\theta}\left\{Mr^2 + I_z - 2Mr(\overline{x}\sin\alpha + \overline{y}\cos\alpha)\right\} = -Mgl\sin\theta
\end{aligned}
\tag{2.19}
$$

Example 4 [51, pp. 150–151]

The following example will illustrate the calculation of kinetic energy, as well as the use of the d'Alembert–Lagrange formulation for the derivation of the equations of motion.

The marble slab, of Fig. 2.6, is free to translate and rotate in the $X_1 - Y_1$ plane under known, given forces F_1, F_2. The origin O is located at any arbitrary point and axes X, Y, Z are attached to the lamina. The variables x, y, θ may be shown

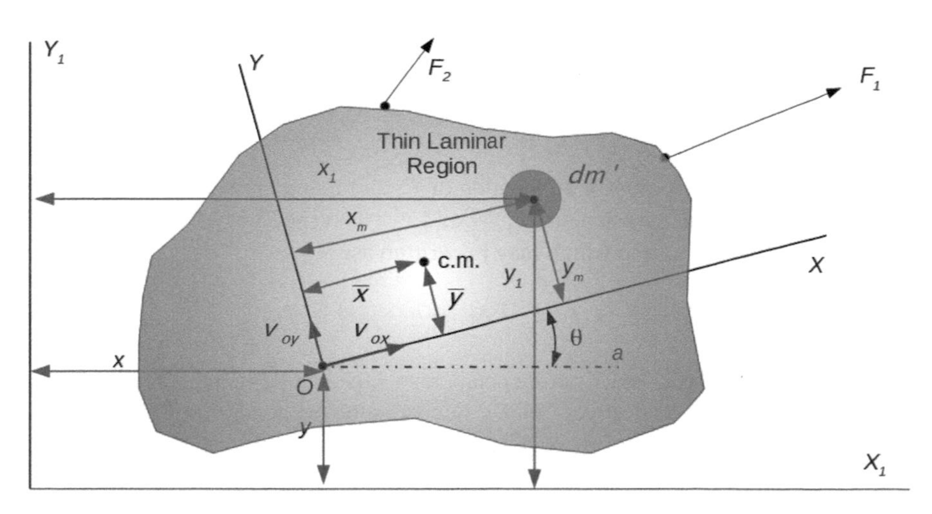

Fig. 2.6 Lamina with two applied forces—Lagrange's equations

to be suitable independent generalized coordinates, thus resulting in a 3 *d*egrees *o*f *f*reedom system. The angular velocities are: $\omega_x = \omega_y = 0$, $\omega_z = \dot{\theta}$, while the linear velocities of the origin are: $v_{ox} = \dot{x}\cos\theta + \dot{y}\sin\theta$, $v_{oy} = \dot{y}\cos\theta - \dot{x}\sin\theta$, $v_{oz} = 0$. The velocities v_{ox}, v_{oy} are components of the velocity of O which are projected onto the instantaneous positions of X and Y. Hence Eq. 2.16 results in:

$$T = \frac{1}{2}M(\dot{x}^2 + \dot{y}^2) + \frac{1}{2}I_z\dot{\theta}^2 + M\dot{\theta}[(\dot{y}\overline{x} - \dot{x}\overline{y})\cos\theta - (\dot{x}\overline{x} + \dot{y}\overline{y})\sin\theta)] \tag{2.20}$$

The equations of motion corresponding to x, y, θ are easily calculated. For example, the θ equation is derived as follows:

$$\begin{aligned}
\frac{d}{dt}\left(\frac{\partial T}{\partial \dot{\theta}}\right) - \frac{\partial T}{\partial \theta} &= F_\theta \\
\frac{\partial T}{\partial \dot{\theta}} &= I_z\dot{\theta} + M[(\dot{y}\overline{x} - \dot{x}\overline{y})\cos\theta - (\dot{x}\overline{x} + \dot{y}\overline{y})\sin\theta)] \\
-\frac{\partial T}{\partial \theta} &= M\dot{\theta}[(\dot{y}\overline{x} - \dot{x}\overline{y})\sin\theta + (\dot{x}\overline{x} + \dot{y}\overline{y})\cos\theta)] \\
\Rightarrow \frac{d}{dt}\left(\frac{\partial T}{\partial \dot{\theta}}\right) &= I_z\ddot{\theta} + M[(\ddot{y}\overline{x} - \ddot{x}\overline{y})\cos\theta - (\ddot{x}\overline{x} + \ddot{y}\overline{y})\sin\theta)] \\
&\quad - M[\dot{\theta}(\dot{y}\overline{x} - \dot{x}\overline{y})\sin\theta + \dot{\theta}(\dot{x}\overline{x} + \dot{y}\overline{y})\cos\theta] \\
\Rightarrow \frac{d}{dt}\left(\frac{\partial T}{\partial \dot{\theta}}\right) - \frac{\partial T}{\partial \theta} &= I_z\ddot{\theta} + M[(\ddot{y}\overline{x} - \ddot{x}\overline{y})\cos\theta - (\ddot{x}\overline{x} + \ddot{y}\overline{y})\sin\theta)] \\
&\quad - M[\dot{\theta}(\dot{y}\overline{x} + \dot{x}\overline{y})\sin\theta + \dot{\theta}(\dot{x}\overline{x} + \dot{y}\overline{y})\cos\theta] \\
&\quad + M\dot{\theta}[(\dot{y}\overline{x} - \dot{x}\overline{y})\sin\theta + (\dot{x}\overline{x} + \dot{y}\overline{y})\cos\theta)] \\
&= I_z\ddot{\theta} + M[(\ddot{y}\overline{x} - \ddot{x}\overline{y})\cos\theta - (\ddot{x}\overline{x} + \ddot{y}\overline{y})\sin\theta)] = F_\theta
\end{aligned} \tag{2.21}$$

The coordinates of the points of application of F_1 and F_2 on the lamina with respect to the X, Y, Z coordinate system are x_1', y_1' and x_2', y_2', respectively. The generalized force F_θ is actually a torque which may be written as:

$$F_\theta = \tau_\theta = f_{1y}x_1' - f_{1x}y_1' + f_{2y}x_2' - f_{2x}y_2' \tag{2.22}$$

where f_{1x}, f_{1y} are the X and Y components, respectively, of F_1 and f_{2x}, f_{2y} are the X and Y components, respectively, of F_2. Accordingly, the generalized forces in the X and Y directions are $F_x = f_{1x} + f_{2x}$, $F_y = f_{1y} + f_{2y}$.

Example 5 [51, pp. 151–152]

The sheet metal lamina depicted in Fig. 2.7(1) is hung up by a string of constant length r and is free to swing as a "double pendulum" in a vertical plane.

1. (a) The center of mass is offset from the origin by $\overline{x}, \overline{y}$ with respect to the X, Y, Z coordinate frame.
   (b) With body-fixed axes X, Y, Z as shown in Fig. 2.7(1), ϕ and θ may be shown to be a set of suitable generalized coordinates.
   (c) As is apparent, the angular rates in the X, Y, Z coordinate system are: $\omega_x = \omega_y = 0$
   (d) $\omega_z = \dot{\phi}$, $(\omega_z \neq \dot{\phi} + \dot{\theta})$. Since ϕ is an independent generalized coordinate, and $\dot{\phi}$ is measured relative to the vertical about the Z axis, then the angular rate about the Z axis is simply $\omega_z = \dot{\phi}$. The angle ϕ is not measured relative to the angle θ.
   (e) The velocity at the origin O is: $v_o = r\dot{\theta}$. The components of v_o along the X and Y axes are v_{ox} and v_{oy}, respectively, and they are: $v_{ox} = r\dot{\theta}\sin(\phi - \theta)$; $v_{oy} = r\dot{\theta}\cos(\phi - \theta)$.

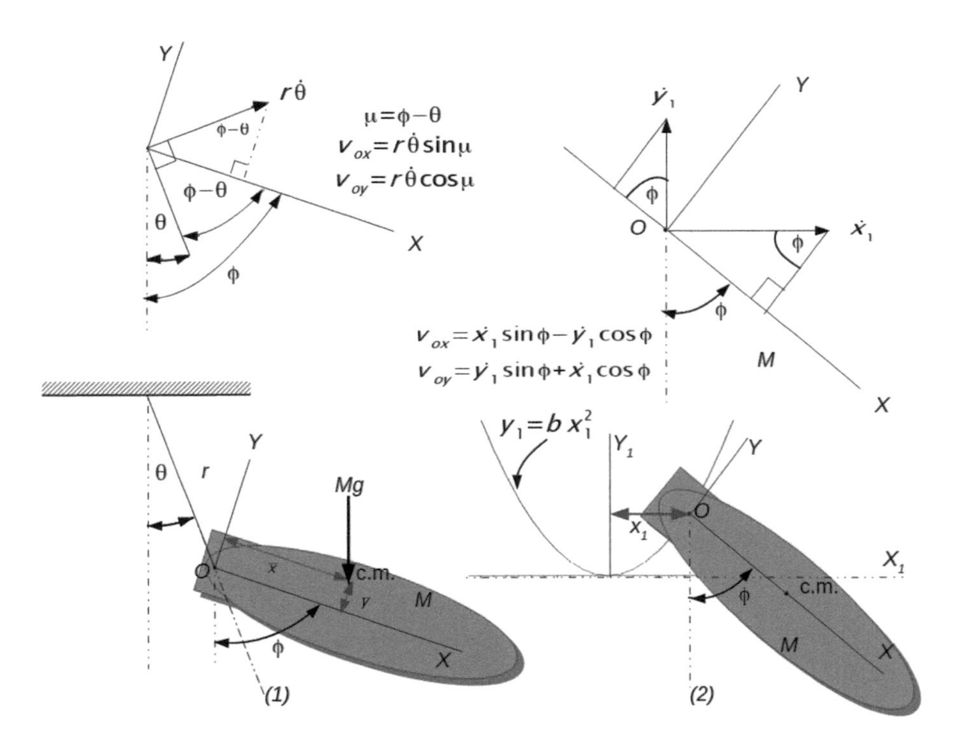

Fig. 2.7 Double "pendulum"

(f) The kinetic energy is calculated by means of Eq. 2.16, repeated below for convenience:

$$T = \frac{1}{2}Mv_o^2 + \frac{1}{2}\Big[\omega_x^2 I_x + \omega_y^2 I_y + \omega_z^2 I_z - 2\omega_x\omega_y I_{xy} - 2\omega_x\omega_z I_{xz} - 2\omega_y\omega_z I_{yz}\Big] + M\left[v_{ox}(\omega_y\overline{z} - \omega_z\overline{y}) + v_{oy}(\omega_z\overline{x} - \omega_x\overline{z}) + v_{oz}(\omega_x\overline{y} - \omega_y\overline{x})\right] \tag{2.23}$$

The kinetic energy then becomes:

$$T = \frac{1}{2}Mr^2\dot{\theta}^2 + \frac{1}{2}I_z\dot{\phi}^2 + Mr\dot{\phi}\dot{\theta}[\overline{x}\cos(\phi - \theta) - \overline{y}\sin(\phi - \theta)] \tag{2.24}$$

Note that I_z, in this example, is calculated with respect to the origin O of the X, Y, Z coordinate frame. I_z could also have been calculated with respect to the center of mass of the lamina, and it would then not have been necessary to include the $\bar{x}$, $\bar{y}$ terms in the resulting kinetic energy expression (see Nielsen et al. [16, p. 4] and Example 8 on page 113).

(g) Expressions for F_θ and F_ϕ are obtained as follows (see Fig. 2.8):

$$\begin{aligned} V &= -Mgh; \quad h = r\cos\theta - \overline{y}\sin\phi + \overline{x}\cos\phi \\ \Rightarrow V &= -Mg(r\cos\theta - \overline{y}\sin\phi + \overline{x}\cos\phi) \\ F_\theta &= -\frac{\partial V}{\partial\theta} = -Mgr\sin\theta \\ F_\phi &= -\frac{\partial V}{\partial\phi} = -Mg(\overline{x}\sin\phi + \overline{y}\cos\phi) \end{aligned} \tag{2.25}$$

2. Suppose that point O on the slab can slide along the smooth parabolic line $y_1 = bx_1^2$ as shown in Fig. 2.7(2). Then, the velocities along the X and Y axes, v_{ox} and v_{oy}, respectively, are (note that $v_{oz} = 0$):

$$\begin{aligned} v_o^2 &= \dot{x}_1^2 + \dot{y}_1^2; \quad v_{ox} = \dot{x}_1\sin\phi - \dot{y}_1\cos\phi \\ v_{oy} &= \dot{y}_1\sin\phi + \dot{x}_1\cos\phi; \quad v_{oz} = 0 \end{aligned} \tag{2.26}$$

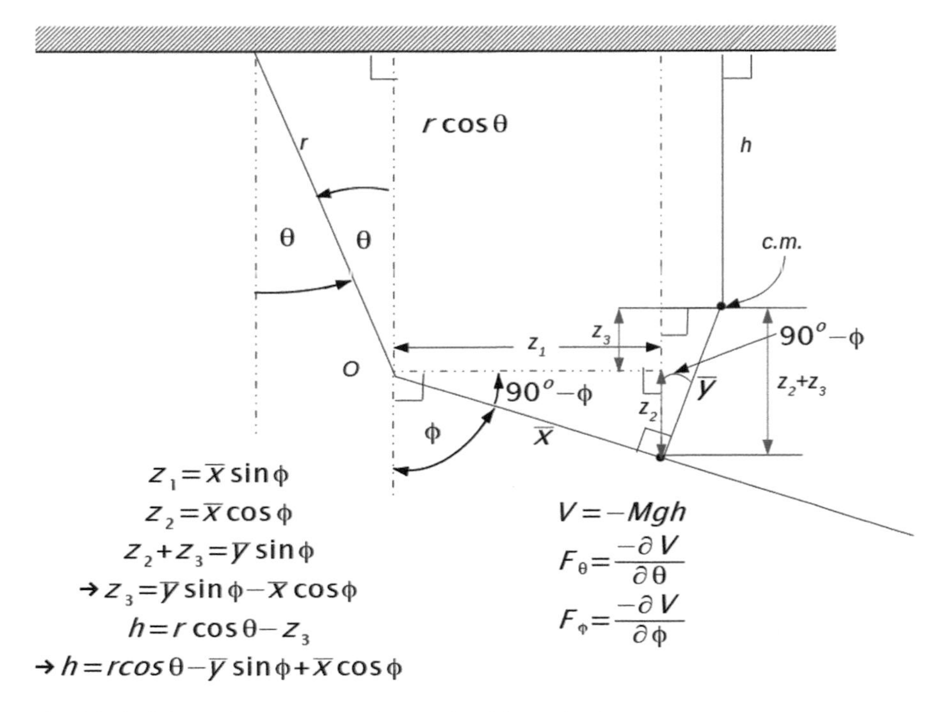

Fig. 2.8 Definition of potential V and generalized forces F_θ, F_ϕ

Since the velocity at point O is: $v_o^2 = \dot{x}_1^2 + \dot{y}_1^2$, and the angular velocities are: $\omega_x = 0, \omega_y = 0, \omega_z = \dot{\phi}$, Eq. 2.16, repeated below becomes:

$$
\begin{aligned}
T &= \frac{1}{2}Mv_o^2 \\
&+ \frac{1}{2}\left[\omega_x^2 I_x + \omega_y^2 I_y + \omega_z^2 I_z - 2\omega_x\omega_y I_{xy} - 2\omega_x\omega_z I_{xz} - 2\omega_y\omega_z I_{yz}\right] \\
&+ M\left[v_{ox}(\omega_y\overline{z} - \omega_z\overline{y}) + v_{oy}(\omega_z\overline{x} - \omega_x\overline{z}) + v_{oz}(\omega_x\overline{y} - \omega_y\overline{x})\right] \\
&\Rightarrow T = \frac{1}{2}M(\dot{x}_1^2 + \dot{y}_1^2) + \frac{1}{2}\dot{\phi}^2 I_z + M\dot{\phi}\left[v_{oy}\overline{x} - v_{ox}\overline{y}\right] \\
&= \frac{1}{2}M(\dot{x}_1^2 + \dot{y}_1^2) + \frac{1}{2}\dot{\phi}^2 I_z \\
&+ M\dot{\phi}\left[(\dot{y}_1\sin\phi + \dot{x}_1\cos\phi)\overline{x} - (\dot{x}_1\sin\phi - \dot{y}_1\cos\phi)\overline{y}\right] \qquad (2.27)
\end{aligned}
$$

Example 6 [51, p. 152]

A slender rod whose linear mass distribution is ρ per unit length and whose total length is L (see Fig. 2.9) rotates freely with an angle θ_2 in the bearing at O about a

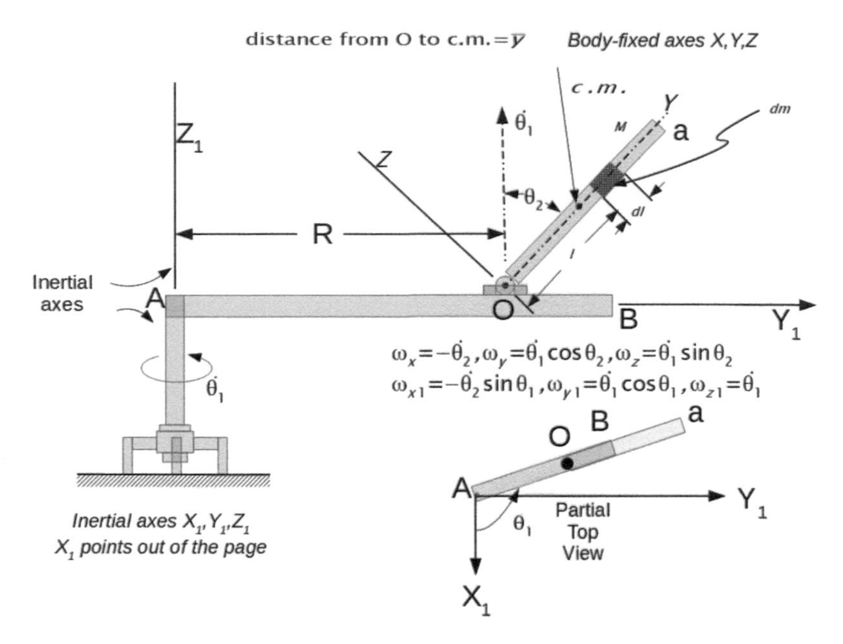

Fig. 2.9 Rotating slender rod

horizontal axis. The bearing at O is attached to the horizontal arm $A-B$, which also rotates with angular rate $d\theta_1/dt$ about the Z_1 axis, through an angle θ_1. The kinetic energy of the rod should be evaluated by integration and the results compared to the following relation:

$$T = \frac{1}{2}Mv_o^2 + \frac{1}{2}\left[\omega_x^2 I_x + \omega_y^2 I_y + \omega_z^2 I_z - 2\omega_x\omega_y I_{xy} - 2\omega_x\omega_z I_{xz} - 2\omega_y\omega_z I_{yz}\right] \\ + M\left[v_{ox}(\omega_y\overline{z} - \omega_z\overline{y}) + v_{oy}(\omega_z\overline{x} - \omega_x\overline{z}) + v_{oz}(\omega_x\overline{y} - \omega_y\overline{x})\right] \tag{2.28}$$

where $\overline{x}$, $\overline{y}$, $\overline{z}$, the centers of gravity of the rigid body are: $\overline{x} = \int xdm/M, \quad \overline{y} = \int ydm/M, \quad \overline{z} = \int zdm/M$.

1. Within the context of this problem, the total mass of the slender rod is $M = \int_0^L \rho dl$. Additionally, the moments of inertia of the slender rod about the instantaneous Z and X axes, respectively, with the origin at O, are evaluated to be: $I_z = I_x = \int_0^L \rho l^2 dl$.
2. From the definition of $\overline{y}$, we have: $M\overline{y} = \int_0^L ydm = \int_0^L \rho l dl$
3. The velocities in the X direction are: $-R\dot{\theta}_1$ and $-\dot{\theta}_1 l \sin\theta_2$, and the velocity in the Z direction is: $l\dot{\theta}_2$. The inertial velocity of point O in X, Y, Z coordinates is: $v_{ox} = -R\dot{\theta}_1, \quad v_{oy} = v_{oz} = 0$.

4. In addition $R\dot{\theta}_1^2 \sin\theta_2 \int_0^L \rho l dl = R\dot{\theta}_1^2 \sin\theta_2 M\overline{y}$. Furthermore, since the angular velocity $\dot{\theta}_2$ takes place around the X axis, the term $\frac{1}{2}\dot{\theta}_2^2 \int_0^L \rho l^2 dl$ becomes $\frac{1}{2}I_x\dot{\theta}_2^2$. Similarly the angular rotation around the Z axis is $\dot{\theta}_1 \sin\theta_2$ and hence the term $\frac{1}{2}\dot{\theta}_1^2 \sin^2\theta_2 \int_0^L \rho l^2 dl$ may be written as $\frac{1}{2}I_z\dot{\theta}_1^2 \sin^2\theta_2$. And finally $\frac{1}{2}R^2\dot{\theta}_1^2 \int_0^L \rho dl = \frac{1}{2}MR^2\dot{\theta}_1^2$.
5. The defining equation $T = 1/2 \int v^2 dm$ can therefore be written as:

$$
\begin{aligned}
T &= \frac{1}{2}\int_0^L [(R + l\sin\theta_2)^2\dot{\theta}_1^2 + l^2\dot{\theta}_2^2]\rho dl \\
\Rightarrow T &= \frac{1}{2}R^2\dot{\theta}_1^2 \int_0^L \rho dl + \frac{1}{2}\dot{\theta}_1^2 \sin^2\theta_2 \int_0^L \rho l^2 dl \\
&\quad + R\dot{\theta}_1^2 \sin\theta_2 \int_0^L \rho l dl + \frac{1}{2}\dot{\theta}_2^2 \int_0^L \rho l^2 dl \\
\Rightarrow T &= \frac{1}{2}MR^2\dot{\theta}_1^2 + \frac{1}{2}I_z\dot{\theta}_1^2 \sin^2\theta_2 + M\overline{y}R\dot{\theta}_1^2 \sin\theta_2 + \frac{1}{2}I_x\dot{\theta}_2^2
\end{aligned} \tag{2.29}
$$

6. Using the formula in Eq. 2.28, where $v_{ox} = -R\dot{\theta}_1$; $v_{oy} = 0$; $v_{oz} = 0$; $v_o = v_{ox}$, $\omega_x = -\dot{\theta}_2$; $\omega_y = \dot{\theta}_1 \cos\theta_2$; $\omega_z = \dot{\theta}_1 \sin\theta_2$, $I_{xy} = I_{xz} = I_{yz} = I_y = 0$ and $\overline{x} = 0$; $\overline{y} \neq 0$; $\overline{z} = 0$, the equation becomes:

$$
\begin{aligned}
T &= \frac{1}{2}Mv_o^2 + \frac{1}{2}\left[\omega_x^2 I_x + \omega_y^2 I_y + \omega_z^2 I_z - 2\omega_x\omega_y I_{xy} - 2\omega_x\omega_z I_{xz} - 2\omega_y\omega_z I_{yz}\right] \\
&\quad + M\left[v_{ox}(\omega_y\overline{z} - \omega_z\overline{y}) + v_{oy}(\omega_z\overline{x} - \omega_x\overline{z}) + v_{oz}(\omega_x\overline{y} - \omega_y\overline{x})\right] \\
&= \frac{1}{2}Mv_o^2 + \frac{1}{2}\left[\omega_x^2 I_x + \omega_z^2 I_z\right] + M\left[v_{ox}(\omega_y\overline{z} - \omega_z\overline{y}) + v_{oz}(\omega_x\overline{y} - \omega_y\overline{x})\right] \\
&= \frac{1}{2}MR^2\dot{\theta}_1^2 + \frac{1}{2}[\dot{\theta}_2^2 I_x + \dot{\theta}_1^2 \sin^2\theta_2 I_z] + MR\overline{y}\dot{\theta}_1^2 \sin\theta_2
\end{aligned} \tag{2.30}
$$

2.6 Notation System Used in This Book

Lecture on Notation Systems: Video Times—00:00–06:01

Two notation systems were used in the 2-003SC course on Engineering Dynamics, given at MIT (see Vandiver [48]), in order to specify position vectors, velocity vectors, and vectors of any kind that might be associated with translating and rotating reference frames (see Fig. 2.10).

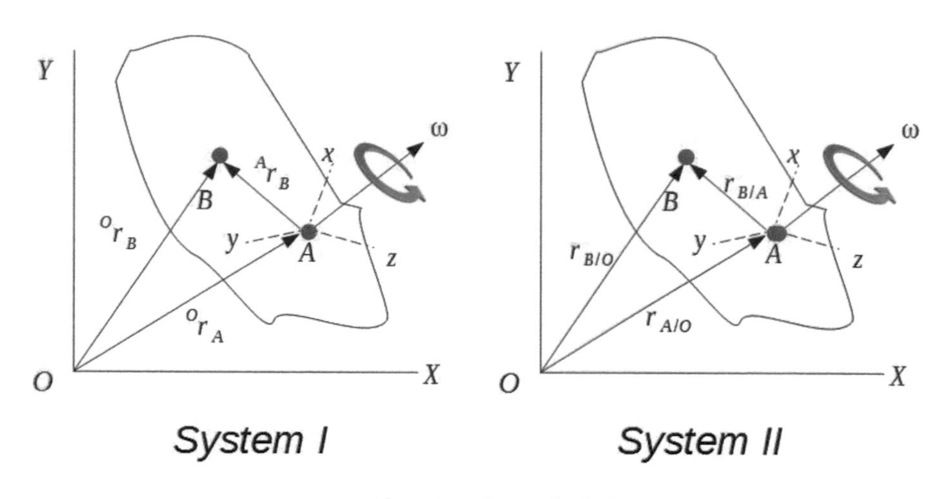

Fig. 2.10 Illustration of notation systems

Figure contains a rigid body, and a reference frame, the $Axyz$ reference frame with origin at A, which is attached to and moves with the rigid body. The whole system is translating and rotating in an inertial frame O with coordinates X, Y, Z. We need to be able to describe the position and the velocity of this rigid body, and any point on this rigid body, say point B, which might actually be moving with respect to the rigid body. The position vector from the origin O to the $Axyz$ reference frame in notation system I is designated as ${}^{O}R_A$. The O superscript that precedes the R signifies that the vector is in the inertial coordinate system or is measured with respect to the inertial coordinate system. Similarly point B is labeled as ${}^{O}R_B$ and signifies the vector from the origin of the inertial coordinate system to the point B on the rigid body. The vector from point A on the rigid body to point B is labeled as ${}^{A}R_B$ and signifies the position of B with respect to point A on the rigid body. This notation is the superscript version of the vector designation scheme.

Notation System II (which will be used henceforth) uses the subscript format, where for instance, the position vector from the origin O to the $Axyz$ reference frame in System II would be: $R_{A/O}$, which implies that "with respect to O," is just the "/O" symbol. Similarly, in this notation system, point B with respect to O would be $R_{B/O}$. Finally, the vector from point A to point B is $R_{B/A}$. So the two notation systems are exactly equivalent. When the context is clear, the subscript format "with respect to O," is just the "O" symbol, for example, the moment of inertia with respect to O may be written as I_O. This method is the preferred one and will be used frequently, although not exclusively, throughout the text.

It would be desirable to be able to take the time derivative of the vector at point B and use it to derive expressions for velocities in a rotating and translating frame.

The position vector $R_{B/O}$ is simply a vector sum which may be written as: $R_{B/O} = R_{A/O} + R_{B/A}$. The time derivatives of $R_{B/O}$ in both notation systems are as follows:

$$v_{B/O} = \frac{dR_{B/O}}{dt} = \underbrace{\frac{dR_{A/O}}{dt}}_{v_{A/O}} + \frac{d}{dt}\left(R_{B/A}\right)_{/O}$$

$$= v_{A/O} + \left(\frac{\partial R_{B/A}}{\partial t}\right)_{/A} + \omega_{/O} \times R_{B/A}$$

$$v_{B/O} = \frac{dR_{B/O}}{dt} = \underbrace{\frac{dR_{A/O}}{dt}}_{v_{A/O}} + \frac{d}{dt}\left(R_{B/A}\right)_{O}$$

$$= v_{A/O} + \left(\frac{\partial R_{B/A}}{\partial t}\right)_{A} + \omega_{O} \times R_{B/A}$$

$$\frac{d^{O}R_B}{dt} = {}^{O}v_B = {}^{O}v_A + \frac{{}^{O}d}{dt}\left({}^{A}R_B\right) = {}^{O}v_A + \frac{{}^{A}\partial}{\partial t}\left({}^{A}R_B\right) + {}^{O}\omega \times {}^{A}R_B \tag{2.31}$$

where $\omega_{/O}$, ω_O, and ${}^{O}\omega$ all signify the rotation of the rigid body with respect to O. The term $\left(\frac{\partial R_{B/A}}{\partial t}\right)_{/A}$ is used to determine if there are any changes in the vector within the rotating $Axyz$ frame. It is as though an observer was situated on the rotating $Axyz$ coordinate frame and was measuring any changes over time of the $R_{B/A}$ vector. The difference between system I and system II notation lies in the fact that the $/O$ subscripts become superscripts. The term ${}^{O}\omega \times {}^{A}R_B$ is the contribution to the velocity, as seen in the inertial frame and caused by the rotation of the rigid body.

2.7 Angular Momentum of a Mass Particle

Lecture 5: Video Times—35:35–55:03

The material in this section was adapted from the MIT Lecture Series on Dynamics (see Vandiver 2.003SC Engineering Dynamics. - Video of Lecture 5: Angular Momentum [45]).

Figure 2.11 depicts a particle with mass m within an inertial coordinate system $OXYZ$. The particle is located at point B and has a total force acting upon it of $F_B = \sum f_B$. The vector from the origin O of the inertial coordinate system to point B is $r_{B/O}$. The particle is traveling with a velocity of v_B at some instant of time and therefore has a linear momentum with respect to the inertial coordinates of $P_{B/O} = mv_{B/O}$.

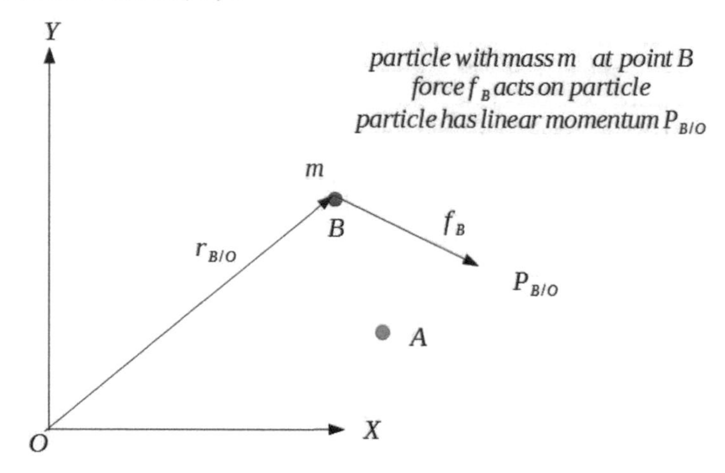

Fig. 2.11 Illustration of linear momentum $P_{B/O}$

The definition of angular momentum for the particle is:

$$h_{B/O} = r_{B/O} \times P_{B/O} = r_{B/O} \times m v_{B/O} \tag{2.32}$$

Lowercase h is used for particles and uppercase H indicates a rigid body. The derivative with respect to time of the angular momentum is the sum of the torques applied to the particle, with respect to the coordinate system in which the angular momentum is being calculated, that is: $\frac{dh_{B/O}}{dt} = \sum \tau_{B/O}$, and the torque at B is with respect to the origin O in the inertial frame $OXYZ$. In planar motion this equation would be written as $I\ddot{\theta} = \sum \tau_{B/O}$. The angular momentum of the particle at B with respect to any other arbitrary point, say, point A, is *by definition*:

$$r_{B/O} = r_{A/O} + r_{B/A};\ h_{B/O} = r_{B/O} \times P_{B/O} = r_{A/O} \times P_{B/O} + r_{B/A} \times P_{B/O}$$
$$= h_{A/O} + h_{B/A}$$

$$\boxed{\Rightarrow h_{B/A} = r_{B/A} \times P_{B/O}} \tag{2.33}$$

The momentum must *always be computed with respect to the inertial frame*. The formula for the torque on the particle at B with respect to point A (which will be proven in the sequel) is as follows:

$$\tau_{B/A} = \frac{dh_{B/A}}{dt} + v_{A/O} \times P_{B/O} \tag{2.34}$$

Fig. 2.12 Two link robot

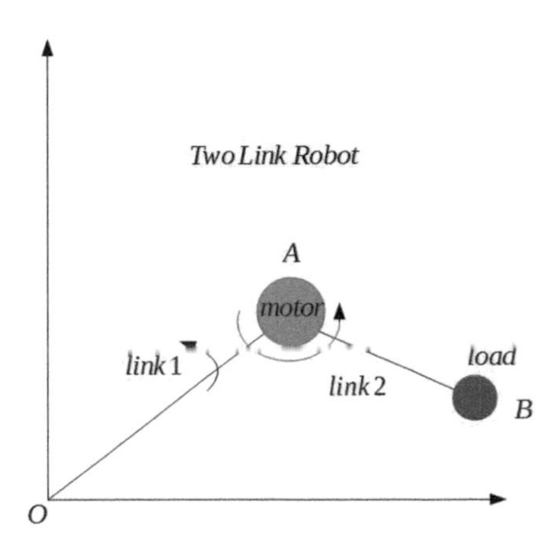

An interesting example would be a two link robot arm (see Fig. 2.12). The two links can rotate. The origin is at O while the load is at point B and the motor is located at point A. The motor rotates the link with the load and has to be able to supply the right amount of torque in order to be able to move the load. The amount of torque required is what needs to be calculated.

The sum of the external forces at B yields the time rate of change of the momentum at B with respect to O, that is: $\sum f_B = F_B = \frac{dP_{B/O}}{dt} = m\frac{dv_{B/O}}{dt}$. This is the momentum vector of our particle. The torque of B with regard to A is:

$$\tau_{B/A} = r_{B/A} \times F_B = r_{B/A} \times \frac{dP_{B/O}}{dt} \tag{2.35}$$

where F_B is the sum of all the forces acting on the particle at point B. A useful vector identity is the following:

$$\frac{d}{dt}(x \times y) = x \times \frac{dy}{dt} + \frac{dx}{dt} \times y \Rightarrow x \times \frac{dy}{dt} = \frac{d}{dt}(x \times y) - \frac{dx}{dt} \times y \tag{2.36}$$

where both x and y are vectors. Applying this latter identity to $r_{B/A} \times \frac{dP_{B/O}}{dt}$ by letting $r_{B/A} = x$; $\frac{dP_{B/O}}{dt} = \frac{dy}{dt}$ results in:

$$\tau_{B/A} = r_{B/A} \times \frac{dP_{B/O}}{dt} = \frac{d}{dt}\underbrace{\left(r_{B/A} \times P_{B/O}\right)}_{h_{B/A}} - \frac{dr_{B/A}}{dt} \times P_{B/O} \tag{2.37}$$

However, $r_{B/A} = r_{B/O} - r_{A/O}$ and so the time derivative of $r_{B/A}$ becomes:

$$\frac{dr_{B/A}}{dt} = \frac{dr_{B/O}}{dt} - \frac{dr_{A/O}}{dt} = v_{B/O} - v_{A/O} \tag{2.38}$$

Now $\left(r_{B/A} \times P_{B/O}\right) = h_{B/A}$ is the angular momentum of the particle at B with respect to A, as previously defined, so that the expression for the torque $\tau_{B/A}$ can be rewritten in the following way:

$$\tau_{B/A} = \frac{dh_{B/A}}{dt} - \left(v_{B/O} - v_{A/O}\right) \times P_{B/O} \tag{2.39}$$

Since $P_{B/O} = mv_{B/O}$, it turns out that:

$$\left(v_{B/O} - v_{A/O}\right) \times P_{B/O} = \underbrace{v_{B/O} \times mv_{B/O}}_{=0} - v_{A/O} \times mv_{B/O} \tag{2.40}$$

Hence the torque $\tau_{B/A}$ turns out to be:

$$\boxed{\tau_{B/A} = \frac{dh_{B/A}}{dt} + v_{A/O} \times mv_{B/O} = \frac{dh_{B/A}}{dt} + v_{A/O} \times P_{B/O}} \tag{2.41}$$

The term $v_{A/O} \times P_{B/O}$ can sometimes be irksome and it would be of advantage not to have to calculate it but revert to the usual formula which states that the torque is the time rate of change (derivative) of the angular momentum, that is, $\tau = \frac{dh}{dt}$. There are two obvious situations in which $v_{A/O} \times P_{B/O} = 0$ and they are:

1. $v_{A/O} = 0 \Rightarrow \tau_{B/A} = \frac{dh_{B/A}}{dt}$. This is a significant result because the point A can be arbitrarily located anywhere in the plane. As long as the point A isn't in motion ($v_{A/O} = 0$), the torque may be calculated with respect to that arbitrary stationary point (you can have a fixed set of axes at point A). This allows for consideration of rotations around a fixed axis which is not necessarily located at the center of mass. The same result as in Eq. 2.41, with $v_{A/O} = 0$ holds for rigid bodies. For example, it is possible for a rigid body to rotate at one end and not necessarily around its center of mass. The formula applies as long as the velocity of the axis about which it's rotating, and about which the torque is being computed, is not in motion (fixed and stationary).
2. If the velocity vector $v_{A/O}$ is parallel to the momentum vector $P_{B/O}$, then $v_{A/O} \times P_{B/O} = 0$. If the point A is at the center of mass, then its momentum is defined as $P_{B/O} = mv_{A/O} \Rightarrow v_{A/O} \times P_{B/O} = v_{A/O} \times mv_{A/O} = 0$, even for rigid bodies. The point A is labeled as the point G (the center of mass).
3. This result is very useful in problems of dynamics. The formula for torque with respect to the center of mass may be written as the time rate of change of h with respect to G, that is: $\tau_{/G} = \frac{dh_{/G}}{dt}$

Although the above center of mass proof is valid for particles, it can easily be extended to rigid bodies by summing or integrating over all of the individual particles in the rigid body.

2.8 Rigid Body Angular Momentum

Lecture 5: Video Times—55:03–59:18
The material in this section was adapted from the MIT Lecture Series on Dynamics (see Vandiver 2.003SC Engineering Dynamics. - Video of Lecture 5: Angular Momentum [45]).

The torque for a rigid body with respect to a point A is:

$$\tau_{/A} = \dot{H}_{/A} + v_{A/O} \times P_{G/O} \tag{2.42}$$

Note that the point A can be in motion (even accelerating). The term $v_{A/O} \times P_{G/O}$ can go to zero if $v_{A/O} = 0$, i.e. the point A is stationary and rotation is taking place around the fixed axis at A, or $v_{A/O}$ is parallel to $P_{G/O}$—it's always the case that $v_{A/O} \times P_{G/O} = 0$ when the point A is at the center of mass of the rigid body. For these two cases, $\tau_{/A} = \dot{H}_{/A}$. The condition for fixed axis rotation around an axis at point A is: $v_{A/O} = 0$. When the problem calls for the calculation of a torque required for a rotation around the center of mass, then the formula to use is: $\tau_{/G} = \dot{H}_{/G}$, and for the torque of a body which is pinned and rotates about a fixed axis at a point A, not situated at the center of mass, the formula that is required is: $\tau_{/A} = \dot{H}_{/A}$.

Example 7: Carnival Ride Problem

Lecture 5: Video Times—59:18–1:17:05
The following example was taken from: Vandiver 2.003SC Engineering Dynamics. - Video of Lecture 5: Angular Momentum[45].

The carnival ride is depicted schematically in Fig. 2.13. There is a fixed point A and a fixed axis around which the arm rotates. The arm can also be extended or retracted. The rider sits on the end of the extendable arm.

The path taken might be an inward spiral, similar to the act performed by a figure skater, of decreasing the rotational moment of inertia by retracting both arms to the sides while turning around in a circle. What forces would be felt by the rider on the ride or what accelerations will the rider feel?

Given Conditions: $\dot{\theta} = constant, \dot{r} = constant$ The rider will feel a centripetal and some Coriolis acceleration, because $\dot{r} \neq 0$. Since the length of the arm, which is rotating at constant speed, is changing, there will also be a change in linear as well as angular momentum, because $r \times P$ is changing (as the length of the arm extends or retracts). Any change in angular momentum requires a torque. The formula developed prior to the example will aid in the calculation of the torques needed and the forces on the rider as the length of the arm becomes longer or shorter. The rider at B can be regarded as a 'particle'. Then $h_{B/O}$, the angular momentum of the rider at B with respect to O, is going to be $h_{B/O} = r_{B/O} \times P_{B/O}$. Since this

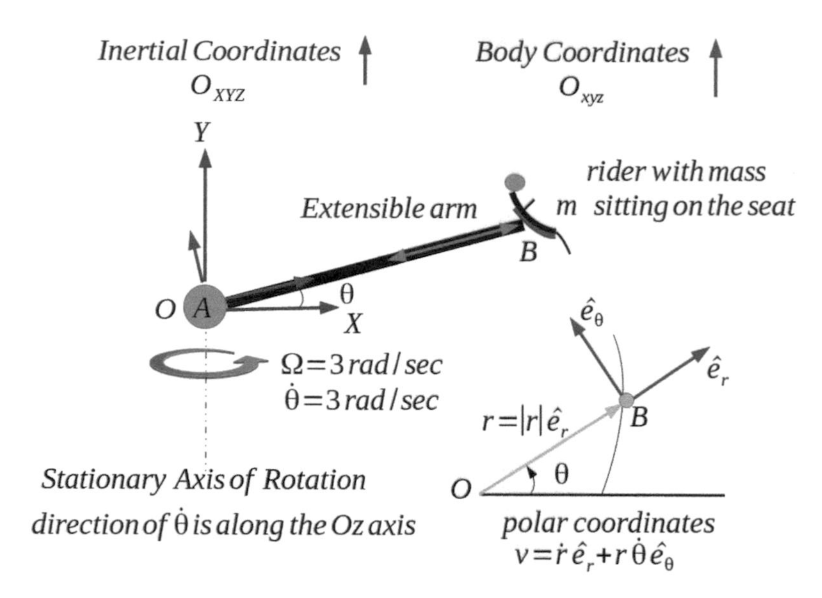

Fig. 2.13 Carnival ride problem

is a planar motion problem, confined to the $X - Y$ plane, with rotation of the arm $\dot{\theta}$ taking place around the Z axis, it is convenient to use polar coordinates, which means that the angular momentum may be rewritten in the form of:

1. Angular Momentum:

$$h_{B/O} = r_{B/O} \times P_{B/O} = r\hat{e}_r \times m \underbrace{(\dot{r}\hat{e}_r + r\dot{\theta}\hat{e}_\theta)}_{v_{B/O}} = r\hat{e}_r \times mr\dot{\theta}\hat{e}_\theta = r^2 m\dot{\theta}\hat{k} \tag{2.43}$$

2. Torque:

$$\tau_{B/O} = \frac{dh_{B/O}}{dt} = 2mr\dot{r}\dot{\theta}\hat{k} + mr^2\ddot{\theta}\hat{k} \tag{2.44}$$

The term $\dot{r}\hat{e}_r$ is the extension rate but there is also a component of velocity in the $\hat{e}_\theta$ direction of $r\dot{\theta}\hat{e}_\theta$. In the equation for torque, the term $v_{A/O} \times P_{B/O} = 0$, since the point A is a fixed point around whose axis, the rotation takes place. In fact the point A and the point O are the same points. The constant terms in the expression for the torque are $\hat{k}, m, \dot{\theta}$ and these terms need not be differentiated with respect to time. Since it was assumed that $\dot{\theta} = constant$, the torque then becomes:

$$\tau_{B/O} = \frac{dh_{B/O}}{dt} = 2mr\dot{r}\dot{\theta}\hat{k} \tag{2.45}$$

The term $2mr\dot{r}\dot{\theta}\hat{k}$ is the Coriolis acceleration and $mr^2\ddot{\theta}\hat{k}$ is referred to as the "Eulerian" acceleration. It's possible to write the torque as $r \times F$, that is:

$$\tau_{B/O} = r\hat{e}_r \times \underbrace{\left(2m\dot{r}\dot{\theta} + rm\ddot{\theta}\right)\hat{e}_\theta}_{\text{“}F\text{”}} \tag{2.46}$$

The term $rm\ddot{\theta}\hat{e}_\theta$ is the force required if the arm (with the rider seated on the end) is accelerating in the $\hat{e}_\theta$ direction. Assuming that $\dot{\theta} = constant$, then the torque may be written as:

$$\tau_{B/O} = r\hat{e}_r \times \hat{e}_\theta \underbrace{\left(2m\dot{r}\dot{\theta}\right)}_{\text{“}F\text{”}} = 2mr\dot{r}\dot{\theta}\hat{k} = r \times F_{Coriolis} \tag{2.47}$$

For this problem we are given the following data: $m = 100\,\text{kg}$, $r = 5\,\text{m}$, $\dot{r} = 0.4\,\text{m/s}$, $\dot{\theta} = 3\,\text{rad/s}$. The rotation rate is slightly less than half a revolution per second. Hence, the Coriolis force $F_{Coriolis} = \left(2m\dot{r}\dot{\theta}\right)$ is 240 N. The torque $\tau_{B/O} = 2mr\dot{r}\dot{\theta} = 240 \times 5 = 1200\,\text{N-m}$. The acceleration $a_{B/O}$ is: $a_{B/O} = F_{Coriolis}/m = \frac{240}{100} = 2.4$ meters per second*second $= 2.4/9.81 \approx 0.25\times g$.

Assuming that the system is spinning but the arm is not moving in and out, that is $\dot{r} = 0$, what force would you expect the rider to feel? The rider would feel a centripetal acceleration in the direction of O (in the inwards direction). If the rider's seat could swing around so that the rider would be facing inwards, towards O, the rider would feel that he is being pushed out of his seat. If his seat is facing outwards (the usual situation), the rider would be pushed backwards into his seat. If the arm is moving outwards at 0.4 m/s, there will be an additional $1/4\times g$'s of acceleration in the $\hat{e}_\theta$ direction, i.e. in the direction of increasing θ, which is perpendicular to the arm. The value of the centripetal acceleration would be: $a_{centripetal} = \Omega^2 r = 3^2 * 5 = 45 = 4.5\times g$. The Coriolis would be insignificant compared to the centripetal acceleration.

2.9 Linear and Angular Momenta and Their Derivatives

This section was adapted from the book on Dynamics by Beer et al. [4, pp. 721–722]. The linear momentum vector L of a particle is defined to be:

$$L = mv \tag{2.48}$$

where v is the particle's velocity vector and m is its mass. Taking the time derivative of L (assuming that the mass is constant) results in:

$$\dot{L} = m\dot{v} = ma \tag{2.49}$$

where a is the particle's acceleration vector. However from Newton's second law of motion:

$$F = ma = m\frac{dv}{dt} = \dot{L} \tag{2.50}$$

where F is the force applied to the particle. If more than one force is applied to the particle simultaneously, then the forces must be summed as follows:

$$\sum F = ma = m\frac{dv}{dt} = \dot{L} \tag{2.51}$$

The following assumptions have been made: A particle P of mass m is moving with velocity v with respect to an inertial frame $Oxyz$, where O is the origin of the inertial frame. The moment about O of the vector mv is termed the *moment of momentum* or the *angular momentum* of the particle P about O, at the instant under consideration and is designated by H_O. From the definition of the moment of a vector, and labeling by r, the position vector of P from the origin of the inertial coordinates, H_O is defined to be:

$$H_O = r \times mv \tag{2.52}$$

The time derivative of H_O is simply:

$$\dot{H}_O = \dot{r} \times mv + r \times m\dot{v} = \underbrace{v \times mv}_{=0} + r \times ma \tag{2.53}$$

Vectors v and mv are collinear, hence the first term of $\dot{H}_O$ is zero. Noticing that $r \times ma = r \times F$ represents the moment about O of the force F, the second term is designated as M_O. If more than one force acts simultaneously on the particle P, then we have:

$$r \times \sum F = \sum M_O = \dot{H}_O \tag{2.54}$$

2.10 Work and Calculation of Kinetic and Potential Energies

Work of a Weight W

The contents of this section have been taken from Beer et al. [4, pp. 756–763]

The work performed by the force F which results in a displacement vector dr is defined as the quantity:

$$dU = F \cdot dr \tag{2.55}$$

where both F ad dr are vector quantities. In Cartesian coordinates the above equation may be written in the form:

$$dU = F_x d_x + F_y d_y + F_z d_z \tag{2.56}$$

Work is a scalar quantity, meaning that it has a sign and a magnitude but no direction.

- Work is positive if the angle between the force vector F and the displacement vector dr is acute (less than 90°)
- Work is negative if the angle between the force vector F and the displacement vector dr is obtuse (between 90° and 270°)
- Zero if the force vector and the displacement vector are perpendicular

The above implies that the total work over a finite path from point A_1 to point A_2 is:

$$U_{1\to 2} = \int_{A_1}^{A_2} (F_x d_x + F_y d_y + F_z d_z) \tag{2.57}$$

If the y axis points upward, for a weight W we have: $F_x = 0$, $F_z = 0$, $F_y = -W$. This implies that:

$$dU = -W dy$$

$$U_{1\to 2} = -\int_{y_1}^{y_2} W dy$$

$$\text{or } U_{1\to 2} = -W(y_2 - y_1) = -W\Delta y \tag{2.58}$$

where Δy is the vertical shift as the particle moves from A_1 to A_2. The work of the weight W equals *the product of W and the vertical displacement of the center of gravity of the body*. From the foregoing, it is apparent that the work W is positive if Δy is negative, that is, when the body moves down.

Work of the Force Exerted by a Spring

Consider a body attached to a fixed point (such as a wall) by a spring. When the spring is undeformed, the body is at A_0. From experimental evidence we know that the force F exerted by the spring on body is linearly proportional to the deflection x of the spring from its un-stretched position A_0, that is:

$$F = kx \tag{2.59}$$

where k is the spring constant in units of *Newtons per meter. Note that the direction of the force exerted by the spring on the body is in the opposite direction to the deflection of the spring*. The work of the force F by the spring during a finite displacement of the body from $A_1(x = x_1)$ to $A_2(x = x_2)$ is determined to be:

$$dU = -Fdx = -kxdx; \quad U_{1\to 2} = -\int_{x_1}^{x_2} kxdx = \frac{1}{2}kx_1^2 - \frac{1}{2}kx_2^2 \tag{2.60}$$

Kinetic Energy of a Particle

Figure 2.14 illustrates the motion of a particle with mass m subject to a force F and traveling along the curved path from point A_1 to point A_2 along a curved trajectory. Writing Newton's second law in terms of the tangential component of the force F, that is F_t, the result is:

$$F_t = ma_t = m\frac{dv}{dt} \tag{2.61}$$

where v is the velocity tangential to the particle's trajectory. The tangential velocity may be written in the form:

$$v = \frac{ds}{dt} \tag{2.62}$$

where s is the distance the particle travels along the curved path. The acceleration a_t may then be shown to be:

$$F_t = m\frac{dv}{dt} = m\frac{dv}{ds}\frac{ds}{dt} = mv\frac{dv}{ds} \tag{2.63}$$

Integrating F_t from A_1 where $s = s_1$ and $v = v_1$ to A_2 where $s = s_2$ and $v = v_2$:

$$\int_{s_1}^{s_2} F_t ds = m\int_{s_1}^{s_2} v\frac{dv}{ds}ds = m\int_{v_1}^{v_2} vdv = \frac{1}{2}mv_2^2 - \frac{1}{2}mv_1^2 \tag{2.64}$$

Note that $\int_{s_1}^{s_2} F_t ds = U_{1\to 2}$, where $U_{1\to 2}$ is the work done by the force on the particle as it is displaced from A_1 to A_2. The entity $\frac{1}{2}mv^2$ is a scalar quantity and

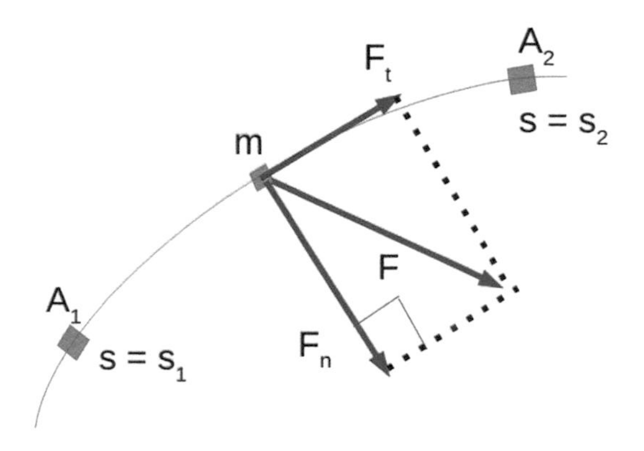

Fig. 2.14 Motion of a particle with mass m subject to a force F and traveling along the curved path from point A_1 to point A_2

it is defined to be the kinetic energy of the particle with the symbol T. This implies that:

$$U_{1\to 2} = \frac{1}{2}mv_2^2 - \frac{1}{2}mv_1^2 = T_2 - T_1 \tag{2.65}$$

The above equation expresses the fact that when a particle moves from A_1 to A_2 while subject to the action of a force F, *the work performed by the force F equals the change in kinetic energy of the particle* and is known as the *principle of work and energy*. Rearranging the terms in Eq. 2.65 of $U_{1\to 2} = T_2 - T_1$, we have: $U_{1\to 2} + T_1 = T_2$. Hence, *the kinetic energy of the particle at A_2 can be calculated by the addition to its kinetic energy at A_1, the work performed on the particle by the force F during its displacement from A_1 to A_2*. In a comparable fashion to Newton's second law, from which it was derived, the principle of work and energy applies only with respect to a Newtonian or inertial frame of reference. The points listed below should be borne in mind concerning the kinetic energy of a particle:

- Both work and kinetic energy are scalar quantities.
- Their sum can be computed as an ordinary algebraic sum, the work $U_{1\to 2}$ being considered as positive or negative in accordance with the direction of F.
- When several forces act on the particle, the expression $U_{1\to 2}$ represents the total work of the forces acting on the particle.
- $U_{1\to 2}$ is obtained by adding together the work of the various forces algebraically. This is possible since the kinetic energy of a particle is a scalar quantity.
- The speed v used to determine the kinetic energy T should be measured with respect to a Newtonian or inertial frame of reference. The kinetic energy is always positive regardless of the direction of motion of the particle.

Conservation of Energy (Beer et al. [4, pp. 785–786])

We know from Eq. 3.33 that the work of a conservative force such as the weight of a particle or the force exerted by a spring may be expressed in the form:

$$U_{1\to 2} = V_1 - V_2 \tag{2.66}$$

Similarly the kinetic energy is:

$$U_{1\to 2} = \frac{1}{2}mv_2^2 - \frac{1}{2}mv_1^2 = T_2 - T_1 \tag{2.67}$$

Equating the two forms of energy for a conservative force, we have:

$$\begin{aligned} U_{1\to 2}(\text{potential energy}) &= U_{1\to 2}(\text{kinetic energy}) \\ \text{or}\quad V_1 - V_2 = T_2 - T_1 &\Rightarrow V_1 + T_1 = V_2 + T_2 \end{aligned} \tag{2.68}$$

The above equation (Eq. 2.68) states that when a particle moves under the action of conservative forces, *the sum of the kinetic and potential energies of the particle*

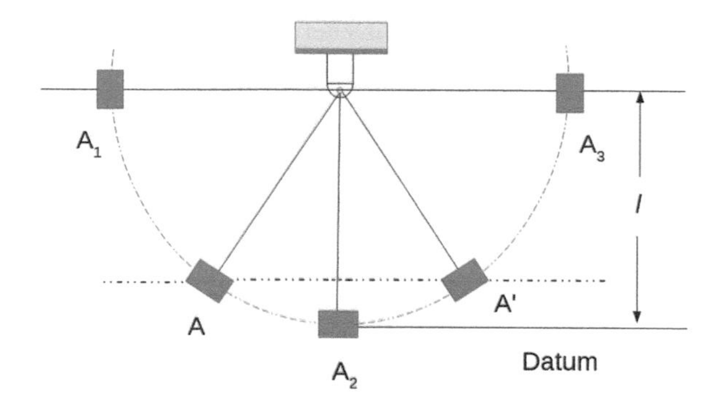

Fig. 2.15 Motion of a pendulum bob

remains constant. The sum $T+V$ is the total energy of the particle and is symbolized by E. For example, the pendulum bob with weight W is released with zero velocity at A_1 and is allowed to swing vertically (see Fig. 2.15). The potential, kinetic, and total energy at A_1 with A_2 as the reference datum is:

$$V_1 = Wl; \quad T_1 = 0; \quad T_1 + V_1 = Wl \tag{2.69}$$

The kinetic energy at A_1 where the velocity of the pendulum bob is zero is: $T_1 = 0$. Similarly at A_2 the kinetic energy is $\frac{1}{2}(W/g)v_2^2$. The speed of the pendulum at A_2 is determined from the principle of work and energy as follows:

$$\begin{aligned} &T_1 = 0; \quad V_1 = Wl; \quad T_2 = \frac{1}{2}\frac{W}{g}v_2^2; \quad V_2 = 0 \\ &T_1 + V_1 = T_2 + V_2 \Rightarrow 0 + Wl = \frac{1}{2}\frac{W}{g}v_2^2 + 0 \\ &\Rightarrow v_2 = \sqrt{2gl} \end{aligned} \tag{2.70}$$

Hence at A_2, with speed of the bob equal to $v_2 = \sqrt{2gl}$, the kinetic energy is:

$$\begin{aligned} &T_2 = \frac{1}{2}mv_2^2 = \frac{1}{2}\frac{W}{g}2gl = Wl; \; V_2 = 0 \\ &T_2 + V_2 = Wl \end{aligned} \tag{2.71}$$

The total mechanical energy $E = T + V$ at A_1 and at A_2 remains the same. At A_1 the energy is entirely potential energy while at A_2 it is entirely kinetic energy.

2.11 Systems of Particles

This section was derived from Beer et al. [4, pp. 856–862]. A system is composed of n particles; each particle P_i has mass m_i and acceleration a_i with respect to an inertial or Newtonian frame of reference. The force on particle P_i, brought to bear by particle P_j, is labeled as f_{ij} and is termed an *internal force*. The sum of all the internal forces exerted on P_i by all of the system's particles is: $\sum_{j=1}^{n} f_{ij}$. Note that f_{ii} has no meaning and is assumed to equal zero. The resultant of all the external forces acting on P_i is labeled as F_i, and thus Newton's second law may be written as:

$$F_i + \sum_{j=1}^{n} f_{ij} = m_i a_i \tag{2.72}$$

The position vector of particle P_i with respect to the origin O is of an inertial frame is r_i. Taking moments about O, we have:

$$r_i \times F_i + \sum_{j=1}^{n} (r_i \times f_{ij}) = r_i \times m a_i \tag{2.73}$$

Note that the force exerted on particle P_i by particle P_j, that is f_{ij} is equal in magnitude but opposite in sign to the force exerted by particle P_j on particle P_i, that is $f_{ij} = -f_{ji}$. This is true because of Newton's third law (for every action, there is an equal and opposite reaction). In fact these two forces come in pairs and have the same line of action but act in opposite directions thus canceling each other out. The sum of the moments about O of forces f_{ij} and f_{ji} are:

$$r_i \times f_{ij} + r_j \times f_{ji} = r_i \times \underbrace{(f_{ij} + f_{ji})}_{=0} + \underbrace{(r_j - r_i) \times f_{ji}}_{=0} = 0 \tag{2.74}$$

Notice from Fig. 2.16 that the vectors $r_j - r_i$ and f_{ji} are collinear and hence $(r_j - r_i) \times f_{ji} = 0$. Adding together all of the system's internal forces and summing the moments due to the internal forces about O, the result is:

$$\sum_{i=1}^{n} \sum_{j=1}^{n} f_{ij} = 0; \quad \sum_{i=1}^{n} \sum_{j=1}^{n} r_i \times f_{ij} = 0 \tag{2.75}$$

In other words, the sum of all of the internal forces and the sum of the moments about O due to the internal forces are all zero. Summing up all of the external

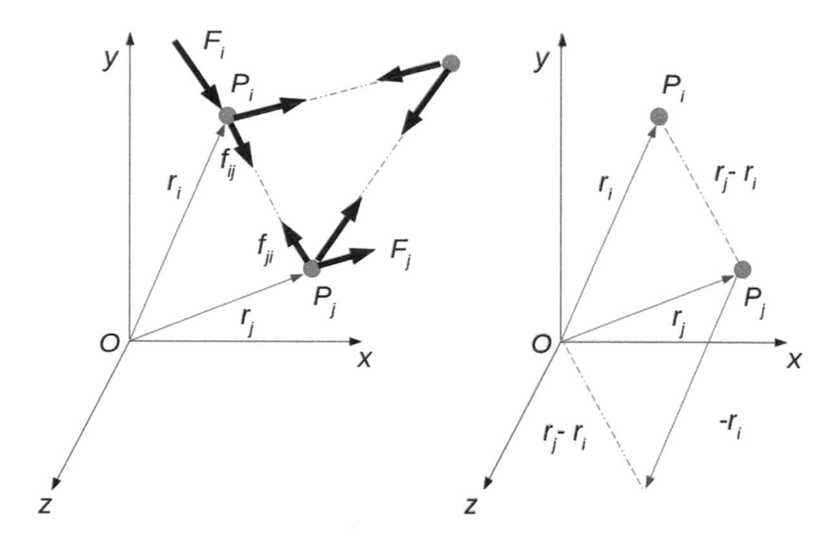

Fig. 2.16 Internal forces cancel each other out

forces and the moments about O due to the external forces leads to:

$$\sum_{i=1}^{n} F_i = \sum_{i=1}^{n} m_i a_i \tag{2.76}$$

$$\sum_{i=1}^{n} (r_i \times F_i) = \sum_{i=1}^{n} (r_i \times m_i a_i) \tag{2.77}$$

The above equation states the fact that the system of the external forces F_i and the system of effective forces $m_i a_i$ have the same resultant and the same moment resultant.

Linear and Angular Momentum of a System Composed of Particles
The linear momentum of the system is defined to be the sum of the linear momenta of all the particles of the system, that is:

$$L = \sum_{i=1}^{n} m_i v_i \tag{2.78}$$

The angular momentum H_o about the origin O of the system of particles may be defined as:

$$H_o = \sum_{i=1}^{n} (r_i \times m_i v_i) \tag{2.79}$$

Differentiating both the linear and angular momenta of Eqs. 2.78 and 2.79, respectively, with respect to time t results in:

$$\dot{L} = \sum_{i=1}^{n} m_i \dot{v}_i = \sum_{i=1}^{n} m_i a_i \tag{2.80}$$

$$\begin{aligned}
\dot{H}_o &= \sum_{i=1}^{n} (\dot{r}_i \times m_i v_i) + \sum_{i=1}^{n} (r_i \times m_i \dot{v}_i) \\
&= \underbrace{\sum_{i=1}^{n} (v_i \times m_i v_i)}_{=0} + \sum_{i=1}^{n} (r_i \times m_i \dot{v}_i) \\
\Rightarrow \dot{H}_o &= \sum_{i=1}^{n} (r_i \times m_i \dot{v}_i) = \sum_{i=1}^{n} (r_i \times m_i a_i)
\end{aligned} \tag{2.81}$$

The right-hand members of Eqs. 2.76 and 2.80 are the same. The same is true of Eqs. 2.77 and 2.81 and so it must be true that:

$$\sum_{i=1}^{n} F_i = \dot{L}; \quad \sum_{i=1}^{n} Mo_i = \dot{H}_o \tag{2.82}$$

These equations state the fact that *the resultant of all of the external forces and the moment of the resultant of all of the external forces about the fixed point O are equal, respectively, to the time derivatives of the linear and angular momenta about O of the system of particles.*

Motion of the Center of Mass of a System of Particles

Denoting by $\overline{r}$ the position vector of the mass center G of the system of particles with respect to the origin O of an inertial frame, the relation between the mass center and the masses and locations of the particles may be written as:

$$m\overline{r} = \sum_{i=1}^{n} m_i r_i; \quad m = \sum_{i=1}^{n} m_i \tag{2.83}$$

Converting the position vectors $\overline{r}$ and r_i, respectively, into rectangular components, Eq. 2.83 becomes:

$$m\overline{x} = \sum_{i=1}^{n} m_i x_i; \quad m\overline{y} = \sum_{i=1}^{n} m_i y_i; \quad m\overline{z} = \sum_{i=1}^{n} m_i z_i \tag{2.84}$$

Taking the time derivatives of both sides of Eq. 2.83:

$$m\dot{\overline{r}} = \sum_{i=1}^{n} m_i \dot{r}_i \Rightarrow m\overline{v} = \sum_{i=1}^{n} m_i v_i \tag{2.85}$$

where the velocity of the mass center G, with respect to the inertial coordinate frame centered at O, is written as $\overline{v}$. Recognizing that the right-hand side of Eq. 2.85 is the same as the linear momentum L (Eq. 2.78), we therefore have:

$$L = m\overline{v} \tag{2.86}$$

If we differentiate both sides of the above equation with respect to t, we find that:

$$\dot{L} = m\dot{\overline{v}} = m\overline{a} \tag{2.87}$$

where the acceleration of the mass center G with respect to the inertial coordinate frame centered at O is denoted by $\overline{a}$. Substituting for $\dot{L} = \sum_{i=1}^{n} F_i$ from Eq. 2.82 into Eq. 2.87 results in:

$$\sum_{i=1}^{n} F_i = m\overline{a} \tag{2.88}$$

Equation 2.88 is identical with the result which would be attained for a particle of mass m, which equals the total mass of the particles of the system, acted upon by the sum of all of the external forces. In other words, *the center of mass G of an aggregate system of particles translates as if the entire mass of the system and all of the external forces acting upon the disparate particles were concentrated at that point.*

Angular Momentum of an Aggregate of Particles About Its Mass Center

There are applications where it is advantageous to consider the movement of the particles of the system with respect to a centroidal (mass center) frame $Gx'y'z'$, which translates with respect to the Newtonian frame $Oxyz$. It will be demonstrated that $\sum_{i=1}^{n} Mo_i = \dot{H}_o$ still holds when the frame $Oxyz$ is replaced by $Gx'y'z'$. Denoting by r_i' and v_i', the position and velocity vector, respectively, of the point P_i relative to the moving axes $Gx'y'z'$, the angular momentum H_G' of the system of particles about the mass center G is as follows:

$$H_G' = \sum_{i=1}^{n} r_i' \times m_i v_i' \tag{2.89}$$

Differentiating both sides of Eq. 2.89 with respect to t we get (see Eq. 2.81):

$$\dot{H}'_G = \sum_{i=1}^{n} r'_i \times m_i a'_i \tag{2.90}$$

where a'_i stands for the acceleration of P_i with respect to $Gx'y'z'$. The acceleration of particle P_i with respect to the inertial coordinate frame is:

$$a_i = \overline{a} + a'_i \Rightarrow a'_i = a_i - \overline{a} \tag{2.91}$$

where a_i and $\overline{a}$, respectively, symbolize the accelerations of particle P_i and the center of mass G with respect to the inertial frame $Oxyz$. Inserting the expression for a'_i above into Eq. 2.90 results in:

$$\dot{H}'_G = \sum_{i=1}^{n} (r'_i \times m_i a_i) - \underbrace{\left(\sum_{i=1}^{n} m_i r'_i\right)}_{=0} \times \overline{a} \tag{2.92}$$

The term $\sum_{i=1}^{n} m_i r'_i = m\overline{r}' = 0$, since the position vector $\overline{r}'$ of G relative to the frame $Gx'y'z'$ is zero. Equation 2.92 becomes:

$$\dot{H}'_G = \sum_{i=1}^{n} (r'_i \times m_i a_i) = \sum_{i=1}^{n} (r'_i \times F_i) \tag{2.93}$$

The sum in Eq. 2.93 therefore reduces to the moment about G of the external forces acting on the particles of the system which can be written as:

$$\sum_{i=1}^{n} M_{G_i} = \dot{H}'_G \tag{2.94}$$

The above equation expresses the fact that *the moment about the point G of all of the external forces is equal to the time derivative of the angular momentum about G of the aggregate system of particles*. In Eq. 2.89 the angular momentum H'_G was defined as the sum of the moments about G of all of the particles $m_i v'_i$ in *their movements with respect to the centroidal frame of reference* $Gx'y'z'$. Sometimes it may be desirable to compute the sum H_G of the moments about G of the particles $m_i v_i$ with regard to the inertial frame of reference $Oxyz$, that is:

$$H_G = \sum_{i=1}^{n} (r'_i \times m_i v_i) \tag{2.95}$$

Since $v_i = \overline{v} + v_i'$, Eq. 2.95 becomes:

$$H_G = \left(\sum_{i=1}^{n} m_i r_i'\right) \times \overline{v} + \sum_{i=1}^{n} m_i r_i' \times v_i' \tag{2.96}$$

As observed earlier, the first term in Eq. 2.96 is equal to zero since $\left(\sum_{i=1}^{n} m_i r_i'\right) = m\overline{r'} = 0$. The location of $\overline{r'}$ is at the center of mass of the particles or at the origin of the $Gx'y'z'$ coordinate frame. Hence, the angular momenta H_G' and H_G are identical. This implies that their respective time derivatives are equal as well, or:

$$\dot{H}_G = \dot{H}_G' \tag{2.97}$$

Summarizing, the angular momentum H_G can be computed by forming the moments about G of the particles in their motion with respect to either the Newtonian frame $Oxyz$ or the centroidal frame $Gx'y'z'$:

$$H_G = \sum_{i=1}^{n} (r_i' \times m_i v_i) = \sum_{i=1}^{n} (r_i' \times m_i v_i') \tag{2.98}$$

Kinetic Energy of an Aggregate (System) of Particles (Beer et al. [4, pp. 872–873])

The kinetic energy T of a system made up of many individual particles is defined to be the sum of the kinetic energies of the individual particles of the system and may be written as:

$$T = \frac{1}{2} \sum_{i=1}^{n} (m_i v_i^2) \tag{2.99}$$

It is often convenient, when computing the kinetic energy of a system comprised of a large number of particles (as is the case for a rigid body) to consider the motion of the mass center G of the system and the motion of the system relative to a moving frame attached to G, separately. The velocity v_i of particle P_i relative to the inertial coordinate frame $Oxyz$ may be expressed in the form of: $v_i = \overline{v} + v_i'$, where $\overline{v}$ is the velocity of the mass center G relative to the inertial frame and v_i' is the velocity of particle P_i relative to the frame $Gx'y'z'$. Since $v_i^2 = v_i \cdot v_i$, the kinetic energy of Eq. 2.99 becomes:

$$\begin{aligned} T &= \frac{1}{2} \sum_{i=1}^{n} (m_i v_i \cdot v_i) = \frac{1}{2} \sum_{i=1}^{n} m_i (\overline{v} + v_i') \cdot (\overline{v} + v_i') \\ &= \frac{\overline{v}^2}{2} \left(\sum_{i=1}^{n} (m_i)\right) + \overline{v} \cdot \sum_{i=1}^{n} m_i v_i' + \frac{1}{2} \sum_{i=1}^{n} m_i v_i'^2 \end{aligned} \tag{2.100}$$

In the above equation (Eq. 2.100) the first sum represents the total mass m of the system. The second sum is equal to $m\overline{v}' = 0$ since $\overline{v}'$ represents the velocity of the center of mass G relative to the centroidal frame $Gx'y'z'$ (which is clearly zero). Hence Eq. 2.100 may be written as:

$$T = \frac{1}{2}m\overline{v}^2 + \frac{1}{2}\sum_{i=1}^{n} m_i v_i'^2 \tag{2.101}$$

The above equation demonstrates that the kinetic energy T of a system of particles can be obtained by adding the kinetic energy of the mass center G and the kinetic energy of the system of particles as it moves relative to the frame $Gx'y'z'$.

2.12 Principle of Work and Energy for a Rigid Body

The concept of work and energy for a rigid body is based upon chapters 10 and 17 in the book by Beer et al. [4, pp. 560–562, 1082–1085].
Assume that the rigid body is composed of a very large number n of mass particles each with weight Δm_i. The mathematical expression of the principle of work and energy is: $T_1 + U_{1\rightarrow 2} = T_2$, where T_1 and T_2 are the initial and final values of the kinetic energy of the particles forming the rigid body and $U_{1\rightarrow 2}$ is the work of all the forces acting on the assorted particles of the body. The total kinetic energy is then:

$$T = \frac{1}{2}\sum_{i=1}^{n} \Delta m_i v_i^2 \tag{2.102}$$

The expression $U_{1\rightarrow 2}$ symbolizes the work of all the forces, either internal or external, acting on the complete set of particles which compose the rigid body. However, the total work of the internal forces which hold together the disparate particles of a rigid body is zero. Consider any two particles A and B of a rigid body and the two equal and opposing forces F and $-F$ which they bring to bear on each other (see Fig. 2.17). In general, small shifts dr and dr' of the two particles are different, however, the components of these displacements along the line connecting the two particles, the line AB, must be equal, otherwise, the distance between the particles would change and the body could not be considered to be rigid. Therefore, the work done by the force F is equal in magnitude to the work of $-F$, and their sum is zero. Hence it may be safely assumed that the total work of the internal forces acting on the particles of a rigid body is zero, and the expression for the work $U_{1\rightarrow 2}$, in the equation of the principle of work and energy, $T_1 + U_{1\rightarrow 2} = T_2$ includes only the work of the external forces acting on the body during the displacement under consideration.

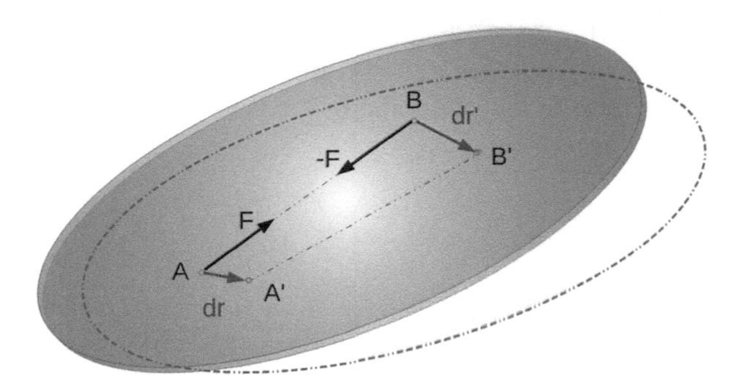

Fig. 2.17 Small displacements dr and dr' of the two particles differ but the components of these displacements along $A - B$ are equal—no net internal forces

Work Performed by Forces Acting on a Rigid Body

Previously it was noted that the work of a force F during a shift of its point of application from B_1 to B_2 is:

$$U_{1\to 2} = \int_{B_1}^{B_2} F \cdot dr \tag{2.103}$$

or

$$U_{1\to 2} = \int_{s_1}^{s_2} (|F| \cos\alpha) ds \tag{2.104}$$

where $|F|$ is the magnitude of the force, α is the angle between the direction of motion of the rigid body and the direction of the applied force at the point of application B, and s is the variable of integration which measures the distance traveled by B along its path.

When calculating the work of the external forces acting on a rigid body, it is advantageous to determine the work of a couple without considering the work of each of the two forces forming the couple separately. The two forces F and $-F$ form a couple of moment M and act on the rigid body (see Fig. 2.18). Any small movement of the rigid body which transports A and B into A' and B'', respectively, can be separated into two parts: in the first part, points A and B undergo matching shifts dr_1; in the second, A' remains in place while B' moves into point B'' through a distance of dr_2 whose magnitude is: $|dr_2| = ds_2 = rd\theta \quad r = |B' - A'|$. During the first phase of the motion of the rigid body, the work of F is commensurate in magnitude and opposite in sign to the work of $-F$ and their sum is therefore zero. In the second phase, only force F performs useful work which is: $dU = Fds_2 = Frd\theta$. Note that the product Fr is equal to the magnitude M, the moment of the applied couple. Hence, the work of a moment M acting on a rigid body is simply:

$$dU = Md\theta \tag{2.105}$$

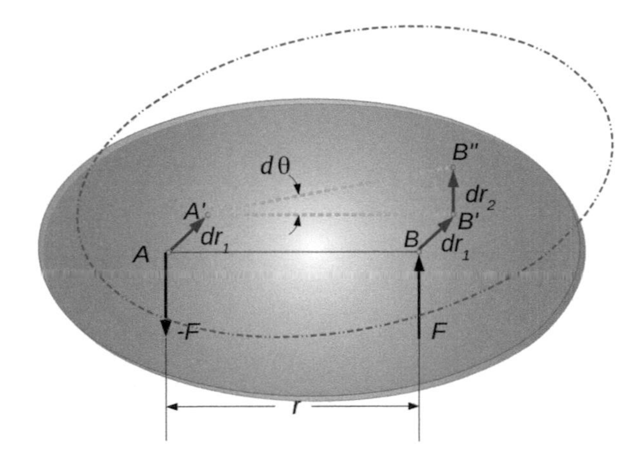

Fig. 2.18 Work done by couple $M = dU = Fds_2 = Frd\theta$

where $d\theta$ is the small angle through which the rigid body pivots (around point A'). The work performed by the couple during a finite rotation of the rigid body is calculated by integrating both sides of Eq. 2.105 from the initial value θ_1 of the angle θ to its final amount θ_2. We write this expression as follows:

$$U_{1\rightarrow 2} = \int_{\theta_1}^{\theta_2} M d\theta \tag{2.106}$$

If the moment M of the couple is constant, the above formula 2.106 becomes:

$$U_{1\rightarrow 2} = M(\theta_2 - \theta_1) \tag{2.107}$$

There exist a number of forces which arise in problems of kinetics which do no work. These include forces applied to fixed points where there is no subsequent motion involved or which act in a direction orthogonal to the displacement of their application point. Among the forces which do not perform any work, the following have been included:

1. the reaction force at a frictionless pin when the body being supported rotates about the pin
2. the reaction force of a frictionless surface when the body in contact moves along the surface
3. the weight of a body when its center of gravity moves horizontally.
4. when a rigid body rolls without slipping on a fixed surface, the friction force F at the point of contact C doesn't perform any work since the velocity v_C at the point of contact C is zero. The work of the friction force F, during a small movement of the rigid body is therefore: $dU = Fds_C = F(v_C dt) = 0$.

Kinetic Energy of a Rigid Body in Planar Motion

Assume that a rigid body of mass m is undergoing planar motion (see Fig. 2.19). If the absolute velocity v_i of each mass particle P_i of the body is written as the sum of

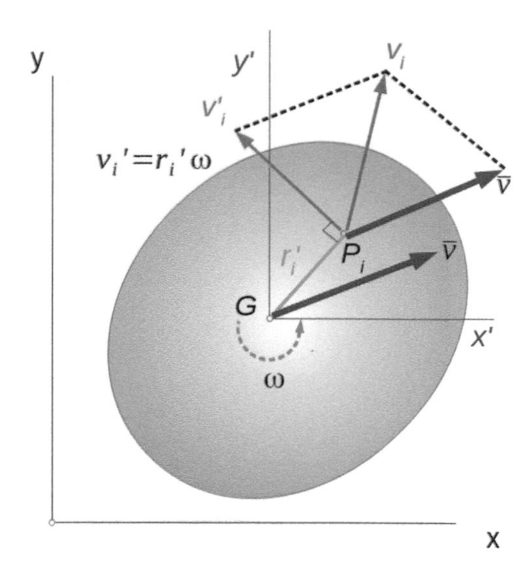

Fig. 2.19 Rigid body in plane motion—kinetic energy

the velocity $\overline{v}$ of the center of mass G and of the velocity v_i' of the particle relative to a non-rotating frame $Gx'y'$ which is rigidly attached to G, then from Eq. 2.101, the kinetic energy of the system of particles forming the rigid body may be written as:

$$T = \frac{1}{2}m\overline{v}^2 + \frac{1}{2}\sum_{i=1}^{n}(\Delta m_i)v_i'^2 \tag{2.108}$$

where Δm_i is the mass of any particle of which the rigid body is made up of. However, the magnitude v_i' equals the product $r_i'\omega$, where r_i' is the distance of P_i from the axis of rotation (which passes through G and is perpendicular to the plane of motion), and of the magnitude ω of the angular velocity of the body at the instant under consideration. Substituting the expression $v_i' = r_i'\omega$ into Eq. 2.108 results in:

$$T = \frac{1}{2}m\overline{v}^2 + \frac{1}{2}\left(\sum_{i=1}^{n} r_i'^2 \Delta m_i\right)\omega^2 \tag{2.109}$$

Since the sum in Eq. 2.109 equals the moment of inertia $\overline{I}$ of the rigid body about the axis of rotation (which passes through G and is perpendicular to the plane), Eq. 2.109 becomes:

$$T = \frac{1}{2}m\overline{v}^2 + \frac{1}{2}\overline{I}\omega^2 \tag{2.110}$$

When a rigid body is in translation only, that is $\omega = 0$, then the above expression is reduced to $\frac{1}{2}m\overline{v}^2$, while in the case of a rotation only about the body's c.g., without translation, that is when $\overline{v} = 0$, Eq. 2.110 becomes: $\frac{1}{2}\overline{I}\omega^2$. In conclusion, it becomes apparent that the kinetic energy of a rigid body in planar motion can be divided into two parts:

1. the kinetic energy $\frac{1}{2}m\overline{v}^2$ associated with the linear translatory motion of the center of mass point G of the body
2. the kinetic energy $\frac{1}{2}\overline{I}\omega^2$ which is due to the rotation of the body about G.

Kinetic Energy of a Rigid Body in Planar Motion: Non-centroidal Rotation
Equation 2.110 is valid for any type of planar motion. Hence it can be used to calculate the kinetic energy of a rigid body which is rotating about a fixed axis through O (see Fig. 2.20) with an angular velocity ω. The kinetic energy of the body can be written in a more direct fashion by noting that the speed v_i of the particle P_i equals the product $r_i\omega$, where r_i is the distance from the fixed axis to the point P_i and ω is the magnitude of the instantaneous angular velocity of the rigid body. Substituting this expression into Eq. 2.102 results in:

$$T = \frac{1}{2}\sum_{i=1}^{n}(\Delta m_i)(r_i\omega)^2 = \frac{1}{2}\left(\sum_{i=1}^{n} r_i^2\Delta m_i\right)\omega^2 \tag{2.111}$$

Since the terms within the brackets on the right-hand side of Eq. 2.111 is defined to be the moment of inertia of the rigid body about the fixed axis through O, that is $I_O = \sum_{i=1}^{n} r_i^2\Delta m_i$, then Eq. 2.111 becomes:

$$T = \frac{1}{2}I_O\omega^2 \tag{2.112}$$

It should be noted that the above results are not limited to the motion of plane slabs or to the motion of bodies which are symmetrical with respect to the reference plane, but can be applied to the investigation of planar motion of any rigid body, regardless of its shape.

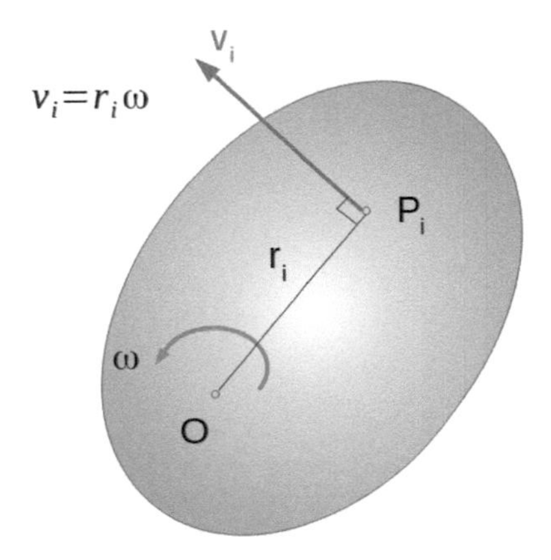

Fig. 2.20 Rigid body in plane motion—noncentroidal rotation

2.13 Angular Momentum of a Rigid Body in Three Dimensions

We now turn to the dynamics of rigid bodies in three dimensions (see Beer et al. [4, pp. 1147–1148]).

H_G, the angular momentum of a body about its mass center G can be calculated from the angular rate ω of the body when three-dimensional motion is considered. The angular momentum of the body about G may be determined in accordance with Eq. 2.98, as follows:

$$H_G = \sum_{i=1}^{n}(r_i' \times \Delta m_i v_i) = \sum_{i=1}^{n}(r_i' \times \Delta m_i v_i') \tag{2.113}$$

where r_i' and v_i' denote the position and velocity vectors, respectively, of the particle P_i, of mass Δm_i, relative to the body fixed non-rotating ("centroidal frame") frame $Gxyz$, whose origin is located at the body's c.g. point (see Fig. 2.21). However, $v_i' = \omega \times r_i'$, where ω is the instantaneous angular velocity of the body. Substitution of this latter result into Eq. 2.113 results in:

$$H_G = \sum_{i=1}^{n}(r_i' \times \Delta m_i v_i') = \sum_{i=1}^{n}[r_i' \times (\omega \times r_i')\Delta m_i] \tag{2.114}$$

Assuming that the angular velocity vector of the body at the instant considered, ω, may be decomposed into three orthogonal components, $\omega = [\omega_x, \omega_y, \omega_z]$, which are the terms along the x, y, and z coordinates, respectively, of the $Gxyz$ coordinate system and further assuming that the position vector r_i' may be also accordingly decomposed, that is $r_i' = [x_i, y_i, z_i]$, we have:

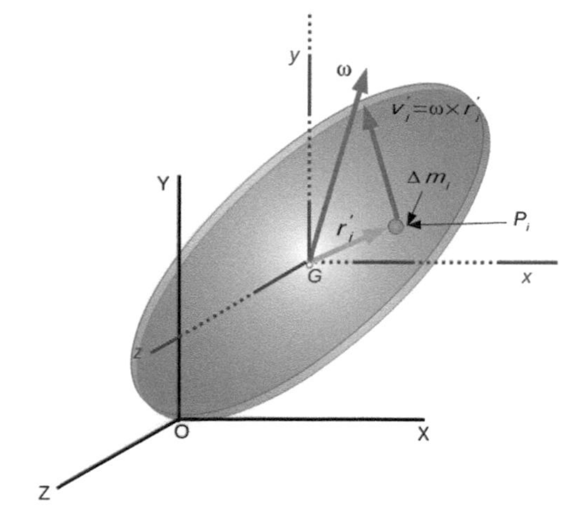

Fig. 2.21 Rigid body angular momentum in three dimensions

$$\omega \times r_i' = \begin{vmatrix} \hat{i} & \hat{j} & \hat{k} \\ \omega_x & \omega_y & \omega_z \\ x_i & y_i & z_i \end{vmatrix} = \hat{i}(\omega_y z_i - y_i\omega_z) + \hat{j}(\omega_z x_i - z_i\omega_x) + \hat{k}(\omega_x y_i - x_i\omega_y) \tag{2.115}$$

Proceeding to calculate the vector product $r_i' \times (\omega \times r_i')$, based upon the result of $\omega \times r_i'$ from Eq. 2.115, we have

$$\begin{aligned} r_i' \times (\omega \times r_i') &= \begin{vmatrix} \hat{i} & \hat{j} & \hat{k} \\ x_i & y_i & z_i \\ (\omega_y z_i - y_i\omega_z) & (\omega_z x_i - z_i\omega_x) & (\omega_x y_i - x_i\omega_y) \end{vmatrix} \\ &= \hat{i}[y_i(\omega_x y_i - x_i\omega_y) - z_i(\omega_z x_i - z_i\omega_x)] \\ &\quad + \hat{j}[z_i(\omega_y z_i - y_i\omega_z) - x_i(\omega_x y_i - x_i\omega_y)] \\ &\quad + \hat{k}[x_i(\omega_z x_i - z_i\omega_x) - y_i(\omega_y z_i - y_i\omega_z)] \\ &= \hat{i}[\omega_x(y_i^2 + z_i^2) - \omega_y x_i y_i - \omega_z x_i z_i] \\ &\quad + \hat{j}[\omega_y(x_i^2 + z_i^2) - \omega_x x_i y_i - \omega_z y_i z_i] \\ &\quad + \hat{k}[\omega_z(x_i^2 + y_i^2) - \omega_x x_i z_i - \omega_y y_i z_i] \end{aligned} \tag{2.116}$$

Adding Δm_i to Eq. 2.116 and summing, the result is:

$$\begin{aligned} H_G &= \hat{i}\sum_{i=1}^{n}[\omega_x(y_i^2 + z_i^2) - \omega_y x_i y_i - \omega_z x_i z_i]\Delta m_i \\ &\quad + \hat{j}\sum_{i=1}^{n}[\omega_y(x_i^2 + z_i^2) - \omega_x x_i y_i - \omega_z y_i z_i]\Delta m_i \\ &\quad + \hat{k}\sum_{i=1}^{n}[\omega_z(x_i^2 + y_i^2) - \omega_x x_i z_i - \omega_y y_i z_i]\Delta m_i \end{aligned} \tag{2.117}$$

Letting the masses Δm_i become infinitesimally small, replacing the summation sign with integrals and letting the components of H_G be $H_G = [H_x, H_y, H_z]$ leads to:

$$
\begin{aligned}
H_x &= \omega_x \int (y^2 + z^2)dm - \omega_y \int xy\,dm - \omega_z \int zx\,dm \\
H_y &= \omega_y \int (x^2 + z^2)dm - \omega_x \int yx\,dm - \omega_z \int yz\,dm \\
H_z &= \omega_z \int (x^2 + y^2)dm - \omega_x \int zx\,dm - \omega_y \int yz\,dm
\end{aligned} \tag{2.118}
$$

It should be noted that the integrals containing squared terms signify the "centroidal" mass moments of inertia of the rigid body about the x, y, and z axes, respectively, and the integrals containing products of the mutually orthogonal coordinates represent the "centroidal" mass products of inertia of the rigid body. Defining the entities within the integrals as follows, we have:

1. $I_x = \int (y^2 + z^2)dm; \quad I_y = \int (x^2 + z^2)dm; \quad I_z = \int (x^2 + y^2)dm$
2. $I_{xy} = \int xy\,dm; \quad I_{yz} = \int yz\,dm; \quad I_{zx} = \int zx\,dm$
3. $I_{xy} = I_{yx}; \quad I_{zy} = I_{yz}; \quad I_{xz} = I_{zx}$

Rewriting Eq. 2.118 with the above definitions results in:

$$
\begin{aligned}
H_x &= \omega_x I_x - \omega_y I_{xy} - \omega_z I_{xz} \\
H_y &= \omega_y I_y - \omega_x I_{yx} - \omega_z I_{yz} \\
H_z &= \omega_z I_z - \omega_x I_{zx} - \omega_y I_{zy}
\end{aligned} \tag{2.119}
$$

Equations 2.119 demonstrate that the operation which transforms the vector ω into the vector H_G consists of the array of moments and products of inertia. In matrix form this relationship is as follows:

$$
H_G = \begin{bmatrix} I_x & -I_{xy} & -I_{xz} \\ -I_{yx} & I_y & -I_{yz} \\ -I_{zx} & -I_{zy} & I_z \end{bmatrix} \begin{bmatrix} \omega_x \\ \omega_y \\ \omega_z \end{bmatrix} \tag{2.120}
$$

It should be noted that, having obtained the linear momentum $m\overline{v}$ and the angular momentum H_G of a rigid body, the angular momentum about a point O, H_O can be easily determined by addition of H_G, the angular momentum at G, with the angular

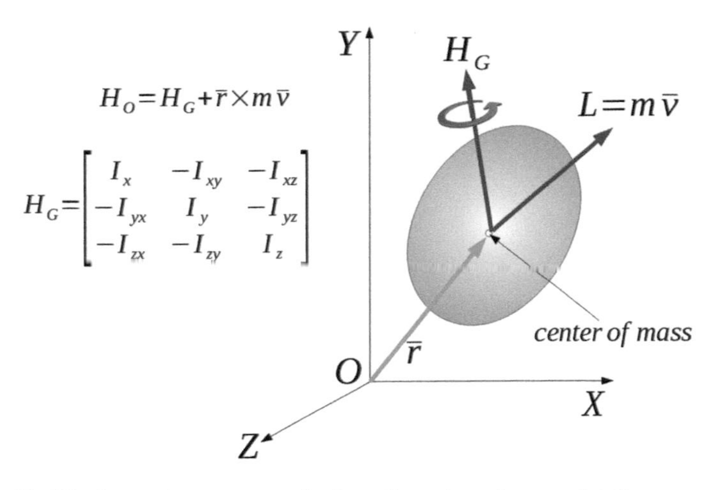

Fig. 2.22 Rigid body angular momentum in three dimensions about a point O

momentum about O of vector $m\overline{v}$ (see Fig. 2.22 and Beer et al. [4, p. 1150]) as follows:

$$H_O = H_G + \overline{r} \times m\overline{v} \tag{2.121}$$

where m is the mass of the body, $\overline{r}$ is the distance from point O to the center of gravity of the body, and $\overline{v}$ is the linear velocity at the c.g. point of the body.

Kinetic Energy of a Rigid Body in Three Dimensions (Beer et al. [4, pp. 1152–1153])

Examining a rigid body of mass m in three-dimensional motion, and recalling that if the absolute velocity v_i of each particle P_i of the rigid body is written as the sum of the velocity $\overline{v}$ of the mass center G of the body and of the velocity v_i' of each of the particles relative to a frame $Gxyz$ attached to G and of fixed orientation (see Fig. 2.23), the kinetic energy of the aggregate of the particles forming the rigid body can be expressed as:

$$T = \frac{1}{2}m\overline{v}^2 + \frac{1}{2}\sum_{i=1}^{n}(\Delta m_i)v_i'^2 \tag{2.122}$$

where the term on the right signifies the kinetic energy T' of the body relative to the centroidal non-rotating frame $Gxyz$. Since $v_i' = \omega \times r_i'$, T' of Eq. 2.122 may be expressed in the following form:

$$T' = \frac{1}{2}\sum_{i=1}^{n}(\omega \times r_i')^2 \Delta m_i \tag{2.123}$$

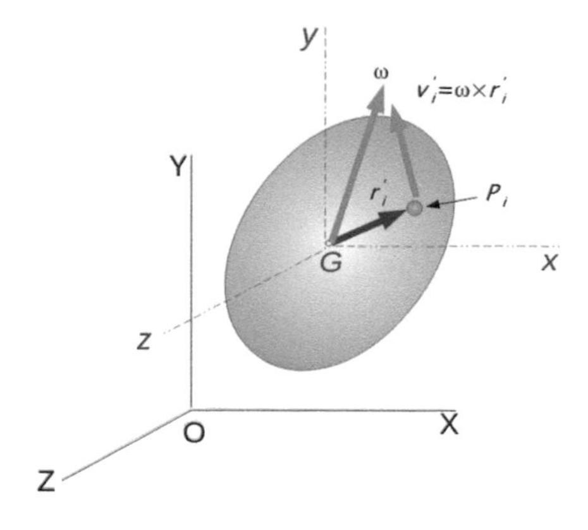

Fig. 2.23 Kinetic energy of a rigid body in three dimensions

From Eq. 2.115 we have: $\omega \times r_i' = \hat{i}(\omega_y z_i - y_i\omega_z) + \hat{j}(\omega_z x_i - z_i\omega_x) + \hat{k}(\omega_x y_i - x_i\omega_y)$, and thus $(\omega \times r_i') \cdot (\omega \times r_i')$ becomes:

$$\begin{aligned}(\omega \times r_i') \cdot (\omega \times r_i') &= (\omega_y z_i - y_i\omega_z)^2 + (\omega_z x_i - z_i\omega_x)^2 + (\omega_x y_i - x_i\omega_y)^2 \\ &= \omega_x^2(z_i^2 + y_i^2) + \omega_y^2(x_i^2 + z_i^2) + \omega_z^2(x_i^2 + y_i^2) \\ &\quad - 2\omega_x\omega_y x_i y_i - 2\omega_x\omega_z x_i z_i - 2\omega_y\omega_z z_i y_i\end{aligned} \tag{2.124}$$

The expression for T', Eq. 2.123 then becomes:

$$\begin{aligned}T' = \frac{1}{2}\sum_{i=1}^{n}\Big[&\omega_x^2(z_i^2 + y_i^2) + \omega_y^2(x_i^2 + z_i^2) + \omega_z^2(x_i^2 + y_i^2) \\ &- 2\omega_x\omega_y x_i y_i - 2\omega_x\omega_z x_i z_i - 2\omega_y\omega_z z_i y_i\Big]\Delta m_i\end{aligned} \tag{2.125}$$

Letting the masses Δm_i become infinitesimally small and replacing the summation sign with integrals results in:

$$\begin{aligned}T' = &\frac{1}{2}\omega_x^2\int(z^2 + y^2)dm + \frac{1}{2}\omega_y^2\int(x^2 + z^2)dm + \frac{1}{2}\omega_z^2\int(x^2 + y^2)dm \\ &-2\omega_x\omega_y\frac{1}{2}\int xydm - 2\omega_x\omega_z\frac{1}{2}\int xzdm - 2\omega_y\omega_z\frac{1}{2}\int zydm\end{aligned} \tag{2.126}$$

The expressions of Eq. 2.126 within the integral signs are the same as those of Eq. 2.118 and so T' may be written in the form:

$$T' = \frac{1}{2}\left[\omega_x^2 I_x + \omega_y^2 I_y + \omega_z^2 I_z - 2\omega_x\omega_y I_{xy} - 2\omega_x\omega_z I_{xz} - 2\omega_y\omega_z I_{yz}\right] \tag{2.127}$$

Substituting Eq. 2.127 into the expression for the kinetic energy of the body relative to centroidal axes, Eq. 2.122, the result is:

$$T = \frac{1}{2}m\overline{v}^2 + \frac{1}{2}\left[\omega_x^2 I_x + \omega_y^2 I_y + \omega_z^2 I_z - 2\omega_x\omega_y I_{xy} - 2\omega_x\omega_z I_{xz} - 2\omega_y\omega_z I_{yz}\right] \tag{2.128}$$

If the coordinate frame is chosen so that it coincides instantaneously with the principal axes x', y', z' of the body, that is the axis system wherein the products of inertia I_{xy}, I_{xz}, I_{yz} are all identically zero, the relation obtained in Eq. 2.128 reduces to:

$$T = \frac{1}{2}m\overline{v}^2 + \frac{1}{2}\left[\omega_{x'}^2 I_{x'} + \omega_{y'}^2 I_{y'} + \omega_{z'}^2 I_{z'}\right] \tag{2.129}$$

Chapter 3
Lagrangian Dynamics

In this chapter, the fundamental ideas which make up the main body of the theory of classical Lagrangian dynamics are presented. These include among others, the notion of generalized coordinates and degrees of freedom, generalized forces, configuration and velocity constraints, Pfaffian forms, and the definitions of holonomic and non-holonomic systems. The concept of virtual work for static systems and the principle of d'Alembert are used for the formulation of the virtual work principle for dynamical systems. These two aforementioned principles are combined into the extended Hamilton's principle which is the precursor to the derivation of the d'Alembert–Lagrange equations of motion. Since conservative forces play such a central role in classical mechanics, these types of forces are investigated and the equivalence between the work performed by a conservative force and the change in the potential energy due to the former is established. Finally, the d'Alembert–Lagrange procedure for the derivation of equations of motion is formulated with the aid of the extended Hamilton's principle. One method by which to deal with the problem of constraints within the framework of the d'Alembert–Lagrange formalism in mechanics is by means of Lagrangian multipliers which are covered in the present chapter. The generalized forces encountered in practice are usually of the non-conservative type and a systematic procedure employing the concept of virtual work and illustrated through examples is presented in order to deal with these forces. The final section of this chapter deals with the notions of Hamiltonians and the connection between the Lagrangian and the Hamiltonian via the Legendre transformation.

A. W. Pila, *Introduction To Lagrangian Dynamics*,
https://doi.org/10.1007/978-3-030-22378-6_3

3.1 Definitions Required for the Study of Lagrangian Dynamics

The following definitions, taken from the lectures by MIT Profs. J. Vandiver and David Gossard (see 2.003SC Engineering Dynamics—An Introduction to Lagrange Equations—Lecture Notes [49]), and additional material taken from Professor Haim Baruh's book [3, pp. 216–220, 223–224], shall be used henceforth:

Generalized Coordinates Consider an inertial coordinate system and let $\mathbf{r}_i(x_i, y_i, z_i)$ represent the location of the ith particle within this coordinate frame. The vector $\mathbf{r}_i$ may be expressed in Cartesian coordinates $\mathbf{i}, \mathbf{j}, \mathbf{k}$ as:

$$\mathbf{r}_i = x_i\mathbf{i} + y_i\mathbf{j} + z_i\mathbf{k} \quad i = 1, 2, \ldots, N \tag{3.1}$$

In many cases, it is more advantageous to use a different set of variables, other than the physical coordinates to describe the motion of the system. A set of variables $q_1, q_2, \ldots, q_n, \ n = 3N$, related to the physical coordinates $x_1, y_1, z_1, \ldots, x_N, y_N, z_N$ may be defined as follows:

$$\begin{aligned}
x_1 &= x_1\left(q_1, q_2, \ldots, q_n\right) \\
y_1 &= y_1\left(q_1, q_2, \ldots, q_n\right) \\
z_1 &= z_1\left(q_1, q_2, \ldots, q_n\right) \\
x_2 &= x_2\left(q_1, q_2, \ldots, q_n\right) \\
&\vdots \\
x_N &= x_N\left(q_1, q_2, \ldots, q_n\right) \\
y_N &= y_N\left(q_1, q_2, \ldots, q_n\right) \\
z_N &= z_N\left(q_1, q_2, \ldots, q_n\right)
\end{aligned} \tag{3.2}$$

The generalized coordinates of a mechanical system are therefore the minimum group of parameters which can completely and unambiguously define the configuration of that system. They are called generalized because they are not restricted to being Cartesian coordinates and are not even required to be measured from an inertial reference frame. However, they are used to express the kinetic and potential energies of the rigid bodies which make up the dynamical system. The kinetic and potential energy must be computed with respect to an inertial reference frame. Therefore the generalized coordinates must be able to express the velocity and displacement of rigid bodies with respect to an inertial frame of reference.

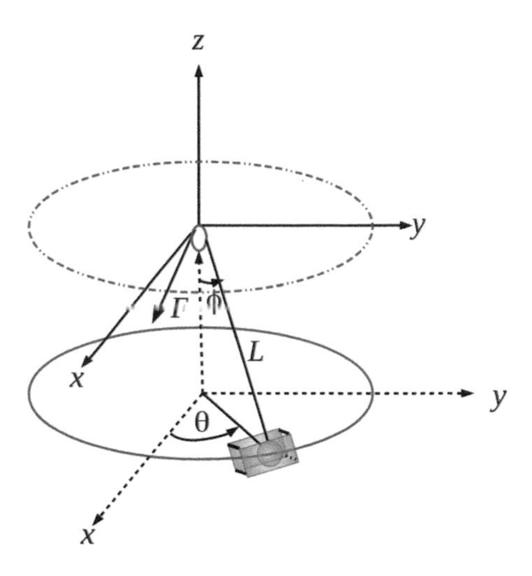

Fig. 3.1 A spherical pendulum whose length L may or may not vary

For example, the spherical pendulum in Fig. 3.1 can be located with Cartesian coordinates x, y, z or by generalized coordinates q_1, q_2, q_3, where $q_1 = L, q_2 = \theta$, $q_3 = \phi$. The two sets of coordinates are related by:

$$\begin{aligned} x &= L\cos\theta\sin\phi = q_1\cos(q_2)\sin(q_3) \\ y &= L\sin\theta\sin\phi = q_1\sin(q_2)\sin(q_3) \\ z &= -L\cos\phi = -q_1\cos(q_3) \end{aligned} \tag{3.3}$$

If the length of the pendulum is constant, that is, $q_1 = L = constant$, then only the variables $q_2 = \theta$ and $q_3 = \phi$ are required. If we decide to use the coordinates x, y, z to describe the motion, they must be related to the constraint equation, that is:

$$x^2 + y^2 + z^2 = L^2 = constant \tag{3.4}$$

Constraint relations such as the latter indicate that the generalized coordinates x, y, z are related to each other and hence it is possible to analyze the system with fewer of them. We therefore need to distinguish between a set of generalized coordinates where each coordinate is independent of all the others and a set where at least one generalized coordinate is not independent, such as in the above example. In general, for a system consisting of N particles with m constraints (m constraint equations) acting on it, the system can be uniquely described by p *independent generalized coordinates* q_k, $(k = 1, 2, \ldots, p)$, where:

$$p = 3N - m = n - m; \quad n = 3N \tag{3.5}$$

in which p signifies the number of degrees of freedom of the system. The number of degrees of freedom is independent of the coordinates used and is a characteristic

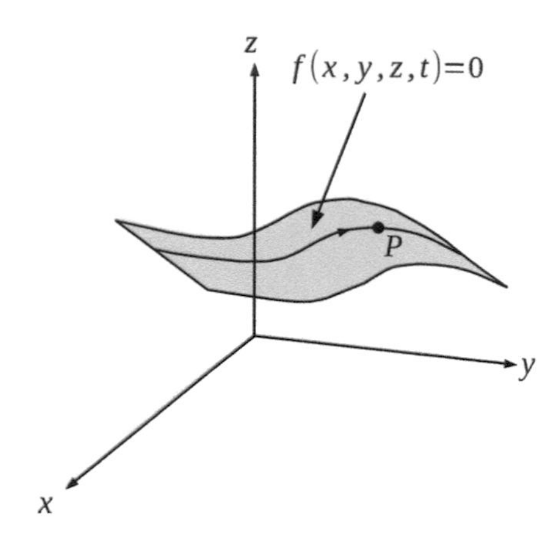

Fig. 3.2 A particle moving on a smooth surface

of the dynamical system. While there exists more than one way of selecting the number and types of generalized coordinates, the number of degrees of freedom $p = n - m$ is invariant.

Equality Constraints In dynamical systems, constraints arise as a result of contact between two or more bodies and their effects are to restrict the motion of the bodies to which the constraints are applied. Each constraint has associated with it a *constraint equation* and a *constraint force*, which may be regarded as a contact or reaction force. The constraint equation is a description of the geometry or kinematics of the contact. Note that equations of constraint may also be written when the observer is situated within a moving reference frame and there are no contacts. The relative motion equation between the observer and the system under observation then becomes the constraint equation. Figure 3.2 depicts a particle P, constrained to move on a smooth surface whose shape function or *configuration constraint* is:

$$f(x, y, z, t) = 0 \tag{3.6}$$

The shape function $f(x, y, z, t) = 0$ is smooth and has both first and second derivatives in all of its variables. For a system described in terms of n generalized coordinates, the corresponding configuration constraint may be written in the form:

$$f(q_1, q_2, \ldots, q_n, t) = 0 \tag{3.7}$$

and the time derivative of f (in terms of its generalized coordinates $q_1, q_2, \ldots, q_n, t$) may be shown to be:

$$\frac{df}{dt} = \frac{\partial f}{\partial q_1}\dot{q}_1 + \frac{\partial f}{\partial q_2}\dot{q}_2 + \cdots + \frac{\partial f}{\partial q_n}\dot{q}_n + \frac{\partial f}{\partial t} = 0 \tag{3.8}$$

The corresponding differential, which expresses the *constraint relations in Pfaffian form*, may be written as:

$$df = \frac{\partial f}{\partial q_1}dq_1 + \frac{\partial f}{\partial q_2}dq_2 + \cdots + \frac{\partial f}{\partial q_n}dq_n + \frac{\partial f}{\partial t}dt = 0$$

The time derivative (Eq. 3.8) of the constraint equation results in *constraint equations in velocity form (or velocity or motion constraints)*, which may be written in terms of the n generalized coordinates as:

$$\frac{df}{dt} = \frac{\partial f}{\partial q_1}\dot{q}_1 + \frac{\partial f}{\partial q_2}\dot{q}_2 + \cdots + \frac{\partial f}{\partial q_n}\dot{q}_n + \frac{\partial f}{\partial t} = 0 \tag{3.9}$$

As becomes apparent, the general form of a velocity constraint in terms of the n generalized coordinates becomes:

$$\sum_{k=1}^{n} a_{jk}\dot{q}_k + a_{j0} = 0 \quad j = 1, 2, \ldots, m \tag{3.10}$$

where a_{jk} and a_{j0} ($j = 1, 2, \ldots, m; k = 1, 2, \ldots, n$) are functions of the generalized coordinates, for example, $a_{jk} = a_{jk}(q_1, q_2, \ldots, q_n, t)$. Assuming that $f(q_1, q_2, \ldots, q_n, t) = 0$ represents the ith constraint, it is possible to relate the partial derivatives of f with respect to the generalized coordinates to the a_{jk} terms and a_{j0} as follows:

$$\frac{df}{dt} = \frac{\partial f}{\partial q_1}\dot{q}_1 + \frac{\partial f}{\partial q_2}\dot{q}_2 + \cdots + \frac{\partial f}{\partial q_n}\dot{q}_n + \frac{\partial f}{\partial t} = 0$$

$$\sum_{k=1}^{n} a_{jk}\dot{q}_k + a_{j0} = 0 \quad j = 1, 2, \ldots, m$$

$$\Rightarrow a_{i1} = \frac{\partial f}{\partial q_1}, a_{i2} = \frac{\partial f}{\partial q_2}, \ldots, a_{in} = \frac{\partial f}{\partial q_n}, a_{i0} = \frac{\partial f}{\partial t} \tag{3.11}$$

It should be noted that when constraints are imposed upon a set of independent generalized coordinates, some of the generalized coordinates lose their independence. A constraint that can be expressed as both a configuration constraint $f(q_1, q_2, \ldots, q_n) = 0$ and a velocity constraint of the type $\sum_{k=1}^{n} a_{jk}\dot{q}_k + a_{j0} = 0$ $j = 1, 2, \ldots, m$ is referred to as a *holonomic constraint* (see Prof. Marghitu [20, pp. 211]). Constraints which do not possess this property are designated as *non-holonomic constraints*. A holonomic constraint which doesn't depend explicitly on time, that is, $f(q_1, q_2, \ldots, q_n) = 0$, is called a *scleronomic constraint*, while a holonomic constraint which is explicitly time dependent is known as a *rheonomic constraint*. Since the majority of constraints encountered in engineering applications are scleronomic, only these types of holonomic constraints will be considered.

Holonomic Systems From the discussion above, it is to be noted that a constraint that can be expressed as both a configuration and a velocity constraint is referred to as a *holonomic constraint*. When there are no constraints which restrict the motion of the system, it becomes apparent that a holonomic system is one in which the number of independent generalized coordinates required to describe the motion of the system equals the number of degrees of freedom, that is, $p = n - m; m = 0$.

Non-Holonomic Systems When the constraint is non-holonomic, there is no integration factor which will allow the differential df to be integrated to become f. Furthermore, the constraint force associated with a non-holonomic constraint cannot be expressed as a force normal to a surface S, i.e., the non-holonomic constraint doesn't define a surface—it is a constraint on the generalized velocities.

Consider a system that originally has n degrees of freedom and is subjected to m holonomic constraints. The constraints reduce the degrees of freedom from n to $n - m$. The resulting system is unconstrained of order $n - m$.

By contrast, a non-holonomic constraint constrains only the generalized velocities without affecting the generalized coordinates. In such systems there are n generalized coordinates and $n - m$ independent generalized velocities.

Figure 3.3 depicts a typical non-holonomic system undergoing planar motion. The configuration coordinates related to point A are X_A, Y_A with respect to the inertial coordinates X, Y and θ, the angle the body makes relative to the inertial X axis. The non-holonomic constraint is associated with the translational velocity of point A, v_A, which, in vector form is:

$$v_A = \dot{X}_A I + \dot{Y}_A J \tag{3.12}$$

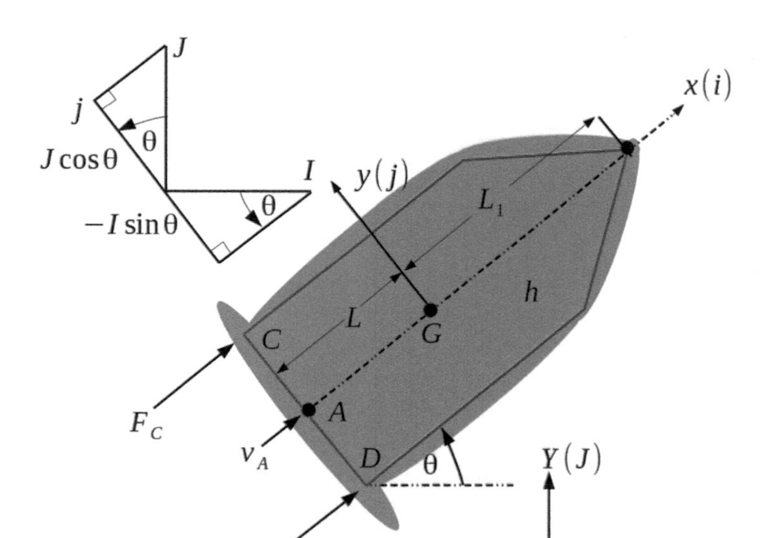

Fig. 3.3 Generic vehicle

Since there is no sideways motion, the constraint may be described by:

$$\begin{aligned} v_A \cdot j &= 0;\ j = \cos\theta J - \sin\theta I \Rightarrow v_A \cdot j \\ &= (\dot{X}_A I + \dot{Y}_A J) \cdot (\cos\theta J - \sin\theta I) \\ &= \dot{Y}_A \cos\theta - \dot{X}_A \sin\theta = 0 \Rightarrow \dot{X}_A - \frac{\dot{Y}_A}{\tan\theta} = 0 \end{aligned} \tag{3.13}$$

The system is non-holonomic since the velocity constraint cannot be integrated to yield a configuration constraint. The associated constraint force is essentially the resistance of point A to any sideways motion or the resistance to any movement perpendicular to the motion in the v_A direction. A different category of constraints restricts a particle or system to a domain bounded by a surface. Such a constraint may be expressed as an inequality of the form: $f(x, y, z, t) \geq 0$, for a bounding surface in motion and is known as an *inequality constraint*. It is also non-holonomic since it cannot be integrated.

Degrees of Freedom The number of degrees of freedom of a system is the number of coordinates that can be independently varied in a small displacement. Stated another way, the number of degrees of freedom is the same as the number of independent directions in which a system can move from any given initial configuration. Consider a system consisting of N rigid bodies in 2D space. Each rigid body has three degrees of freedom: two translational and one rotational. The N-body system has 3N degrees of freedom. Now let's say that there are k kinematic constraints. Then the system has $d = 3N - k$ degrees of freedom. For example, a single uniform stick confined to a vertical, x–y plane has three degrees of freedom, x, y, and rotation, θ about the z axis. It has no additional constraints. If we connect one end of the stick to a pivot, it now has two constraints, one in each of the x and y directions. It is still free to rotate. Therefore $d = 3 - 2 = 1$ dof. The motion of the pinned stick may be completely specified by one coordinate, the angle of rotation.

Completeness and Independence of Generalized Coordinates In choosing which coordinates to use one must obey two requirements:

1. A set of generalized coordinates is complete, if it is capable of locating all parts of the system at all times.
2. A set of generalized coordinates is independent if, when all but one of the generalized coordinates are fixed, there remains a continuous range of values for that one coordinate.

For example, specifying coordinates x_1, y_1 and x_2, y_2 in the double pendulum system depicted in Fig. 3.4, we have four coordinates which fully describe the locations of the masses of the system. However, the system has only two degrees of freedom. It turns out that the specified coordinates are not independent, since fixing x_2, y_2 and y_1 will not allow us to obtain a continuous range of values for x_1 (we assume that the strings are of fixed length). A better choice for the generalized coordinates would be θ_1 and θ_2. These two coordinates are indeed independent.

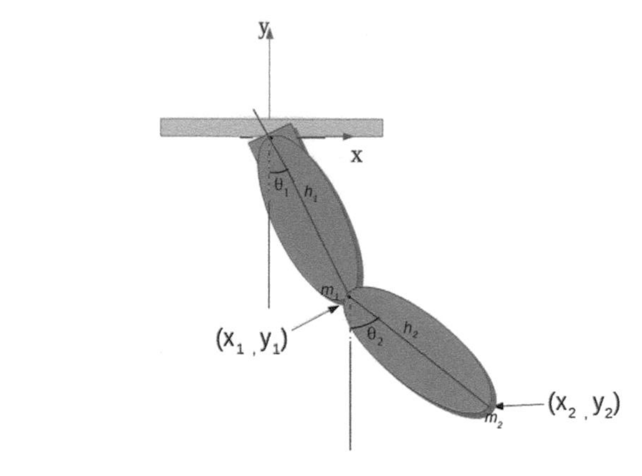

Fig. 3.4 Double pendulum system

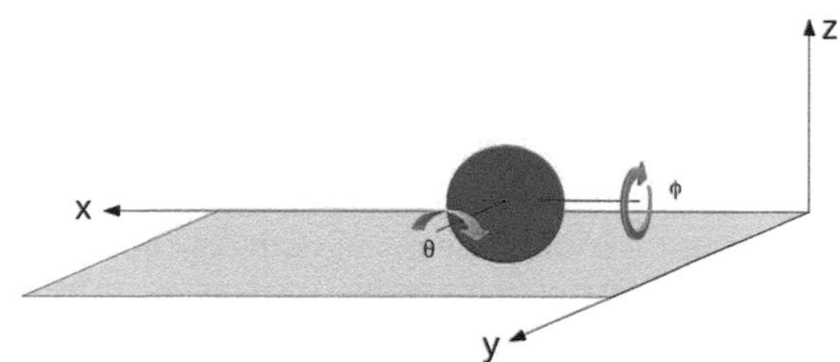

Fig. 3.5 Non-holonomic system—# of degrees of freedom ≤ # coord. needed to fully determine position of the ball

In summary, in order to find the equations of motion of a dynamical system using Lagrange equations, one must first determine the number of degrees of freedom, "d," and then choose a set of "d" generalized coordinates, which make up a complete and independent set. If the system is non-holonomic, and of the following type: $\sum_{e=1}^{n} a_{je}\dot{q}_e + a_{jt}dt = 0;\ j = 1, 2, \ldots, k$, it is possible to solve a modified form of Lagrange's equations in order to obtain the system's dynamics. This topic will be dealt with in greater detail in Chap. 4. An example of a non-holonomic system in which more coordinates are needed than the system's degrees of freedom is described by the following system (see Fig. 3.5).

The ball is allowed to roll about the x and y axes, but not spin around the z axis. The ball without any constraints has six degrees of freedom. When it is constrained to not have any translatory motion in the z direction, it then has five degrees of freedom. Without being allowed to spin along the z axis it is therefore reduced to four degrees of freedom. The ball cannot slip, i.e., it has no translatory motion in the $x - y$ plane and this reduces the number of degrees of freedom from four to two. However, to fully describe the location of the ball and its orientation, four coordinates are required, two translation coordinates in the $x - y$ plane and two angular coordinates θ around the y axis and ϕ around the x axis, respectively. Hence the number of degrees of freedom is smaller than the number of coordinates needed to fully describe its position and the system is therefore non-holonomic (see Prof. Vandiver, Lecture 15 [42]).

3.2 Summary: Holonomic and Non-holonomic Systems

Definitions

- Holonomic Constraints

  1. If the conditions of constraint can be expressed as equations connecting the coordinates of the particle (and possibly the time) having the form: $f(r_1, r_2, \ldots, t) = 0$, then the constraints are said to be *holonomic*. Constraints which are not expressible in the form of the above equation are *non-holonomic* (see Goldstein [28, pp. 12–16]).
  2. A constraint that can be expressed as an equation in just the configuration variables, and possibly time, but independent of the rate variables" (see Mason [21, pp. 25–27]).
  3. Since a holonomic constraint is one that can be expressed as a functional relationship between the coordinates: $f(q_1, q_2, \ldots) = 0$, using such a constraint, by simple substitution, one of the coordinates may be eliminated.

- Non-Holonomic Constraints

  1. If the conditions of constraint can be expressed as follows: $f(r_1, r_2, \ldots, r_n, \dot{r}_1, \dot{r}_2, \ldots, \dot{r}_n, t) = 0$, then the constraints are said to be *non-holonomic*.
  2. For a non-holonomic constraint, either rate variables or inequalities are required (see Mason [21, pp. 25–27]).
  3. A non-holonomic constraint is one that cannot be expressed in terms of the position variables alone, but includes the time derivative of one or more of those variables. These constraint equations cannot be integrated to obtain relationships solely between the joint variables. The most common example in robotic systems arises from the use of a wheel or roller that rolls without slipping on another member (see Springer Handbook of Robotics [33, pp. 25]).
  4. In general, a non-holonomic constraint is a differential equality constraint that cannot be integrated into a constraint that involves no derivatives. A simple example is that of a car which cannot move sideways, thereby making parallel parking more difficult (see Springer Handbook of Robotics [33, pp. 148]).
  5. Three variables are required to locate a car on the plane—two for its position and one for its angle. However a car has only two degrees of freedom at any instant, acceleration/deceleration and steering, yet it can reach any configuration in the plane by judicious maneuvers.

6. In the case of non-holonomic constraints it is not possible to give arbitrary and independent variations to the generalized coordinates without violating the constraint relations. In a non-holonomic system, the variations or changes $\delta q_1, \delta q_2, \ldots, \delta q_k$ are related and are not independent. This implies that (for scleronomic non-holonomic constraints):

$$\sum_{e=1}^{n} a_{je}\delta q_e = 0; \; j = 1, 2, \ldots, \; k; k < n \tag{3.14}$$

The number of degrees of freedom is denoted by n, while k expresses the number of constraint relations. These equations are in differential form and are not integrable (see Greenwood [14, pp. 36–37] and Greiner [15, pp. 301–303]). Non-holonomic constraints cannot be written in closed form (algebraic equation), but instead must be expressed in terms of the differentials of the coordinates (and possibly time), that is:

$$\sum_{e=1}^{n} a_{je} dq_e + a_{jt} dt = 0; \; j = 1, 2, \ldots, k$$

$$a_{je} = \psi\left(q_1, q_2, \ldots, q_n, t\right) \tag{3.15}$$

7. Constraints of this type are non-integrable and restrict the velocities of the system, which implies that:

$$\sum_{e=1}^{n} a_{je}\dot{q}_e + a_{jt} = 0; \; j = 1, 2, \ldots, k \tag{3.16}$$

8. The following example was taken from Lecture 7 of the lecture series by Prof. Deyst and How [9, pp. 15].
9. **Example**: Wheel rolling without slipping on a curved path (see Fig. 3.6). Define ϕ as the angle between the tangent to the path and the X-axis at the instant of interest.
10. The non-holonomic constraints of rolling without slipping are as follows:

$$\dot{x} = v \cos\phi = r\dot{\theta} \cos\phi$$

$$\dot{y} = v \sin\phi = r\dot{\theta} \sin\phi$$

$$dx - r\cos\phi d\theta = 0$$

$$dy - r\sin\phi d\theta = 0 \tag{3.17}$$

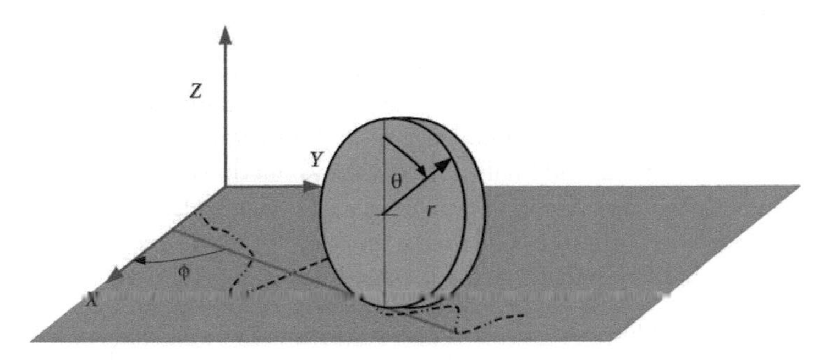

Fig. 3.6 Wheel rolling without slipping on a curved path—classic example of a non-holonomic system

11. There are two differential equations of constraint, neither of which can be integrated without solving the entire problem. Hence the constraints are non-holonomic. The reason is that the change in θ can be related to the change in x, y for a given ϕ, but the absolute value of θ depends on the path taken to get to that point (which is the "solution"). In other words, the absolute value of θ depends on the path taken to get to that point or the angle ϕ, which constitutes the solution to the problem. The problem cannot be solved without knowledge of the solution. Since the angle ϕ is known only after solving the problem, the equations are not integrable. Hence, the problem is non-holonomic.
12. Why study non-holonomic constraints?
    (a) It is fundamental to all of robotics.
    (b) The robot has only a few motors, say k, while the task has many degrees of freedom (DoF), say n. How many independent motions can the robot produce? At most k. How many degrees of freedom in the task does the robot wish to control? Perhaps all n. The difference between the number of DoFs and the number of motors implies a non-holonomic constraint (see Mason Lecture 5 [22, pp. 8]).
    (c) A holonomic constraint is a constraint on configuration: it says there are places you cannot go. It implies a reduction in freedom of motion, while a non-holonomic constraint is a constraint on velocity: there are directions you cannot go in, but you can still get to wherever you desire (see Mason Lecture 5 [22, pp. 25]).
13. Non-holonomic Constraints: Constraints that constrain the velocities of particles but not their positions (see Meam [18, pp. 7]).
14. With non-holonomic systems, the generalized coordinates are not independent of each other (see Fantoni and Lozano [10, pp. 17]).
15. A non-holonomic system is one in which there is a continuous closed circuit of the governing parameters, by which the system may be transformed from any given state to any other state. Because the final state of the system

depends on the intermediate values of its trajectory through parameter space, the system cannot be represented by a conservative potential function such as, for example, the inverse square law of the gravitational force. This latter is an example of a holonomic system: path integrals in the system depend only upon the initial and final states of the system (positions in the potential), completely independent of the trajectory of transition between those states. The system is therefore said to be *integrable*, while the non-holonomic system is said to be *non-integrable* (see Wikipedia [1]).

For systems with holonomic constraints, the dependent coordinates can be eliminated by introducing generalized coordinates. If the constraints are non-holonomic, this approach does not work. There is no general method for treating non-holonomic problems. Only for those special non-holonomic constraints that can be given in differential form can one eliminate the dependent equations by the method of Lagrange multipliers or quasi-velocities. The approach taken closely mirrors that of Meirovitch [23, pp. 157–164], Cameron and Book [8, pp 47–59], and Vepa[50] and will be presented in the next chapter. There exists a more advanced approach to the subject of non-holonomic systems, their characterization, identification, and control based on the following topics found in differential geometry and related subject matter such as Lie groups, Lie algebras, etc. A partial list includes the following topics:

1. Manifolds, differentiable manifolds, manifolds and maps
2. Tangent vectors, spaces, vector fields
3. Fiber bundles
4. Differential k-forms
5. Exterior derivatives
6. Jacobi–Lie brackets, Lie groups
7. Vector fields and flows
8. Lie brackets and Frobenius' theorem, the Lie algebra associated with a Lie group, actions of Lie groups, Canonical coordinates on a Lie group
9. Tangent spaces and tangent maps
10. Cotangent spaces and cotangent maps
11. Differential forms
12. The exponential map
13. The geometry of the Euclidean group, metric properties of SE(3), volume forms on SE(3)
14. Lie groups and robot kinematics

The interested reader is encouraged to pursue these topics in greater detail by referring to the following works by Murray et al. (A Mathematical Introduction to Robotic Manipulation [25]), Selig (Geometrical Methods in Robotics [32]), Bullo and Lewis (Geometric Control of Mechanical Systems Modeling, Analysis, and Design for Simple Mechanical Control Systems [7]), Bloch et al. (Nonholonomic Mechanics and Control [5]), Soltakhanov et al. (Mechanics of Non-Holonomic systems [35]), to name but a few.

3.3 Virtual Work for Static Systems Only

The approach taken in the sequel is that outlined in the book by Meirovitch (see [24, pp. 263–267]). Consider a system of N particles residing within a three-dimensional enclosure and let the *virtual displacements*

$$\delta x_1,\ \delta y_1,\ \delta z_1,\ \delta x_2,\ \delta y_2, \delta z_2, \ldots, \delta x_N,\ \delta y_N,\ \delta z_N$$

be defined as *infinitesimal changes* in the Cartesian coordinates of $x_1, y_1, z_1, x_2, y_2, z_2, \ldots, x_N, y_N, z_N$. Although the virtual displacements are arbitrary, they must however *comply with the system constraints*. For example, if a particle is confined to a surface, then the virtual displacement must be along that surface, since a real particle can neither penetrate nor leave the surface. The virtual displacements are small (imaginary) variations in the coordinates, achieved by envisaging that the system's position has been ever so slightly displaced. The underlying assumption is that the virtual displacement process is instantaneous, and so $\delta t = 0$. Since the virtual displacements are infinitesimal they obey the rules of differential calculus. A key presupposition is that *on every particle in the system there is a resultant force* R_i which is the sum of an externally applied force such as gravity, aerodynamic lift, drag, etc. and a constraining force, such as a force which confines the motion of the particle to the surface upon which it moves. Mathematically, we have:

$$R_i = F_i + f_i\ , i = 1, 2, \ldots, N$$

where F_i is an externally applied force and f_i is a constraining force. When the system is in equilibrium, each and every particle must be at rest, which implies that the resultant force on each particle must be nullified, or $R_i = F_i + f_i = 0\ , i = 1, 2, \ldots, N$. The virtual work on the ith particle is defined to be:

$$\overline{\delta W_i} = R_i \cdot \delta r_i = 0\ , i = 1, 2, \ldots, N$$

since $R_i = 0\ , i = 1, 2, \ldots, N$. The above scalar products $R_i \cdot \delta r_i = 0\ , i = 1, 2, \ldots, N$ represent the *virtual* work carried out by the resultant force R_i due to the virtual displacement vector δr_i of the ith particle. Summing up the virtual work for all of the N particles, we have:

$$\overline{\delta W} = \sum_{i=1}^{N} R_i \cdot \delta r_i = 0$$

$$\Rightarrow \overline{\delta W} = \sum_{i=1}^{N} F_i \cdot \delta r_i + \sum_{i=1}^{N} f_i \cdot \delta r_i = 0 \tag{3.18}$$

We limit ourselves to systems for which the constraint forces act in directions which are perpendicular or normal (or conjugate) to the virtual motions, such as a particle confined to a smooth surface. *This implies that the virtual work due to constraint forces are zero.* Conversely, if a particle is confined to a jagged, uneven surface, there is a tangential component of the constraint force due to friction, in addition to the component in the normal direction, which results in positive virtual work due to the tangential constraint force. Thus, we rule out any constraining forces, such as friction, etc., which gives rise to a non-zero virtual work value. With the virtual work due to forces of constraint nullified, the sum of the virtual work due to constraint forces and external forces may be written as:

$$\sum_{i=1}^{N} f_i \cdot \delta r_i = 0 \Rightarrow \overline{\delta W} = \sum_{i=1}^{N} F_i \cdot \delta r_i = 0 \tag{3.19}$$

When the virtual displacements δr_i are all independent and comply with system constraints, we can invoke the arbitrariness of the virtual displacements and conclude that

$$\overline{\delta W} = \sum_{i=1}^{N} F_i \cdot \delta r_i = 0 \tag{3.20}$$

which can be satisfied for all possible values of δr_i only if

$$F_i = 0, i = 1, 2, \ldots, N$$

The preceding statement is known as *the principle of virtual work*. However, when the coordinates $r_i (i = 1, 2, \ldots, N)$ are related by constraint equations, i.e., they are not independent, it is more convenient to use "generalized coordinates" $q_1, q_2, \ldots, q_n$, which may or may not be Cartesian coordinates. In vector form, we have:

$$r_i = r_i (q_1, q_2, \ldots, q_n) \, , i = 1, 2, \ldots, N \tag{3.21}$$

The rules of differential calculus must now be invoked in order to obtain the virtual displacements:

$$\delta r_i = \frac{\partial r_i}{\partial q_1} \delta q_1 + \frac{\partial r_i}{\partial q_2} \delta q_2 + \cdots + \frac{\partial r_i}{\partial q_n} \delta q_n = \sum_{k=1}^{n} \frac{\partial r_i}{\partial q_k} \delta q_k \ \ , i = 1, 2, \ldots, N \tag{3.22}$$

where $\delta q_1, \delta q_2, \ldots, \delta q_n$ are all independent virtual generalized displacements, as opposed to $\delta r_i (i = 1, 2, \ldots, N)$ which are not independent. Replacing δr_i in

Eq. (3.20), with the value determined in Eq. (3.22) and changing the order of the summations, we have:

$$\overline{\delta W} = \sum_{i=1}^{N} F_i \cdot \delta r_i = \sum_{i=1}^{N} F_i \cdot \sum_{k=1}^{n} \frac{\partial r_i}{\partial q_k} \delta q_k = \sum_{k=1}^{n} \left(\sum_{i=1}^{N} F_i \cdot \frac{\partial r_i}{\partial q_k} \right) \delta q_k$$
$$= \sum_{k=1}^{n} Q_k \delta q_k = 0 \qquad (3.23)$$

where

$$Q_k = \sum_{i=1}^{N} F_i \cdot \frac{\partial r_i}{\partial q_k}, k = 1, 2, \ldots, n \qquad (3.24)$$

are designated as the *generalized forces*. Since all of the δq_k are now independent by definition and entirely arbitrary, it is possible to assign values to them according to our whim. Letting $\delta q_1 = 1, \delta q_2 = \delta q_3 =, \ldots, = \delta q_n = 0$, it becomes apparent that only $Q_1 = 0$ can satisfy Eq. (3.23). Repeating the same procedure, but with $k = 2, 3, \ldots, n$, in sequence, we obtain the conditions for the static equilibrium of the system of particles:

$$Q_k = 0, k = 1, 2, \ldots, n$$
$$\delta q_1 = \delta q_2 = \cdots = \delta q_{k-1} = \delta q_{k+1} = \ldots, \delta q_n = 0, \ \delta q_k = 1 \qquad (3.25)$$

3.4 The Principle of d'Alembert for Dynamical Systems

This section was adapted from Meirovitch [23, pp. 65–66] and Meirovitch [24, pp. 267–268].

Since the principle of virtual work was developed for the static equilibrium of systems, it cannot be used directly to derive the equations of motion of systems with dynamics. The principle of virtual work may however be extended to cover dynamical systems by employing a quasi-equilibrium tool, known as *d'Alembert's principle for dynamical systems*. Returning to the definition of the resultant force $R_i = F_i + f_i, i = 1, 2, \ldots, N$, assume that m_i, a typical mass particle in a system of particles ($i = 1, 2, \ldots, N$), is subjected to the applied force F_i and the constraint force f_i, while any internal forces are miniscule and may be neglected. Newton's second law for particle of mass m_i then takes the form:

$$F_i + f_i - m_i \ddot{r}_i = 0, i = 1, 2, \ldots, N \qquad (3.26)$$

where $-m_i\ddot{r}_i$ may be looked upon as an *inertial force*. Recall that an inertial force resists a change in the velocity of an object and is equal to and in the opposite direction of an applied force, as well as a resistive force. The inertial force $-m_i\ddot{r}_i$ is the negative of the time derivative of the momentum vector $p_i = m_i\dot{r}_i$. Equation (3.26) is the mathematical statement of d'Alembert's principle. The beauty of d'Alembert's principle is that it allows for the treatment of problems of dynamics in the same manner as static problems. Applying the principle of virtual work in Eq. (3.20) to the mass particle m_i and using d'Alembert's principle, Eq. (3.26), yields:

$$(F_i + f_i - m_i\ddot{r}_i) \cdot \delta r_i = 0\,, i = 1, 2, \ldots, N \tag{3.27}$$

Once again limiting the discussion to constraint forces f_i with zero virtual work and taking the sum over all of the mass particles leads to:

$$\sum_{i=1}^{N}(F_i - m_i\ddot{r}_i) \cdot \delta r_i = 0 \tag{3.28}$$

Equation (3.28) incorporates both the virtual work principle (of statics) and d'Alembert's principle for dynamical systems and is termed as the *generalized principle of d'Alembert* or as the *Lagrange version of d'Alembert's principle*. The sum of the applied and the inertial force, that is, $F_i - m_i\ddot{r}_i$, is sometimes designated as the *effective force* acting on mass particle m_i. The generalized principle of d'Alembert may be stated as follows: *The virtual work carried out by the effective forces over infinitesimal virtual displacements, which are compatible with system constraints, is zero.*

If the position vectors $r_i(i = 1, 2, \ldots, N)$ are all independent, d'Alembert's principle, Eq. (3.28), may be used to derive all of the equations of motion of the system. Otherwise, coordinate transformations, from the dependent coordinates $r_i(i = 1, 2, \ldots, N)$ to the independent generalized coordinates $q_k(k = 1, 2, \ldots, n)$, must be carried out as indicated by the following equation:

$$r_i = r_i(q_1, q_2, \ldots, q_n)\,, i = 1, 2, \ldots, N \tag{3.29}$$

3.5 The Mathematics of Conservative Forces

This section is based upon Meirovitch [23, pp. 15–17], Ginsburg [13, pp. 213–214], and Baruh [3, pp. 44].

A very important class of forces is the class of conservative forces for which *the work depends only on the initial position r_1 and the final position r_2* and not on the path taken from r_1 to r_2. Denoting two distinct paths from r_1 to r_2 by I and II, respectively, the non-dependence of the work on the specific path may be stated mathematically as follows:

$$\underbrace{\int_{r_1}^{r_2} F \cdot dr}_{PathI} = \underbrace{\int_{r_1}^{r_2} F \cdot dr}_{PathII} \tag{3.30}$$

The above equation may also be rewritten in the form:

$$\underbrace{\int_{r_1}^{r_2} F \cdot dr}_{PathI} - \underbrace{\int_{r_1}^{r_2} F \cdot dr}_{PathII} = \underbrace{\int_{r_1}^{r_2} F \cdot dr}_{PathI} + \underbrace{\int_{r_2}^{r_1} F \cdot dr}_{PathII} = \oint F \cdot dr = 0 \tag{3.31}$$

in which $\oint$ denotes an integral over a closed path. In view of Eq. (3.31) it is possible to state that *the work performed by conservative forces over a closed path is zero.* In the sequel, conservative forces will be labeled with the subscript c and non-conservative forces with the superscript *NC* in order to distinguish between them.

Consider a conservative force F_c and choose a path from r_1 to r_2 which passes through the reference position r_{ref}. Define the potential energy as the work performed by the conservative force in moving a particle from position r along the path to the reference position r_{ref}, or

$$V(r) = \int_{r}^{r_{ref}} F_c \cdot dr \tag{3.32}$$

where $V(r)$ is a scalar function depending on r alone, since r_{ref} is arbitrary. The work performed by conservative forces in moving a particle from position r_1 to position r_2 may be expressed in the form:

$$\begin{aligned}\int_{r1}^{r2} F_c \cdot dr = \int_{r1}^{r_{ref}} F_c \cdot dr + \int_{r_{ref}}^{r2} F_c \cdot dr = \int_{r1}^{r_{ref}} F_c \cdot dr - \int_{r2}^{r_{ref}} F_c \cdot dr \\ = V(r_1) - V(r_2) = -(V_2 - V_1)\end{aligned} \tag{3.33}$$

where $V_i = V(r_i)(i = 1, 2)$. Equation (3.33) states that *the work performed by conservative forces in moving a particle from r_1 to r_2 is equal to the negative of the change in potential energy from V_1 to V_2.*

In general, forces can be divided into conservative and non-conservative classes with the latter being denoted by a superscript of NC, that is, F^{NC}. In accordance with the above, work may be expressed as the sum of the work due to the conservative and non-conservative forces, that is:

$$\int_{r1}^{r2} F \cdot dr = \int_{r1}^{r2} F_c \cdot dr + \int_{r1}^{r2} F^{NC} \cdot dr$$
$$= T_2 - T_1 = -(V_2 - V_1) + \int_{r1}^{r2} F^{NC} \cdot dr \qquad (3.34)$$

where T_i is the kinetic energy related to the coordinate of position $r_i (i = 1, 2)$. When dW is a *perfect differential*, the integral of Eq. (3.33) may be written in differential form as:

$$dW = F(r) \cdot dr = -dV(r)$$

where the potential function V is a function of r only. In order for dW to be a perfect differential, the force must depend only on the position vector, which implies that the force must be conservative

3.6 The Extended Hamilton's Principle

Beginning with the case where all of the position vectors $r_i (i = 1, 2, \ldots, N)$ are independent, the virtual work of all of the applied forces including conservative and non-conservative forces may be written as:

$$\sum_{i=1}^{N} F_i \cdot \delta r_i = \overline{\delta W} \qquad (3.35)$$

We now consider $m_i \ddot{r}_i$ which we would like to modify to a more suitable form:

$$\frac{d}{dt}(m_i \dot{r}_i \cdot \delta r_i) = m_i \ddot{r}_i \cdot \delta r_i + m_i \dot{r}_i \cdot \delta \dot{r}_i$$
$$= m_i \ddot{r}_i \cdot \delta r_i + \delta \left(\frac{1}{2} m_i \dot{r}_i \cdot \dot{r}_i \right) = m_i \ddot{r}_i \cdot \delta r_i + \delta T_i \qquad (3.36)$$

where δT_i is the kinetic energy of particle m_i. Note that the variation δ obeys the rules of differential calculus and hence $\delta(\frac{1}{2} m_i \dot{r}_i \cdot \dot{r}_i)$ may be written in the form:

$$\delta \left(\frac{1}{2} m_i \dot{r}_i \cdot \dot{r}_i \right) = \frac{1}{2} m_i \delta \dot{r}_i \cdot \dot{r}_i + \frac{1}{2} m_i \dot{r}_i \cdot \delta \dot{r}_i = m_i \delta \dot{r}_i \cdot \dot{r}_i \qquad (3.37)$$

Rearranging $\frac{d}{dt}(m_i\dot{r}_i \cdot \delta r_i) = m_i\ddot{r}_i \cdot \delta r_i + \delta T_i$ and integrating with respect to time over the interval $t_1 \leq t \leq t_2$ leaves us with:

$$\begin{aligned} -\int_{t_1}^{t_2} m_i\ddot{r}_i \cdot \delta r_i dt &= \int_{t_1}^{t_2} \delta T_i dt - \int_{t_1}^{t_2} \frac{d}{dt}(m_i\dot{r}_i \cdot \delta r_i)dt \\ &= \int_{t_1}^{t_2} \delta T_i dt - m_i\dot{r}_i \cdot \delta r_i|_{t_1}^{t_2} \end{aligned} \tag{3.38}$$

Since the virtual displacements are arbitrary, they may be chosen so as to satisfy $\delta r_i = 0$ at $t = t_1$ and $t = t_2$. This being the case, Eq. (3.38) reduces to:

$$-\int_{t_1}^{t_2} m_i\ddot{r}_i \cdot \delta r_i dt = \int_{t_1}^{t_2} \delta T_i dt; \delta r_i = 0, t = t_1, t_2; i = 1, 2, \ldots, N \tag{3.39}$$

Evaluating the sum taken over i and integrating with respect to t over the interval $t_1 \leq t \leq t_2$, we have:

$$-\int_{t_1}^{t_2} \sum_{i=1}^{N} m_i\ddot{r}_i \cdot \delta r_i dt = \int_{t_1}^{t_2} \delta T dt; \delta r_i = 0,\ i = 1, 2, \ldots, N\ , t = t_1, t_2 \tag{3.40}$$

where T is the total of the kinetic energies of all of the mass particles or the system's kinetic energy. Integrating Eq. (3.28) with respect to time over the interval $t_1 \leq t \leq t_2$ and using the results of Eqs. (3.35) and (3.40), the ensuing equation is:

$$\begin{aligned} \int_{t_1}^{t_2} \left(\sum_{i=1}^{N} (F_i - m_i\ddot{r}_i) \cdot \delta r_i \right) dt &= \int_{t_1}^{t_2} (\overline{\delta W} + \delta T) dt = 0 \\ \delta r_i = 0,\ i = 1, 2, \ldots, N\ , t = t_1, t_2 \end{aligned} \tag{3.41}$$

Equation (3.41) represents the mathematical expression of the *extended Hamilton's principle*. The virtual work $\overline{\delta W}$ may be divided into two parts, the first due to conservative forces and the second due to non-conservative forces. We then have:

$$\overline{\delta W} = \overline{\delta W}_c + \overline{\delta W}^{NC} \tag{3.42}$$

Recalling from Eq. (3.33) that $\overline{\delta W}_c = -\delta V$, Eq. (3.42) may be written as:

$$\overline{\delta W} = \overline{\delta W}_c + \overline{\delta W}^{NC} = -\delta V + \overline{\delta W}^{NC} \tag{3.43}$$

The extended Hamilton's principle, as expressed in Eq. (3.41) becomes:

$$\int_{t_1}^{t_2} \left(\overline{\delta W}^{NC} - \delta V + \delta T\right) dt = 0$$
$$\delta r_i = 0, \ i = 1, 2, \ldots, N \ , t = t_1, t_2 \tag{3.44}$$

We assumed that the position vectors r_i were independent. If, however, this is not the case, then the position vectors must be transformed to an independent set of generalized coordinates. It is therefore expedient to use an independent set of generalized coordinates $q_k (k = 1, 2, \ldots, n)$ at the outset in order to calculate potential and kinetic energies. The energies are the same in any coordinate system. Recalling that $\delta r_i = 0$ $(i = 1, 2, \ldots, N)$ when the δr_i are independent, by analogy $\delta q_k = 0$ $(k = 1, 2, \ldots, n)$, and the extended Hamilton's principle for generalized coordinates may be rewritten in the form:

$$\int_{t_1}^{t_2} \left(\overline{\delta W}^{NC} \underbrace{-\delta V + \delta T}_{\delta \mathcal{L}}\right) dt = 0$$
$$\delta q_k = 0, \ i = 1, 2, \ldots, n \ , t = t_1, t_2 \tag{3.45}$$

Letting $\mathcal{L} = T - V$, and assuming that we have only conservative forces operating on the system, that is, $\overline{\delta W}^{NC} = 0$, Eq. (3.45) becomes:

$$\int_{t_1}^{t_2} (\delta \mathcal{L}) dt = \delta \int_{t_1}^{t_2} \mathcal{L} dt = 0$$
$$\delta q_k = 0, \ i = 1, 2, \ldots, n \ , t = t_1, t_2 \tag{3.46}$$

The expression $\mathcal{L} = T - V$ is known as the Lagrangian and Eq. (3.46) is referred to as the extended Hamilton's principle (see Meirovitch [23, pp. 68] and Baruh [3, pp. 249–251]).

3.7 Lagrange's Equations and Lagrangian Dynamics

The extended Hamilton's principle is not the most effective method for the derivation of equations of motion since it requires certain operations which must be performed, such as integration by parts. The d'Alembert–Lagrange equations, which are derived from the extended Hamilton's principle is a more expedient and efficient method for obtaining equations of motion (see Baruh [3, pp. 253–254]). The kinetic energy of a general dynamic system can be written in terms of generalized coordinates (displacements) and their derivatives (generalized velocities) in the following functional form:

$$T = T(q_1, q_2, \ldots, q_n, \dot{q}_1, \dot{q}_2, \ldots, \dot{q}_n) \tag{3.47}$$

The variation in kinetic energy is simply stated as:

$$\delta T = \sum_{k=1}^{n} \left(\frac{\partial T}{\partial q_k} \delta q_k + \frac{\partial T}{\partial \dot{q}_k} \delta \dot{q}_k \right) \tag{3.48}$$

Similarly, the potential energy V has the form:

$$V = V\,(q_1, q_2, \ldots, q_n) \tag{3.49}$$

and the variation of V may be written as:

$$\delta V = \sum_{k=1}^{n} \left(\frac{\partial V}{\partial q_k} \delta q_k \right) \tag{3.50}$$

In addition, the virtual work of the non-conservative forces, from Eq. (3.23) takes on the form:

$$\overline{\delta W}^{NC} = \sum_{k=1}^{n} Q_k \delta q_k \tag{3.51}$$

where Q_k $(k = 1, 2, \ldots, n)$ are the generalized non-conservative forces. Substituting the expressions for T (Eq. 3.49) and V (Eq. 3.50) into the extended Hamilton's principle (Eq. 3.45), we arrive at:

$$\int_{t_1}^{t_2} (\overline{\delta W}^{NC} - \delta V + \delta T) dt = \int_{t_1}^{t_2} \sum_{k=1}^{n} \left[\left(\frac{\partial T}{\partial q_k} - \frac{\partial V}{\partial q_k} + Q_k \right) \delta q_k + \frac{\partial T}{\partial \delta \dot{q}_k} \delta \dot{q}_k \right] dt = 0$$

$$\delta q_k = 0, \;\; k = 1, 2, \ldots, n \;\; , t = t_1, t_2 \tag{3.52}$$

We now integrate the following term by parts as follows:

$$\int_{t_1}^{t_2} \frac{\partial T}{\partial \delta \dot{q}_k} \delta \dot{q}_k dt = \int_{t_1}^{t_2} \underbrace{\frac{\partial T}{\partial \delta \dot{q}_k}}_{u} \underbrace{\frac{d}{dt} \delta q_k \, dt}_{dv}$$
$$= \underbrace{\frac{\partial T}{\partial \delta \dot{q}_k}}_{u} \underbrace{\delta q_k}_{v} \Big|_{t_1}^{t_2} - \int_{t_1}^{t_2} \underbrace{\frac{d}{dt}\left(\frac{\partial T}{\partial \dot{q}_k}\right)}_{du} \underbrace{\delta q_k}_{v} \, dt$$
$$= - \int_{t_1}^{t_2} \frac{d}{dt}\left(\frac{\partial T}{\partial \dot{q}_k}\right) \delta q_k dt$$
$$\delta q_k = 0, \; k = 1, 2, \ldots, n \;, t = t_1, t_2 \tag{3.53}$$

Notice that we utilized the conditions $\delta q_k = 0, t = t_1, t = t_2$ in the integration by parts procedure above. These latter conditions are due to the fact that the end points must vanish by definition of the variation. Introducing this latter result into Eq. (3.52), we then have:

$$\int_{t_1}^{t_2} \sum_{k=1}^{n} \left[\frac{\partial T}{\partial q_k} - \frac{\partial V}{\partial q_k} + Q_k - \frac{d}{dt}\left(\frac{\partial T}{\partial \dot{q}_k}\right) \right] \delta q_k dt = 0 \tag{3.54}$$

Since the virtual displacement values δq_k $(k = 1, 2, \ldots, n)$ are arbitrary, we can assign $\delta q_1 = 1$ while setting all of the other virtual displacements $\delta q_k = 0$ $(k = 2, 3, \ldots, n)$. Under these circumstances, Eq. (3.54) can be satisfied only if the coefficient of δq_1 is zero, that is:

$$\frac{\partial T}{\partial q_1} - \frac{\partial V}{\partial q_1} + Q_1 - \frac{d}{dt}\left(\frac{\partial T}{\partial \dot{q}_1}\right) = 0 \Rightarrow \frac{d}{dt}\left(\frac{\partial T}{\partial \dot{q}_1}\right) - \frac{\partial T}{\partial q_1} + \frac{\partial V}{\partial q_1} = Q_1 \tag{3.55}$$

The same procedure as above is carried out for all of the virtual displacements and the resulting Lagrange equations are:

$$\frac{d}{dt}\left(\frac{\partial T}{\partial \dot{q}_k}\right) - \frac{\partial T}{\partial q_k} + \frac{\partial V}{\partial q_k} = Q_k; \, k = 1, 2, \ldots, n \tag{3.56}$$

Non-conservative forces which have hitherto not been dealt with explicitly are those which are proportional to the velocity of a particle and which resist the motion, in that, they act in the direction opposite to the particle's velocity (see Meirovitch [23, pp. 73 , 88–91]). Due to the fact that the system loses energy when such forces come into play, they are termed as *dissipative forces*, i.e., some of the system's energy is dissipated. Assume that the dissipative forces acting on particle i are written in Cartesian coordinates and are of the form:

$$F_{x_i} = -c_{x_i}\dot{x}_i;\; F_{y_i} = -c_{y_i}\dot{y}_i;\; F_{z_i} = -c_{z_i}\dot{z}_i;\; i = 1, 2, \ldots, N$$

Note that the parameters c_{x_i}, c_{y_i}, and c_{z_i} are functions which depend only on the coordinates and not on the velocities. The virtual work may be shown to be:

$$\sum_{i=1}^{N} F_i \cdot \delta r_i = \sum_{i=1}^{N} \left(F_{x_i}\delta x_i + F_{y_i}\delta y_i + F_{z_i}\delta z_i\right)$$

$$= -\sum_{i=1}^{N} \left(c_{x_i}\dot{x}_i\delta x_i + c_{y_i}\dot{y}_i\delta y_i + c_{z_i}\dot{z}_i\delta z_i\right)$$

Using the fact that $\delta r_i = \delta r_i(q_1, q_2, \ldots, q_n)$, and $\partial r_i/\partial q_k = \partial \dot{r}_i/\partial \dot{q}_k$, δr_i takes the form:

$$\delta r_i = \sum_{k=1}^{n} \frac{\partial r_i}{\partial q_k}\delta q_k = \sum_{k=1}^{n} \frac{\partial \dot{r}_i}{\partial \dot{q}_k}\delta q_k$$

This implies that:

$$\delta x_i = \sum_{k=1}^{n} \frac{\partial \dot{x}_i}{\partial \dot{q}_k};\; \delta y_i = \sum_{k=1}^{n} \frac{\partial \dot{y}_i}{\partial \dot{q}_k};\; \delta z_i = \sum_{k=1}^{n} \frac{\partial \dot{z}_i}{\partial \dot{q}_k}$$

Applying this latter result to the equation for virtual work results in the virtual work becoming:

$$\sum_{i=1}^{N} F_i \cdot \delta r_i = -\sum_{i=1}^{N} \left(c_{x_i}\dot{x}_i\delta x_i + c_{y_i}\dot{y}_i\delta y_i + c_{z_i}\dot{z}_i\delta z_i\right)$$

$$= -\sum_{k=1}^{n} \left[\sum_{i=1}^{N} c_{x_i}\dot{x}_i\frac{\partial \dot{x}_i}{\partial \dot{q}_k} + c_{y_i}\dot{y}_i\frac{\partial \dot{y}_i}{\partial \dot{q}_k} + c_{z_i}\dot{z}_i\frac{\partial \dot{z}_i}{\partial \dot{q}_k}\right]\delta q_k$$

$$= -\sum_{k=1}^{n} \left[\sum_{i=1}^{N} \frac{1}{2}\frac{\partial}{\partial \dot{q}_k}(c_{x_i}\dot{x}_i^2 + c_{y_i}\dot{y}_i^2 + c_{z_i}\dot{z}_i^2)\right]\delta q_k$$

$$= -\sum_{k=1}^{n} \frac{\partial}{\partial \dot{q}_k}\left[\sum_{i=1}^{N} \frac{1}{2}(c_{x_i}\dot{x}_i^2 + c_{y_i}\dot{y}_i^2 + c_{z_i}\dot{z}_i^2)\right]\delta q_k$$

$$= -\sum_{k=1}^{n} \frac{\partial F}{\dot{q}_k}\delta q_k$$

where $F = \sum_{i=1}^{N} \frac{1}{2}(c_{x_i}\dot{x}_i^2 + c_{y_i}\dot{y}_i^2 + c_{z_i}\dot{z}_i^2)$ is termed a *Rayleigh's dissipation function*. Assuming that the only non-conservative forces are of the dissipative type, the virtual work equation becomes:

$$\overline{\delta W}^{NC} = \sum_{k=1}^{n} Q_k \delta q_k = -\sum_{k=1}^{n} \frac{\partial F}{\partial \dot{q}_k} \delta q_k \Rightarrow Q_k = -\frac{\partial F}{\partial \dot{q}_k}, \quad k = 1, 2, \ldots, n$$

It then follows that Eq. (3.56) may be expressed as:

$$\frac{d}{dt}\left(\frac{\partial T}{\partial \dot{q}_k}\right) - \frac{\partial T}{\partial q_k} + \frac{\partial V}{\partial q_k} - Q_k = 0; k = 1, 2, \ldots, n$$

$$\Rightarrow \frac{d}{dt}\left(\frac{\partial T}{\partial \dot{q}_k}\right) - \frac{\partial T}{\partial q_k} + \frac{\partial V}{\partial q_k} + \frac{\partial F}{\partial \dot{q}_k} = 0; k = 1, 2, \ldots, n$$

Examples

The following four examples have been taken from Widnall's MIT OpenCourseWare course (Widnall [53]) on Dynamics, Lecture L-20.

Example 1

Let's consider a simple mass–spring one degree of freedom system (see Fig. 3.7) with one generalized coordinate x_1. It possesses kinetic energy of the form: $T = \frac{1}{2}m\dot{x}^2$ and its potential energy is: $V = \frac{1}{2}kx^2$ (see Eq. 2.60). The Lagrangian is $\mathcal{L} = T - V = \frac{1}{2}m\dot{x}^2 - \frac{1}{2}kx^2$. Application of Eq. (3.56) to the Lagrangian of this simple system results in the familiar differential equation for the mass–spring oscillator, which is: $m\ddot{x} + kx = 0$. It is obtained as follows:

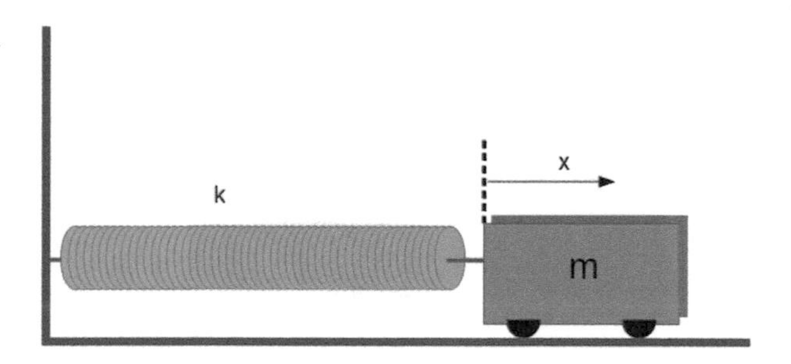

Fig. 3.7 Mass–spring system

$$\frac{d}{dt}\left(\frac{\partial \mathcal{L}}{\partial \dot{x}_1}\right) - \frac{\partial \mathcal{L}}{\partial x_1} = \frac{d}{dt}\left(\frac{\partial T}{\partial \dot{x}_1}\right) - \frac{\partial T}{\partial x_1} + \frac{\partial V}{\partial x_1} = Q_1 = 0$$

$$x_1 = x;\ Q_1 = 0 \Rightarrow \frac{d}{dt}\left(\frac{\partial T}{\partial \dot{x}}\right) - \frac{\partial T}{\partial x} + \frac{\partial V}{\partial x} = 0$$

$$\frac{\partial T}{\partial \dot{x}} = m\dot{x};\ \frac{d}{dt}\left(\frac{\partial T}{\partial \dot{x}}\right) = m\ddot{x};\ \frac{\partial T}{\partial x} = 0;\ \frac{\partial V}{\partial x} = kx$$

$$\frac{d}{dt}\left(\frac{\partial T}{\partial \dot{x}}\right) - \frac{\partial T}{\partial x} + \frac{\partial V}{\partial x} = m\ddot{x} + kx = 0 \tag{3.57}$$

Example 2: Two DOF Mass–Spring System

Figure 3.8 depicts a two degrees of freedom system with three springs, described by two differential equations. The relevant equations can be obtained by direct application of the d'Alembert–Lagrange equation. The expressions for the kinetic and potential energies are, respectively:

$$T = \sum_{i=1}^{2} \frac{1}{2} m_i \dot{x}_i^2;\ V = \frac{1}{2} k_1 x_1^2 + \frac{1}{2} k_2 (x_2 - x_1)^2 + \frac{1}{2} k_3 x_2^2 \tag{3.58}$$

Utilizing the d'Alembert–Lagrange's equation on $\mathcal{L} = T - V =$, we have:

$$\frac{d}{dt}\left(\frac{\partial \mathcal{L}}{\partial \dot{x}_1}\right) - \frac{\partial \mathcal{L}}{\partial x_1} = 0;\ \frac{d}{dt}\left(\frac{\partial \mathcal{L}}{\partial \dot{x}_2}\right) - \frac{\partial \mathcal{L}}{\partial x_2} = 0$$

$$\Rightarrow \frac{d}{dt}\left(\frac{\partial \mathcal{L}}{\partial \dot{x}_1}\right) = m_1 \ddot{x}_1;\ \frac{d}{dt}\left(\frac{\partial \mathcal{L}}{\partial \dot{x}_2}\right) = m_2 \ddot{x}_2$$

$$-\frac{\partial \mathcal{L}}{\partial x_1} = -\frac{\partial V}{\partial x_1} = k_1 x_1 + k_2 (x_1 - x_2);\ -\frac{\partial \mathcal{L}}{\partial x_2} = -\frac{\partial V}{\partial x_2} = k_2 (x_2 - x_1) + k_3 x_2$$

$$\Rightarrow m_1 \ddot{x}_1 = k_2 (x_2 - x_1) - k_1 x_1;\ m_2 \ddot{x}_2 = -k_2 (x_2 - x_1) - k_3 x_2 \tag{3.59}$$

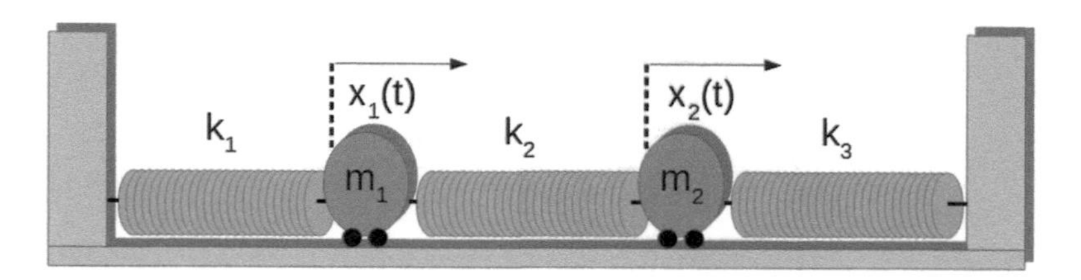

Fig. 3.8 Two degrees of freedom system composed of springs and masses

Example 3: Simple Pendulum

We begin by applying the d'Alembert–Lagrange's equation in order to derive the equations of motion of a simple pendulum. It is expedient to do so in polar coordinates. This is a one degree of freedom system. However, it will prove advantageous, when we shall be required to analyze the double pendulum, to begin with a description of the position of the mass point m_1(located at the bottom of the pendulum) in Cartesian coordinates x_1 and y_1 and then write down the Lagrangian in the polar angle θ_1. Referring to sub-figure (a) of Fig. 3.9, we have:

$$\begin{aligned} x_1 &= h_1 \sin\theta_1;\ \dot{x}_1 = h_1 \cos\theta_1 \dot{\theta}_1 \\ y_1 &= -h_1 \cos\theta_1;\ \dot{y}_1 = h_1 \sin\theta_1 \dot{\theta}_1 \end{aligned} \tag{3.60}$$

and so the kinetic energy is:

$$T = \frac{1}{2} m_1 (\dot{x}_1^2 + \dot{y}_1^2) = \frac{1}{2} m_1 h_1^2 (\dot{\theta}_1)^2 \tag{3.61}$$

The potential energy V is:

$$V = m_1 g y_1 = -m_1 g h_1 \cos\theta_1 \tag{3.62}$$

The Lagrangian $\mathcal{L}$ is:

$$\mathcal{L} = T - V = \frac{1}{2} m_1 h_1^2 (\dot{\theta}_1)^2 + m_1 g h_1 \cos\theta_1 \tag{3.63}$$

From the above equation it becomes apparent that the generalized variable q_1 is: $q_1 = \theta_1$. The entity h_1 is constant, and so the only variable quantity is θ_1, and it naturally becomes the only available generalized variable. Application of 3.56 with $q_1 = \theta_1$ yields the differential equation governing this single pendulum's motion.

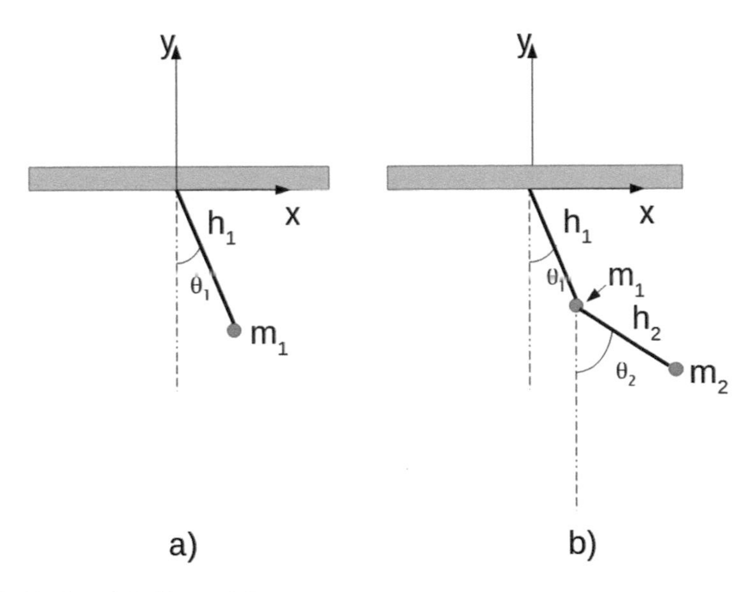

Fig. 3.9 Single and double pendulums

$$
\begin{aligned}
&\frac{d}{dt}\left(\frac{\partial \mathcal{L}}{\partial \dot{\theta}_1}\right) - \frac{\partial T}{\partial \theta_1} + \frac{\partial V}{\partial \theta_1} = 0 \\
&\frac{\partial \mathcal{L}}{\partial \dot{\theta}_1} = m_1 h_1^2 (\dot{\theta}_1); \ \frac{d}{dt}\left(\frac{\partial \mathcal{L}}{\partial \dot{\theta}_1}\right) = m_1 h_1^2 \ddot{\theta}_1 \\
&\frac{\partial V}{\partial \theta_1} - \frac{\partial T}{\partial \theta_1} = m_1 g h_1 \sin\theta_1 \Rightarrow m_1 h_1^2 \ddot{\theta}_1 + m_1 g h_1 \sin\theta_1 = 0
\end{aligned}
\tag{3.64}
$$

Example 4: Double Pendulum

The double pendulum is depicted in sub-figure (b) of Fig. 3.9 and consists of two point masses of mass m_1 and m_2, connected by massless strings of length h_1 and h_2, respectively (the c.g. points of both bodies are situated at the bottom of each pendulum, respectively). The two pendulae are connected to their respective pivot points, about which they swing. This system is endowed with two degrees of freedom: θ_1 and θ_2. The first step in the application of the d'Alembert–Lagrange's equations is the determination of the expressions for the kinetic and the potential energies, as the system rotates through the independent angles $\theta_1$and θ_2. From geometrical considerations, the Cartesian coordinates of the two masses are:

$$
\begin{aligned}
x_1 &= h_1 \sin\theta_1;\ y_1 = -h_1 \cos\theta_1 \\
x_2 &= x_1 + h_2 \sin\theta_2 = h_1 \sin\theta_1 + h_2 \sin\theta_2 \\
y_2 &= y_1 - h_2 \cos\theta_2 = -h_1 \cos\theta_1 - h_2 \cos\theta_2
\end{aligned}
\tag{3.65}
$$

Differentiating the coordinates in Eq. (3.65) with respect to time t, we have:

$$
\begin{aligned}
\dot{x}_1 &= h_1 \cos\theta_1\dot{\theta}_1;\ \dot{y}_1 = h_1 \sin\theta_1\dot{\theta}_1 \\
\dot{x}_2 &= h_1 \cos\theta_1\dot{\theta}_1 + h_2 \cos\theta_2\dot{\theta}_2 \\
\dot{y}_2 &= h_1 \sin\theta_1\dot{\theta}_1 + h_2 \sin\theta_2\dot{\theta}_2
\end{aligned}
\tag{3.66}
$$

Squaring $\dot{x}_1$ and $\dot{y}_1$ and adding the terms together:

$$
\dot{x}_1^2 + \dot{y}_1 = h_1^2 \cos^2\theta_1(\dot{\theta}_1)^2 + h_1^2 \sin^2\theta_1(\dot{\theta}_1)^2 = h_1^2(\dot{\theta}_1)^2 \tag{3.67}
$$

Similarly for $\dot{x}_2$ and $\dot{y}_2$

$$
\begin{aligned}
\dot{x}_2^2 &= (h_1 \cos\theta_1\dot{\theta}_1 + h_2 \cos\theta_2\dot{\theta}_2)^2 \\
&= h_1^2 \cos^2\theta_1(\dot{\theta}_1)^2 + h_2^2 \cos^2\theta_2(\dot{\theta}_2)^2 + 2h_1h_2 \cos\theta_1 \cos\theta_2\dot{\theta}_1\dot{\theta}_2 \\
\dot{y}_2^2 &= (h_1 \sin\theta_1\dot{\theta}_1 + h_2 \sin\theta_2\dot{\theta}_2)^2 \\
&= h_1^2 \sin^2\theta_1(\dot{\theta}_1)^2 + h_2^2 \sin^2\theta_2(\dot{\theta}_2)^2 + 2h_1h_2 \sin\theta_1 \sin\theta_2\dot{\theta}_1\dot{\theta}_2 \\
\Rightarrow \dot{x}_2^2 + \dot{y}_2^2 &= h_1^2(\dot{\theta}_1)^2 + h_2^2(\dot{\theta}_2)^2 + 2h_1h_2\dot{\theta}_1\dot{\theta}_2 \cos(\theta_1 - \theta_2)
\end{aligned}
\tag{3.68}
$$

The kinetic energy is:

$$
\begin{aligned}
T &= \frac{1}{2}m_1(\dot{x}_1^2 + \dot{y}_1^2) + \frac{1}{2}m_2(\dot{x}_2^2 + \dot{y}_2^2) \\
&= \frac{1}{2}m_1h_1^2(\dot{\theta}_1)^2 + \frac{1}{2}m_2[h_1^2(\dot{\theta}_1)^2 + h_2^2(\dot{\theta}_2)^2 + 2h_1h_2\dot{\theta}_1\dot{\theta}_2 \cos(\theta_1 - \theta_2)]
\end{aligned}
\tag{3.69}
$$

The system's potential energy V is:

$$
\begin{aligned}
V &= m_1gy_1 + m_2gy_2 = -m_1gh_1 \cos\theta_1 - m_2g(h_1 \cos\theta_1 + h_2 \cos\theta_2) \\
&= -(m_1 + m_2)gh_1 \cos\theta_1 - m_2gh_2 \cos\theta_2
\end{aligned}
\tag{3.70}
$$

With the preceding information now at our disposal, the Lagrangian $\mathcal{L} = T - V$ may be easily calculated and it turns out to be:

$$\mathcal{L} = T - V = \frac{1}{2} m_1 h_1^2 (\dot{\theta}_1)^2 + \frac{1}{2} m_2 [h_1^2 (\dot{\theta}_1)^2 + h_2^2 (\dot{\theta}_2)^2 + 2 h_1 h_2 \dot{\theta}_1 \dot{\theta}_2 \cos(\theta_1 - \theta_2)] + (m_1 + m_2) g h_1 \cos\theta_1 + m_2 g h_2 \cos\theta_2 \tag{3.71}$$

The terms required for the d'Alembert–Lagrange's equation (Eq. 3.56) are therefore:

$$\begin{aligned}
\frac{\partial \mathcal{L}}{\partial \dot{\theta}_1} &= m_1 h_1^2 \dot{\theta}_1 + m_2 h_1^2 \dot{\theta}_1 + m_2 h_1 h_2 \dot{\theta}_2 \cos(\theta_1 - \theta_2); \\
\frac{\partial \mathcal{L}}{\partial \dot{\theta}_2} &= m_2 h_2^2 \dot{\theta}_2 + m_2 h_1 h_2 \dot{\theta}_1 \cos(\theta_1 - \theta_2) \\
\frac{d}{dt}\left(\frac{\partial \mathcal{L}}{\partial \dot{\theta}_1}\right) &= m_1 h_1^2 \ddot{\theta}_1 + m_2 h_1^2 \ddot{\theta}_1 + m_2 h_1 h_2 \ddot{\theta}_2 \cos(\theta_1 - \theta_2) \\
&\quad + m_2 h_1 h_2 \dot{\theta}_2 \frac{d}{dt}[\cos(\theta_1 - \theta_2)] \\
&= m_1 h_1^2 \ddot{\theta}_1 + m_2 h_1^2 \ddot{\theta}_1 + m_2 h_1 h_2 \ddot{\theta}_2 \cos(\theta_1 - \theta_2) \\
&\quad - m_2 h_1 h_2 \dot{\theta}_2 \sin(\theta_1 - \theta_2)(\dot{\theta}_1 - \dot{\theta}_2) \\
\frac{d}{dt}\left(\frac{\partial \mathcal{L}}{\partial \dot{\theta}_2}\right) &= m_2 h_2 \ddot{\theta}_2 + m_2 h_1 h_2 \ddot{\theta}_1 \cos(\theta_1 - \theta_2) - m_2 h_1 h_2 \dot{\theta}_1 \sin(\theta_1 - \theta_2)(\dot{\theta}_1 - \dot{\theta}_2) \\
\frac{\partial \mathcal{L}}{\partial \theta_1} &= m_2 h_1 h_2 \dot{\theta}_1 \dot{\theta}_2 \frac{\partial \cos(\theta_1 - \theta_2)}{\partial \theta_1} - (m_1 + m_2) g h_1 \sin\theta_1 \\
&= -m_2 h_1 h_2 \dot{\theta}_1 \dot{\theta}_2 \sin(\theta_1 - \theta_2) - (m_1 + m_2) g h_1 \sin\theta_1 \\
\Rightarrow -\frac{\partial \mathcal{L}}{\partial \theta_1} &= m_2 h_1 h_2 \dot{\theta}_1 \dot{\theta}_2 \sin(\theta_1 - \theta_2) + (m_1 + m_2) g h_1 \sin\theta_1 \\
\frac{\partial \mathcal{L}}{\partial \theta_2} &= m_2 h_1 h_2 \dot{\theta}_1 \dot{\theta}_2 \frac{\partial \cos(\theta_1 - \theta_2)}{\partial \theta_2} - m_2 g h_2 \sin\theta_2 \\
&= m_2 h_1 h_2 \dot{\theta}_1 \dot{\theta}_2 \sin(\theta_1 - \theta_2) - m_2 g h_2 \sin\theta_2 \\
\Rightarrow -\frac{\partial \mathcal{L}}{\partial \theta_2} &= -m_2 h_1 h_2 \dot{\theta}_1 \dot{\theta}_2 \sin(\theta_1 - \theta_2) + m_2 g h_2 \sin\theta_2
\end{aligned} \tag{3.72}$$

Applying Eq. (3.56) to $\mathcal{L} = T - V$ with $q_1 = \theta_1$and $q_2 = \theta_2$ and using the results of Eq. (3.72) leads to the following set of differential equations:

$$
\begin{aligned}
&\frac{d}{dt}\left(\frac{\partial \mathcal{L}}{\partial \dot{\theta}_1}\right) - \frac{\partial \mathcal{L}}{\partial \theta_1} = 0 \\
&\quad \Rightarrow (m_1 + m_2)h_1^2\ddot{\theta}_1 + m_2 h_1 h_2 \ddot{\theta}_2 \cos(\theta_1 - \theta_2) - m_2 h_1 h_2 \dot{\theta}_2 \sin(\theta_1 - \theta_2)(\dot{\theta}_1 - \dot{\theta}_2) \\
&\quad + m_2 h_1 h_2 \dot{\theta}_1 \dot{\theta}_2 \sin(\theta_1 - \theta_2) + (m_1 + m_2) g h_1 \sin\theta_1 = 0 \\
&\quad \Rightarrow (m_1 + m_2)h_1^2\ddot{\theta}_1 + m_2 h_1 h_2 \ddot{\theta}_2 \cos(\theta_1 - \theta_2) + m_2 h_1 h_2 \dot{\theta}_2^2 \sin(\theta_1 - \theta_2) \\
&\quad - m_2 h_1 h_2 \dot{\theta}_1 \dot{\theta}_2 \sin(\theta_1 - \theta_2) + m_2 h_1 h_2 \dot{\theta}_1 \dot{\theta}_2 \sin(\theta_1 - \theta_2) \\
&\quad + (m_1 + m_2) g h_1 \sin\theta_1 = 0 \\
&\quad \Rightarrow \frac{d}{dt}\left(\frac{\partial \mathcal{L}}{\partial \dot{\theta}_1}\right) - \frac{\partial \mathcal{L}}{\partial \theta_1} = (m_1 + m_2)h_1^2\ddot{\theta}_1 + m_2 h_1 h_2 \ddot{\theta}_2 \cos(\theta_1 - \theta_2) \\
&\quad + m_2 h_1 h_2 \dot{\theta}_2^2 \sin(\theta_1 - \theta_2) + (m_1 + m_2) g h_1 \sin\theta_1 = 0 \\
&\frac{d}{dt}\left(\frac{\partial \mathcal{L}}{\partial \dot{\theta}_2}\right) - \frac{\partial \mathcal{L}}{\partial \theta_2} = 0 \\
&\quad \Rightarrow m_2 h_2 \ddot{\theta}_2 + m_2 h_1 h_2 \ddot{\theta}_1 \cos(\theta_1 - \theta_2) - m_2 h_1 h_2 \dot{\theta}_1 \sin(\theta_1 - \theta_2)(\dot{\theta}_1 - \dot{\theta}_2) \\
&\quad - m_2 h_1 h_2 \dot{\theta}_1 \dot{\theta}_2 \sin(\theta_1 - \theta_2) + m_2 g h_2 \sin\theta_2 = 0 \\
&\quad \Rightarrow m_2 h_2 \ddot{\theta}_2 + m_2 h_1 h_2 \ddot{\theta}_1 \cos(\theta_1 - \theta_2) - m_2 h_1 h_2 \dot{\theta}_1^2 \sin(\theta_1 - \theta_2) \\
&\quad + m_2 h_1 h_2 \dot{\theta}_1 \dot{\theta}_2 \sin(\theta_1 - \theta_2) \\
&\quad - m_2 h_1 h_2 \dot{\theta}_1 \dot{\theta}_2 \sin(\theta_1 - \theta_2) + m_2 g h_2 \sin\theta_2 = 0 \\
&\quad \Rightarrow m_2 h_2 \ddot{\theta}_2 + m_2 h_1 h_2 \ddot{\theta}_1 \cos(\theta_1 - \theta_2) - m_2 h_1 h_2 \dot{\theta}_1^2 \sin(\theta_1 - \theta_2) \\
&\quad + m_2 g h_2 \sin\theta_2 = 0
\end{aligned}
\tag{3.73}
$$

Example 5: Quad Copter (see Rodolfo et al. [29, pp. 23–34])

Introduction

A helicopter is an airborne system, such that, with the use of rapidly spinning rotors, it is able to push a mass of air in a descending direction through its rotor or rotors, which results in the creation of thrust to support the vehicle in mid-air. The principle by which this is done is similar to the propeller on a conventional aircraft, although the thrust in a conventional aircraft is in the horizontal plane, while that of the helicopter is directed vertically. Conventional helicopters have two rotors, arranged to counter-rotate in the plane of the rotor disk, both of which provide upward thrust. The counter-rotation is required in order to cancel out the reaction torque exerted on the helicopter's body from the rotor's engines. An additional conventional arrangement for countering engine torque on the body is to have a main rotor directed vertically and a tail rotor which produces a lateral thrust, thus creating a moment which counterbalances the torque on the body. The drawback of these two "conventional" schemes is the very complicated mechanism required to synchronize and redirect the rotor thrust to where it is desired. The system charged with the task of synchronization and redirection of thrust is the swash plate. The swash plate controls the angles of attack on the rotor blades and tries to periodically adjust them such that they are at their maximum values when the rotor is moving against the oncoming wind (retreating) in forward flight and at their minimum values when the blade is advancing into the oncoming wind. The swash plate is composed of a very elaborate, costly, intricate, and very robust set of linkages, servomechanisms, dampers, and gears with a constant need for maintenance and repairs.

A quadrotor helicopter or quadcopter, in contrast, has four (or more) equally spaced rotors, situated on the periphery of a square, rectangular, or circular body and arranged in either an "x" or "+" configuration. Having four (or more) motors, each of which is capable of an independent rotor rotational speed adjustment, allows for the elimination of the swash plate. The disadvantage of this arrangement are the resulting losses in energy incurred by speeding up or slowing down the rotors in mid-flight, as compared to a conventional helicopter where a constant rotor speed is maintained throughout all flight phases.

The quadrotor's controls derive from the angular speeds of four electric motors (see Fig. 3.10). Each motor generates a thrust and a torque, whose combination results in the main thrust, the yaw torque, the pitch torque, and the roll torque which in concert act on the quadrotor. In addition, the rotor blades on the quadrotor all have a constant pitch angle and are not adjustable. The quadrotor motors turn in a fixed direction and hence the forces produced by the rotors are all positive. However motors m_1 and m_3 rotate in a clockwise direction (producing counter-clockwise reaction torques), while motors m_2 and m_4 rotate in an anti-clockwise direction (producing clockwise reaction torques).

While gyroscopic effects and aerodynamic torques tend to cancel each other out in trimmed flight, because of the particular spatial arrangement of the motors, their effects will be included as part of the non-conservative generalized forces (torques). The main thrust u is the sum of individual thrusts $(f_1 + f_2 + f_3 + f_4)$ of each motor. The pitch torque is derived from the difference in thrusts $f_1 - f_3$, while the roll torque depends upon the difference between forces f_2 and f_4, that is, $f_2 - f4$. The yaw torque is the sum of the reaction torques of the motors, $\tau_{m_1} - \tau_{m_2} + \tau_{m_3} - \tau_{m_4}$. The reaction torques $\tau_{m_i}, i = 1, 2, 3, 4$ are due to shaft acceleration and blade drag. The explanation contained in Gibiansky's blog [12] is straightforward, very comprehendible, and elegant in its simplicity. The torque which an electric motor produces is directly proportional to the difference between the input current I, when the motor is working under load conditions, and the no load current I_o, that is,

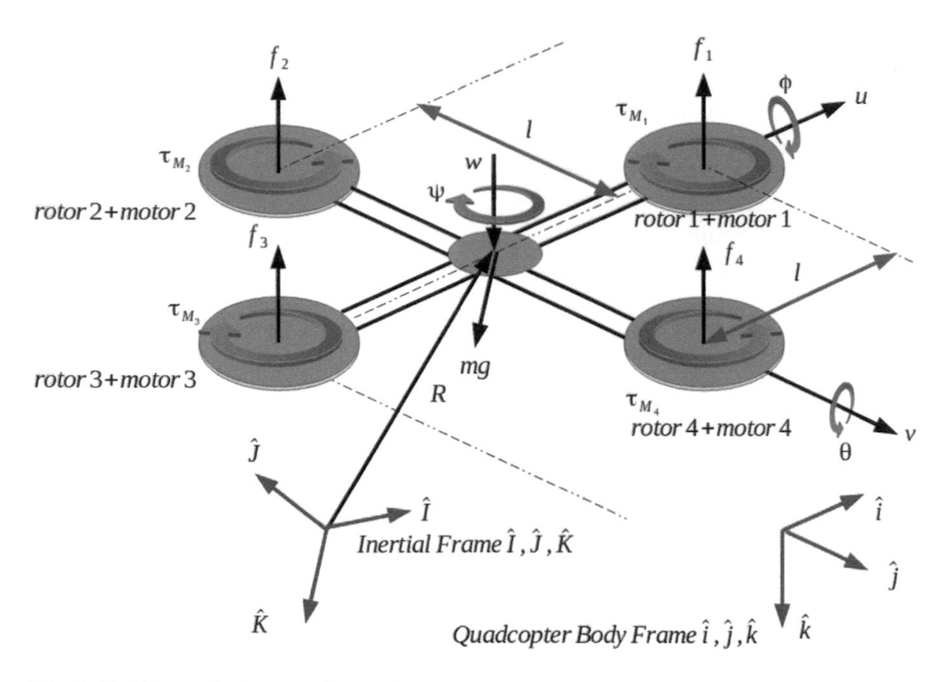

Fig. 3.10 Schematic diagram of a quadcopter

$\tau = K_t(I - I_o)$, where K_t is the current to torque scale factor. This implies that the current I is: $I = \frac{\tau + K_t I_o}{K_t}$. The voltage across the input terminals of the motor may be expressed as: $V = I R_m + K_v \omega$. R_m is the motor's ohmic winding resistance and K_v is its back-emf scale factor with units of volts per radians per second. The back-emf is directly related to its rotational speed ω. The voltage due to the winding resistance is negligible in comparison with the back-emf and hence may be neglected. We also assume that the motor torque τ is much larger than $K_t I_o$, and so the power produced by the electric motor is:

$$P = I V = \left(\frac{\tau + K_t I_o}{K_t}\right)(I R_m + K_v \omega) = \left(\frac{\tau + K_t I_o}{K_t}\right)\left[\left(\frac{\tau + K_t I_o}{K_t}\right) R_m + K_v \omega\right]$$
$$\approx \left(\frac{\tau}{K_t}\right)(K_v \omega)$$

The power is used to generate the lift required to keep the vehicle in the air. By the law of conservation of energy, the energy consumed by the motor, in a given period of time, is equivalent to the force generated on the propeller times the distance that the displaced air (air that the rotor has displaced) moves, that is, $P \cdot dt = F \cdot dx \Rightarrow P = F dx/dt = F v$. If the velocity of the displaced air in hover is v_h and the force equals the weight or thrust T of the vehicle, then $P = T v_h$. From momentum theory (see Seddon et al. [31, pp. 25]), the hover velocity and hence the power are given as:

$$v_h = \sqrt{\frac{T}{2\rho A}};\ P = \frac{K_v \tau}{K_t}\omega = \frac{K_v K_\tau}{K_t} T\omega = \frac{T^{3/2}}{\sqrt{2\rho A}}$$

where A is the area of the rotor disk or the area swept out by the rotor, ρ is the density of the air in which the vehicle is flying, and K_τ is the scale factor relating thrust to torque. Note that torque is proportional to the thrust since, by definition, $\tau = \vec{r} \times \vec{F}$, where $\vec{r}$ is the distance vector from the center of rotation to the point of application of the force $\vec{F}$. Furthermore, K_τ depends upon the rotor blade shape and layout. It turns out that the thrust T may be written as:

$$\frac{T^{3/2}}{\sqrt{2\rho A}} = \frac{K_v K_\tau}{K_t} T\omega \Rightarrow \frac{T^{1/2}}{\sqrt{2\rho A}} = \frac{K_v K_\tau}{K_t}\omega$$

$$\Rightarrow T^{1/2} = \frac{K_v K_\tau}{K_t}\omega\sqrt{2\rho A} \Rightarrow T = \left(\frac{K_v K_\tau}{K_t}\omega\sqrt{2\rho A}\right)^2 = k\omega^2$$

From basic fluid dynamics, the drag force equation describing the frictional force which opposes motion in the direction of the velocity vector of the rotor is:

$$D = \frac{1}{2}\rho C_D A_b v^2$$

where C_D is the drag coefficient of the blade, A_b is the cross sectional area of the rotor blade, and $v = \omega \times R$ is the linear velocity of the blade at its tip (R is the radius of the rotor blade). The torque due to the rotor blade's profile drag is therefore:

$$\tau_{drag_i} = D \times R = \frac{1}{2}\rho C_D A_b v^2 R = \frac{1}{2}\rho C_D A_b (\omega R)^2 R = b\omega^2 \tag{3.74}$$

Since the motor's torque is opposed by the torque due to aerodynamic drag, the torque equation for each motor/rotor combination is:

$$I_{rotor}\dot{\omega} = \tau_{m_i} - \tau_{drag_i} \quad i = 1, 2, 3, 4 \tag{3.75}$$

where I_{rotor} is the moment of inertia of the motor and rotor combination in the direction perpendicular to its rotation (motor's z axis), ω is the rotation rate of the ith motor, and τ_{m_i} is the torque produced by the ith motor. For quasi-stationary maneuvers, ω is constant, which implies that $I_{rotor}\dot{\omega} = 0$ and so the motor's reaction torque equals the drag induced torque or $\tau_{m_i} = \tau_{drag_i}$. In order to maneuver, the quadrotor must adjust the speed of its motors as follows:

1. Forward pitch motion is obtained by increasing the speed of motor m_3 while reducing the speed of motor m_1.
2. Similarly, roll motion is obtained by using motors m_2 and m_4.

3. Yaw motion is obtained by increasing the torque τ_{m_1} and τ_{m_3} of motors m_1 and m_3, respectively, while decreasing the torques τ_{m_2} and τ_{m_4} of motors m_2 and m_4, respectively.

Torques

There are two distinct sources for torques which act on the quadcopter's body (we have neglected aerodynamic torques, except for the blade profile drag torques). The first source is due to the torques exerted by the motors on the body or the motors' reaction torques τ_{m_i}, which were shown to be related directly to the rotor blades' profile drag and are proportional to each individual rotor's rotation rate $b\omega^2$. The sum of all of the four motor reaction torques is the torque which will tend to turn the vehicle about the body's $\hat{k}$ axis (see Fig. 3.10). Mathematically, this may be written as: $\sum_{i=1}^{4}(-1)^{i+1}\tau_{m_i} = \tau_{yaw}$. The torque which gives rise to angular roll is directly related to the imbalance between the forces f_2 and f_2, that is, $(f_2 - f_4)l = \tau_{roll}$. A similar force imbalance situation arises for the angular pitch; however, forces f_2 and f_3 are involved, leading to: $(f_2 - f_4)l = \tau_{pitch}$. The second source of externally applied torques is due to the "gyro effect" of each individual motor and rotor blade combination (see Beer et al. [4, pp. 1185]). The rotating rotor may be looked upon as a gyro, and when it is subject to an angular rotation perpendicular to its axis of rotation or perpendicular to the angular momentum vector which defines its rotation, a torque results which is perpendicular to both the angular momentum vector and the applied angular rate. In mathematical terms, we have:

$$\tau_{gyro} = \Omega \times H$$

$$H\hat{k} = \sum_{i=1}^{4} H_i\hat{k} = \sum_{i=1}^{4} I_{m_i}\omega_i\hat{k};\ \Omega = \begin{bmatrix} p_b \\ q_b \\ r_b \end{bmatrix}$$

$$\tau_{gyro_{roll}} = -q_b\hat{j} \times H\hat{k} = -q_b\sum_{i=1}^{4} I_{m_i}\omega_i\hat{i};$$

$$\tau_{gyro_{pitch}} = -p_b\hat{i} \times H\hat{k} = -p_b\sum_{i=1}^{4} I_{m_i}\omega_i\hat{j}$$

D'Alembert–Lagrangian Dynamics

Defining the generalized coordinates to be: $q = (X, Y, Z, \psi, \theta, \phi)$, where $R = X\hat{I} + Y\hat{J} + Z\hat{K}$ is the position vector from the origin of the inertial coordinate system to the center of mass of the quadrotor (in inertial coordinates) and $\eta = (\psi, \theta, \phi)$

are the Euler angles around the yaw (around the $Z(\hat{K})$), pitch (around the $Y(\hat{J})$), and roll (around the $X(\hat{I})$) axes, respectively, which define the orientation of the quadrotor with respect to the inertial coordinate system. The kinetic energy has a translational component as well as a rotational part. The derivatives of the Euler angles are: $\dot{\eta} = (\dot{\psi}, \dot{\theta}, \dot{\phi})$. The rotational kinetic energy is of the form: $T_{rotate} = \frac{1}{2}\dot{\eta}^T I_{full}\dot{\eta}$. Since the angular rates have not been measured along the quadrotor's body axes, the full inertia tensor must be used in the computation of rotational kinetic energy, and it is:

$$I_{full} = \begin{bmatrix} I_x(t) & -I_{xy}(t) & -I_{xz}(t) \\ -I_{yx}(t) & I_y(t) & -I_{yz}(t) \\ -I_{zx}(t) & -I_{zy}(t) & I_z(t) \end{bmatrix} \tag{3.76}$$

Because of the motion of the quadrotor, the parameters of the inertia tensor are changing at every instant of time and it would be more advantageous to be able to use the principal moments of inertia measured along the body fixed axes, which remain invariant throughout the flight. In body fixed axes, the inertia tensor is constant, diagonal, and doesn't contain any off-diagonal terms. The resulting inertia tensor is:

$$I = \begin{bmatrix} I_x & 0 & 0 \\ 0 & I_y & 0 \\ 0 & 0 & I_z \end{bmatrix} \tag{3.77}$$

The angular rates in body axes are: $\Omega = (p, q, r)$, and the connection between the derivatives of the Euler angles and the body axis angular rates is as follows:

$$\Omega = \begin{bmatrix} p \\ q \\ r \end{bmatrix} = \begin{bmatrix} \dot{\phi} - \dot{\psi}\sin\theta \\ \dot{\theta}\cos\phi + \dot{\psi}\cos\theta\sin\phi \\ \dot{\psi}\cos\theta\cos\phi - \dot{\theta}\sin\phi \end{bmatrix} = \underbrace{\begin{bmatrix} -\sin\theta & 0 & 1 \\ \cos\theta\sin\phi & \cos\phi & 0 \\ \cos\theta\cos\phi & -\sin\phi & 0 \end{bmatrix}}_{W_\eta} \begin{bmatrix} \dot{\psi} \\ \dot{\theta} \\ \dot{\phi} \end{bmatrix} = W_\eta \dot{\eta} \tag{3.78}$$

Rewriting the rotational kinetic energy in terms of the derivatives of the Euler angles results in:

$$T_{rotate} = \frac{1}{2}\Omega^T I \Omega = \frac{1}{2}\dot{\eta}^T \underbrace{W_\eta^T I W_\eta}_{\mathcal{J}} \dot{\eta} = \frac{1}{2}\dot{\eta}^T \mathcal{J} \dot{\eta} \tag{3.79}$$

The matrix $\mathcal{J} = \mathcal{J}(\eta)$ becomes the inertia matrix for the complete rotational kinetic energy of the quadrotor system, expressed directly in terms of the generalized coordinate vector η. The Lagrangian for the quadrotor may be written as follows:

$$\begin{aligned}
\mathcal{L} &= T_{translate} + T_{rotate} - \mathcal{U} \\
\dot{\xi} &= \dot{X}\hat{I} + \dot{Y}\hat{J} + \dot{Z}\hat{K};\ T_{translate} = \frac{1}{2} m_{quad} \dot{\xi}^T \cdot \dot{\xi} \\
T_{rotate} &= \frac{1}{2}\dot{\eta}^T \mathcal{J}\dot{\eta};\ \mathcal{U} = m_{quad} g Z \\
\Rightarrow \mathcal{L} &= \frac{1}{2} m_{quad} \dot{\xi}^T \cdot \dot{\xi} + \frac{1}{2}\dot{\eta}^T \mathcal{J}\dot{\eta} - m_{quad} g Z \\
-\frac{\partial \mathcal{L}}{\partial Z} &= m_{quad} g
\end{aligned} \tag{3.80}$$

where m_{quad} is the total mass of the quadrotor system and $\mathcal{U}$ is the potential energy of the system. The generalized forces are $\begin{bmatrix} F_{\xi} \\ \tau \end{bmatrix}$, where $F_{\xi} = T_{E/B}\hat{F}$. The Euler transformation from body to inertial coordinates $T_{E/B}$ is:

$$\begin{bmatrix} x_{ea} \\ y_{ea} \\ z_{ea} \end{bmatrix} = \begin{bmatrix} \cos\theta\cos\psi & \sin\phi\sin\theta\cos\psi - \cos\phi\sin\psi & \cos\phi\sin\theta\cos\psi + \sin\phi\sin\psi \\ \cos\theta\sin\psi & \sin\phi\sin\theta\sin\psi + \cos\phi\cos\psi & \cos\phi\sin\theta\sin\psi - \sin\phi\cos\psi \\ -\sin\theta & \sin\phi\cos\theta & \cos\phi\cos\theta \end{bmatrix} \tag{3.81}$$

The generalized forces, $F_{\xi} = T_{E/B}\hat{F}$ are the translational external forces applied to the quadrotor due to main the thrust from the four propellers. The generalized torques or moments, τ, represent the yaw, pitch, and roll moments, respectively, and $T_{E/B}$ denotes the Euler rotation matrix, which transforms a vector in body coordinates to a vector in the inertial coordinate frame. The forces in the body frame $\hat{F}$ are transformed into forces in the inertial frame F_{ξ}. The overall thrust in the $-z\hat{k}$ direction is $u = \sum_{i=1}^{4} f_i$ and the vector of forces in body axes is: $\hat{F} = \begin{bmatrix} 0 \\ 0 \\ u \end{bmatrix}$.

The force f_i is the force produced by motor m_i, as shown in Fig. 3.10. Commonly the force produced by each motor is proportional to the square of its angular rate ω_i in the following manner: $f_i = k_i\omega_i^2$, where k_i is a constant and ω_i is the angular speed of the ith motor. The faster the motor turns, the more thrust it generates. The generalized torques may be written as:

$$\begin{aligned}
\tau = \begin{bmatrix} \tau_{\psi} \\ \tau_{\theta} \\ \tau_{\phi} \end{bmatrix} &= \begin{bmatrix} \tau_{yaw} \\ \tau_{pitch} + \tau_{gyro_{pitch}} \\ \tau_{roll} + \tau_{gyro_{roll}} \end{bmatrix} \\
&= \begin{bmatrix} \tau_{m_1} - \tau_{m_2} + \tau_{m_3} - \tau_{m_4} \\ (f_2 - f_4)l \\ (f_3 - f_1)l \end{bmatrix} - \begin{bmatrix} 0 \\ p_b\hat{i} \times H\hat{k} \\ q_b\hat{j} \times H\hat{k} \end{bmatrix}
\end{aligned}$$

$$\Rightarrow \begin{bmatrix} \tau_{\psi} \\ \tau_{\theta} \\ \tau_{\phi} \end{bmatrix} = \begin{bmatrix} \tau_{m_1} - \tau_{m_2} + \tau_{m_3} - \tau_{m_4} \\ (f_2 - f_4)l \\ (f_3 - f_1)l \end{bmatrix} - \begin{bmatrix} 0 \\ p_b \sum_{i=1}^{4} I_{m_i}\omega_i \\ q_b \sum_{i=1}^{4} I_{m_i}\omega_i \end{bmatrix} \tag{3.82}$$

where l is the distance from the center of any motor to the center of gravity of the system. Since the Lagrangian is free of cross terms in the kinetic energy equation (Eq. 3.80) between $\dot{\xi}$ and $\dot{\eta}$, the d'Alembert–Lagrange equations can be separated into the dynamics for the X, Y, Z generalized position coordinates and for the $\eta -(\psi, \theta, \phi)$ generalized Euler angle coordinates. The d'Alembert–Lagrange equations for translation become:

$$\mathcal{L}_{translate} = \frac{1}{2} m_{quad} \dot{\xi}^T \cdot \dot{\xi} - m_{quad} g Z = \frac{1}{2} m_{quad} \left[\dot{X}^2 + \dot{Y}^2 + \dot{Z}^2\right] - m_{quad} g Z$$

$$\left[\frac{\partial \mathcal{L}_{translate}}{\partial \dot{\xi}}\right] = \left[\frac{\partial \frac{1}{2} m_{quad}\left[\dot{X}^2 + \dot{Y}^2 + \dot{Z}^2\right]}{\partial \dot{X}}, \frac{\partial \frac{1}{2} m_{quad}\left[\dot{X}^2 + \dot{Y}^2 + \dot{Z}^2\right]}{\partial \dot{Y}}, \right.$$

$$\left. \frac{\partial \frac{1}{2} m_{quad}\left[\dot{X}^2 + \dot{Y}^2 + \dot{Z}^2\right]}{\partial \dot{Z}}\right]^T$$

$$= m_{quad}\left[\dot{X}, \dot{Y}, \dot{Z}\right]^T = m_{quad}\dot{\xi} \Rightarrow \frac{d(m_{quad}\dot{\xi})}{dt} = m_{quad}\ddot{\xi}$$

$$= m_{quad}\left[\ddot{X}, \ddot{Y}, \ddot{Z}\right]^T$$

$$\frac{\partial \mathcal{L}_{translate}}{\partial \xi} = \left[\frac{\partial \mathcal{L}_{translate}}{\partial X}, \frac{\partial \mathcal{L}_{translate}}{\partial Y}, \frac{\partial \mathcal{L}_{translate}}{\partial Z}\right]^T$$

$$= \left[0, 0, m_{quad} g\right]^T \frac{d}{dt}\left[\frac{\partial \mathcal{L}_{translate}}{\partial \dot{\xi}}\right] - \frac{\partial \mathcal{L}_{translate}}{\partial \xi} = F_{\xi}$$

$$= m_{quad}\left[\ddot{X}, \ddot{Y}, \ddot{Z}\right]^T - \left[0, 0, m_{quad} g\right]^T = \underbrace{T_{E/B}\left[0, 0, u\right]^T}_{F_{\xi}} \tag{3.83}$$

For the generalized angular coordinates, the d'Alembert–Lagrange's equations may be written as:

$$\mathcal{L}_{rotate} = \frac{1}{2}\dot{\eta}^T \mathcal{J}\dot{\eta}; \left[\frac{\partial \mathcal{L}_{rotate}}{\partial \dot{\eta}}\right] = \mathcal{J}\dot{\eta}; \frac{d}{dt}\left[\frac{\partial \mathcal{L}_{rotate}}{\partial \dot{\eta}}\right] = \dot{\mathcal{J}}\dot{\eta} + \mathcal{J}\ddot{\eta}$$

$$\frac{d}{dt}\left[\frac{\partial \mathcal{L}_{rotate}}{\partial \dot{\eta}}\right] - \left[\frac{\partial \mathcal{L}_{rotate}}{\partial \eta}\right] = \dot{\mathcal{J}}\dot{\eta} + \mathcal{J}\ddot{\eta} - \frac{1}{2}\frac{\partial(\dot{\eta}^T \mathcal{J}\dot{\eta})}{\partial \eta} = \tau \tag{3.84}$$

With the following definition of the "Coriolis-centripetal" vector: $\hat{V}(\eta, \dot{\eta}) = \dot{\mathcal{J}}\dot{\eta} - \frac{1}{2}\frac{\partial(\dot{\eta}^T \mathcal{J}\dot{\eta})}{\partial \eta}$, the d'Alembert–Lagrange's equations for the rotational dynamics become:

$$\mathcal{J}\ddot{\eta} + \dot{\mathcal{J}}\dot{\eta} - \frac{1}{2}\frac{\partial\left(\dot{\eta}^T \mathcal{J}\dot{\eta}\right)}{\partial \eta} = \mathcal{J}\ddot{\eta} + \hat{V}(\eta, \dot{\eta}) = \tau \tag{3.85}$$

Factoring out the $\dot{\eta}$ term, the "Coriolis-centripetal" $\hat{V}(\eta, \dot{\eta})$ vector may be expressed as:

$$\hat{V}(\eta, \dot{\eta}) = \left(\dot{\mathcal{J}} - \frac{1}{2}\frac{\partial\left(\dot{\eta}^T \mathcal{J}\right)}{\partial \eta}\right)\dot{\eta} = C(\eta, \dot{\eta})\dot{\eta} \tag{3.86}$$

The term $\left(\frac{1}{2}\frac{\partial(\dot{\eta}^T \mathcal{J})}{\partial \eta}\right)$ is evaluated as follows:

$$\mathcal{J}(\psi, \theta, \phi) = \begin{bmatrix} j_{11} & j_{12} & j_{13} \\ j_{21} & j_{22} & j_{23} \\ j_{31} & j_{32} & j_{33} \end{bmatrix};\ \dot{\eta}^T = \begin{bmatrix} \dot{\psi} & \dot{\theta} & \dot{\phi} \end{bmatrix}$$

$$\dot{\eta}^T \mathcal{J} = \Big[\underbrace{\dot{\psi} j_{11} + \dot{\theta} j_{21} + \dot{\phi} j_{31}}_{\mathcal{P}_{11}}\ \underbrace{\dot{\psi} j_{12} + \dot{\theta} j_{22} + \dot{\phi} j_{32}}_{\mathcal{P}_{12}}\ \underbrace{\dot{\psi} j_{13} + \dot{\theta} j_{23} + \dot{\phi} j_{33}}_{\mathcal{P}_{13}} \Big]$$

$$= \Big[\mathcal{P}_{11}(\psi, \theta, \phi)\ \mathcal{P}_{12}(\psi, \theta, \phi)\ \mathcal{P}_{13}(\psi, \theta, \phi)\Big]$$

$$\mathcal{P}_{11}(\psi, \theta, \phi) = \frac{\partial \mathcal{P}_{11}}{\partial \psi}\Delta\psi + \frac{\partial \mathcal{P}_{11}}{\partial \theta}\Delta\theta + \frac{\partial \mathcal{P}_{11}}{\partial \phi}\Delta\phi$$

$$\mathcal{P}_{12}(\psi, \theta, \phi) = \frac{\partial \mathcal{P}_{12}}{\partial \psi}\Delta\psi + \frac{\partial \mathcal{P}_{12}}{\partial \theta}\Delta\theta + \frac{\partial \mathcal{P}_{12}}{\partial \phi}\Delta\phi$$

$$\mathcal{P}_{13}(\psi, \theta, \phi) = \frac{\partial \mathcal{P}_{13}}{\partial \psi}\Delta\psi + \frac{\partial \mathcal{P}_{13}}{\partial \theta}\Delta\theta + \frac{\partial \mathcal{P}_{13}}{\partial \phi}\Delta\phi$$

$$\Rightarrow \left(\frac{1}{2}\frac{\partial\left(\dot{\eta}^T \mathcal{J}\right)}{\partial \eta}\right) = \mathcal{F}(\psi, \theta, \phi) = \frac{1}{2}\begin{bmatrix} \frac{\partial \mathcal{P}_{11}}{\partial \psi} & \frac{\partial \mathcal{P}_{11}}{\partial \theta} & \frac{\partial \mathcal{P}_{11}}{\partial \phi} \\ \frac{\partial \mathcal{P}_{12}}{\partial \psi} & \frac{\partial \mathcal{P}_{12}}{\partial \theta} & \frac{\partial \mathcal{P}_{12}}{\partial \phi} \\ \frac{\partial \mathcal{P}_{13}}{\partial \psi} & \frac{\partial \mathcal{P}_{13}}{\partial \theta} & \frac{\partial \mathcal{P}_{13}}{\partial \phi} \end{bmatrix}$$

$$\Rightarrow \hat{V}(\eta, \dot{\eta}) = \left(\dot{\mathcal{J}} - \frac{1}{2}\frac{\partial\left(\dot{\eta}^T \mathcal{J}\right)}{\partial \eta}\right)\dot{\eta} = \underbrace{\left(\dot{\mathcal{J}} - \mathcal{F}\right)}_{C(\eta, \dot{\eta})}\dot{\eta}$$

Summarizing the equations of motion, the result is:

$$m_{quad}\begin{bmatrix}\ddot{X}\\ \ddot{Y}\\ \ddot{Z}\end{bmatrix}+\begin{bmatrix}0\\0\\m_{quad}g\end{bmatrix}=F_{\xi}$$
$$\mathcal{J}\ddot{\eta}=\tau-C(\eta,\dot{\eta})\dot{\eta} \tag{3.87}$$

To simplify, define $\tilde{\tau}$ as follows:

$$\tilde{\tau}=\begin{bmatrix}\tilde{\tau}_{\psi}\\ \tilde{\tau}_{\theta}\\ \tilde{\tau}_{\phi}\end{bmatrix}=\mathcal{J}^{-1}\left(\tau-C(\eta,\dot{\eta})\dot{\eta}\right)\Rightarrow\ddot{\eta}=\mathcal{J}^{-1}\left(\tau-C(\eta,\dot{\eta})\dot{\eta}\right)=\begin{bmatrix}\tilde{\tau}_{\psi}\\ \tilde{\tau}_{\theta}\\ \tilde{\tau}_{\phi}\end{bmatrix} \tag{3.88}$$

Finally, the outcome is:

$$\begin{aligned}
&m_{quad}\begin{bmatrix}\ddot{X}\\ \ddot{Y}\\ \ddot{Z}\end{bmatrix}+\begin{bmatrix}0\\0\\m_{quad}g\end{bmatrix}\\
&=\begin{bmatrix}\cos\theta\cos\psi & \sin\phi\sin\theta\cos\psi-\cos\phi\sin\psi & \cos\phi\sin\theta\cos\psi+\sin\phi\sin\psi\\ \cos\theta\sin\psi & \sin\phi\sin\theta\sin\psi+\cos\phi\cos\psi & \cos\phi\sin\theta\sin\psi-\sin\phi\cos\psi\\ -\sin\theta & \sin\phi\cos\theta & \cos\phi\cos\theta\end{bmatrix}\begin{bmatrix}0\\0\\u\end{bmatrix}\\
&\Rightarrow m_{quad}\ddot{X}=u\left(\cos\phi\sin\theta\cos\psi+\sin\phi\sin\psi\right)\\
&m_{quad}\ddot{Y}=u\left(\cos\phi\sin\theta\sin\psi-\sin\phi\cos\psi\right)\\
&m_{quad}\ddot{Z}=u\left(\cos\phi\cos\theta\right)-m_{quad}g\\
&\ddot{\psi}=\tilde{\tau}_{\psi};\ddot{\theta}=\tilde{\tau}_{\theta};\ddot{\phi}=\tilde{\tau}_{\phi}
\end{aligned} \tag{3.89}$$

It is also possible to expand the terms $\tilde{\tau}_{\psi},\tilde{\tau}_{\theta},\tilde{\tau}_{\phi}$.

Numerator of $\tilde{\tau}_{\psi}$

$$\begin{aligned}
&2I_x\tau_{\psi}+2I_x\tau_{\phi}\sin\theta+(2I_y-2I_x)\tau_{\psi}\cos^2\phi\\
&\quad+(2I_y-2I_x)\tau_{\phi}\cos^2\phi\sin\theta+(I_xI_y-\frac{I_xI_z}{2})\dot{\psi}\dot{\theta}\sin 2\theta\\
&\quad+(I_xI_y+I_xI_z-I_x^2)\dot{\phi}\dot{\theta}\cos\theta+(2I_x-2I_y)\tau_{\theta}\cos\phi\cos\theta\sin\phi\\
&\quad+(I_xI_y-I_x^2)\dot{\theta}^2\cos\phi\sin\phi\sin\theta+(I_x^2+I_y^2-2I_xI_y)\dot{\theta}^2\cos^3\phi\sin\phi\sin\theta\\
&\quad+(I_x^2-I_y^2)\dot{\phi}\dot{\psi}\cos\phi\cos^2\theta\sin\phi\\
&\quad+(I_yI_z-I_xI_z+I_x^2-I_y^2)\dot{\phi}\dot{\theta}\cos^2\phi\cos\theta
\end{aligned}$$

$$+ (2I_xI_y - I_x^2 - I_y^2)\dot{\psi}\dot{\theta}\cos^4\phi\cos\theta\sin\theta$$

$$+ (I_xI_z - I_yI_z + I_x^2 - 2I_xI_y + I_y^2)\dot{\psi}\dot{\theta}\cos^2\phi\cos\theta\sin\theta$$

Denominator of $\tilde{\tau}_\psi$

$$2I_xI_y\cos^2\theta$$

Numerator of $\tilde{\tau}_\theta$

$$(I_x - I_y)\tau_\psi\sin 2\phi + 2I_y\tau_\theta\cos\theta + (I_y^2 - I_xI_y)\dot{\phi}\dot{\psi}\cos^2\theta$$

$$+ (2I_x - 2I_y)\tau_\theta\cos^2\phi\cos\theta + (2I_xI_y - I_x^2 - I_y^2)\dot{\theta}^2\cos^2\phi\sin\theta\sin^2\phi$$

$$+ (I_x^2 - I_y^2)\dot{\phi}\dot{\psi}\cos^2\phi\cos^2\theta + (2I_x - 2I_y)\tau_\phi\cos\phi\sin\phi\sin\theta$$

$$- I_y^2\dot{\psi}\dot{\theta}\cos\phi\cos\theta\sin\phi\sin\theta$$

$$+ (I_xI_z - I_yI_z + I_y^2 - I_x^2)\dot{\phi}\dot{\theta}\cos\phi\cos\theta\sin\phi$$

$$+ (I_x^2 + I_y^2 - 2I_xI_y)\dot{\psi}\dot{\theta}\cos^3\phi\cos\theta\sin\phi\sin\theta$$

$$+ (I_xI_y - I_xI_z + I_yI_z)\dot{\psi}\dot{\theta}\cos\phi\cos\theta\sin\phi\sin\theta$$

Denominator of $\tilde{\tau}_\theta$

$$2I_xI_y\cos\theta$$

Numerator of $\tilde{\tau}_\phi$

$$- 2I_xI_z\tau_\phi\cos^2\theta\sin^2\theta - 2I_xI_y\tau_\phi\cos^4\theta + I_xI_z^2\dot{\psi}\dot{\theta}\cos^3\theta\sin^2\theta$$

$$- 2I_xI_z\tau_\psi\cos^2\theta\sin\theta + (2I_xI_z - 2I_yI_z)\tau_\phi\cos^2\phi\cos^2\theta\sin^2\theta$$

$$+ I_x^2I_z\dot{\theta}^2\cos\phi\cos^2\theta\sin^2\theta\sin\phi + (I_x^2I_z - I_xI_z^2)\dot{\phi}\dot{\theta}\cos^3\theta\sin\theta$$

$$- 2I_xI_yI_z\dot{\psi}\dot{\theta}\cos^3\theta + I_xI_yI_z\dot{\psi}\dot{\theta}\cos^5\theta$$

$$+ (I_x^2I_z + I_y^2I_z)\dot{\psi}\dot{\theta}\cos^4\phi\cos^3\theta\sin^2\theta$$

$$+ (I_yI_z^2 - I_y^2I_z - I_xI_z^2 - I_x^2I_z)\dot{\psi}\dot{\theta}\cos^2\phi\cos^3\theta\sin^2\theta$$

$$+ (2I_xI_z - 2I_yI_z)\tau_\psi\cos^2\phi\cos^2\theta\sin\theta$$

$$- I_xI_yI_z\dot{\phi}\dot{\theta}\cos^3\theta\sin\theta$$

$$+ (2I_yI_z - 2I_xI_z)\tau_\theta\cos\phi\cos^3\theta\sin\phi\sin\theta$$

$$+ (2I_xI_yI_z - I_x^2I_z - I_y^2I_z)\dot{\theta}^2\cos^3\phi\cos^2\theta\sin^2\theta\sin\phi$$

$$
\begin{aligned}
&+ (I_x I_z^2 - I_x^2 I_z - I_y I_z^2 + I_y^2 I_z)\dot{\phi}\dot{\theta} \cos^2\phi \cos^3\theta \sin\theta \\
&+ 2 I_x I_y I_z \dot{\psi}\dot{\theta} \cos^2\phi \sin^2\phi \cos^3\theta \sin^2\theta \\
&- I_x I_y I_z \dot{\theta}^2 \cos\phi \cos^2\theta \sin^2\theta \sin\phi \\
&+ (I_y^2 I_z - I_x^2 I_z)\dot{\phi}\dot{\psi} \cos\phi \cos^4\theta \sin\phi \sin\theta
\end{aligned}
$$

Denominator of $\tilde{\tau}_\phi$

$$
-2 I_x I_y I_z (\sin^2\theta - 1)^2
$$

It soon becomes apparent that the terms $\tilde{\tau}_\psi, \tilde{\tau}_\theta, \tilde{\tau}_\phi$ are best calculated numerically. In the example on page 239, the moment equations are left in the body axis system and hence the torque equations are much simpler. If we assume that the vehicle is hovering, that is, $\psi = \theta = \phi = 0°$, the torques become:

$$
\begin{aligned}
\tilde{\tau}_\psi &= \frac{2\tau_\psi + I_x \dot{\phi}\dot{\theta} - I_y \dot{\phi}\dot{\theta} + I_z \dot{\phi}\dot{\theta}}{2 I_x} \\
\tilde{\tau}_\theta &= \frac{2\tau_\theta + I_x \dot{\phi}\dot{\psi} - I_y \dot{\phi}\dot{\psi}}{2 I_y} \\
\tilde{\tau}_\phi &= \frac{2\tau_\phi + I_z \dot{\psi}\dot{\theta}}{2 I_z}
\end{aligned}
$$

Further simplifications result when we use the fact that $I_x = I_y$ (due to symmetry of the $\hat{i}$ and $\hat{j}$ coordinates), in addition to the assumption that the angular rates are zero. We have for the torques:

$$
\tilde{\tau}_\psi = \frac{\tau_\psi}{I_x};\ \tilde{\tau}_\theta = \frac{\tau_\theta}{I_y};\ \tilde{\tau}_\phi = \frac{\tau_\phi}{I_z}
$$

3.8 Recap: Writing d'Alembert–Lagrangian Dynamics

Lecture 15—Video Times—1:24–11:08, 20:26–25:29

This section was adapted from the MIT Lecture Series on Dynamics as presented by Prof. Vandiver (see Vandiver—2.003SC Engineering Dynamics. Video of Lecture 15: Introduction to Lagrange With Examples) [42]. The Lagrangian is: $\mathcal{L} = T - V$, where T is the kinetic energy of the system and all of its component parts and V is the potential energy of the system. In addition, q_j are the j generalized coordinates and Q_j are the j non-conservative generalized forces. The d'Alembert–Lagrangian equation is:

$$\frac{d}{dt}\left(\frac{\partial \mathcal{L}}{\partial \dot{q}_j}\right) - \frac{\partial \mathcal{L}}{\partial q_j} = Q_j$$

$$\Rightarrow \underbrace{\frac{d}{dt}\left(\frac{\partial T}{\partial \dot{q}_j}\right)}_{1} - \underbrace{\frac{\partial T}{\partial q_j}}_{2} - \underbrace{\frac{d}{dt}\left(\frac{\partial V}{\partial \dot{q}_j}\right)}_{=0} + \underbrace{\frac{\partial V}{\partial q_j}}_{3} = \underbrace{Q_j}_{4} \tag{3.90}$$

It should be noted that the potential energy is neither dependent on velocity nor on time t for mechanical systems and hence $\frac{d}{dt}\left(\frac{\partial V}{\partial \dot{q}_j}\right) = 0$. When dealing with electromagnetics, however, the term $\frac{d}{dt}\left(\frac{\partial V}{\partial \dot{q}_j}\right)$ might not equal zero. As may be observed, the left-hand side of the equations of motion contains terms with t and velocity v or $\dot{q}_j$ in them. The right-hand side has these generalized forces which are the non-conservative forces in the system.

The procedure for obtaining the Lagrangian dynamics of a system is outlined as follows:

Left-Hand Side

1. Determine the number of degrees of freedom available and decide upon your set of generalized coordinates, the q_j's.
2. The generalized coordinates need not be Cartesian nor inertial nor orthogonal. The set of selected generalized coordinates must however be independent and complete, and the system (with the chosen generalized coordinates) must be holonomic.
3. Verify for completeness, independence, and holonomicity (see page 69).
4. Calculate the kinetic energy T and the potential energy V for every rigid body in the system.
5. Compute items 1, 2, and 3 in the d'Alembert–Lagrangian equation above for each q_j or degree of freedom. That is, for every generalized coordinate, items 1, 2, and 3 must be computed. The computation of the non-conservative generalized force will be treated in the sequel. The components 1, 2, and 3 for each q_j constitute the left-hand side of the d'Alembert–Lagrangian equation. If there are no external forces, non-conservative forces, and only conservative external forces which act on the system, then items 1 plus 2 plus 3 equal 0. But if there exist non-conservative forces which act upon the system, such as friction, then the right-hand side of the d'Alembert–Lagrangian equation or Q_j must be accounted for.

Right-Hand Side

1. For each q_j, that is, for each generalized coordinate, the generalized non-conservative force Q_j that potentially accompanies the q_j and acts upon the system must be found.
2. Compute the virtual work δW^{NC} for the non-conservative forces. The delta δW_j^{NC} is associated with the virtual displacement δq_j.

   For every generalized coordinate shift the system by a small amount δq_j while keeping all of the other coordinates $q_k, k = 1, \ldots m, k \neq j$ constant and compute the amount of much virtual work performed by this virtual motion. The virtual work δW_j (the superscript NC is dropped for convenience) is: $\delta W_j = Q_j \delta q_j$. The entity being sought after is Q_j and it's going to be a function of all those external non-conservative forces acting through a very small virtual displacement, resulting in a very small amount of virtual work.

Example 6: Mass–Spring Dashpot (Damper) Single Degree of Freedom System

Lecture 15—Video Times—25:40–37:40

This example is originally from the video of Vandiver's Lecture 15 (see Vandiver—2.003SC Engineering Dynamics. Video of Lecture 15: Introduction to Lagrange With Examples) [42]. The mass–spring dashpot (damper) system has a single degree of freedom x and hence only one coordinate is required to describe its motion. The generalized coordinate x is Cartesian so $q_j = q_1 = x$. It is complete, independent, and holonomic.

1. The kinetic energy T is: $T = \frac{1}{2}m\dot{x}^2$.
2. The potential energy V is: $V = \frac{1}{2}mkx^2 - mgx$.
3. The non-conservative forces include the external excitation force and that due to the dashpot and are: $\sum F_1^{NC} = (-b\dot{x} + F(t))\,\hat{i}$.

The force $F(t)$ is an external excitation force and could be non-conservative since it could do work on the system and dissipate energy. It's neither a potential nor a spring. It could be an external vibration. The conservative forces are kx and mg, while the non-conservative forces are $F(t)$ and $b\dot{x}$. Carrying out the operations required in the d'Alembert–Lagrange's equation we have:

$$\frac{d}{dt}\left(\frac{\partial T}{\partial \dot{q}_j}\right) - \frac{\partial T}{\partial q_j} - \frac{d}{dt}\left(\frac{\partial V}{\partial \dot{q}_j}\right) + \frac{\partial V}{\partial q_j} = Q_j$$

$$T = \frac{1}{2}m\dot{x}^2;\ V = \frac{1}{2}mkx^2 - mgx;\ q_1 = x$$

$$\frac{d}{dt}\left(\frac{\partial T}{\partial \dot{q}_j}\right) = \frac{d}{dt}\left(\frac{\partial T}{\partial \dot{x}}\right) = \frac{1}{2}2m\frac{d\dot{x}}{dt} = m\ddot{x}$$

$$-\frac{\partial T}{\partial q_j} = -\frac{\partial T}{\partial x} = 0$$

$$\frac{\partial V}{\partial q_j} = \frac{\partial V}{\partial x} = kx - mg$$

$$\sum F_1^{NC} \cdot \delta_x = (-b\dot{x} + F(t))\,\hat{i} \cdot \delta_x\hat{i} = -b\dot{x}\delta_x + F(t)\delta_x = Q_1\delta_x \tag{3.91}$$

In summary, we have:

1. $\frac{d}{dt}\left(\frac{\partial T}{\partial \dot{q}_j}\right) = \frac{d}{dt}\left(\frac{\partial T}{\partial \dot{x}}\right) = \frac{1}{2}2m\frac{d\dot{x}}{dt} = m\ddot{x}$
2. $-\frac{\partial T}{\partial q_j} = -\frac{\partial T}{\partial x} = 0$ since T is a function of $\dot{x}$ only
3. $\frac{\partial V}{\partial q_j} = \frac{\partial V}{\partial x} = kx - mg$
4. $m\ddot{x} + kx - mg = Q_1$
5. $\left(\sum F_i\right) \cdot dr = (F(t) - b\dot{x})\,\hat{i} \cdot dr$, where F_i and dr are both vectors. dr is the infinitesimal motion and it is actually the same as δx. The virtual motion $dr(\delta_1, \delta_2, \delta_3, \ldots, \delta_j)$ is in general a function of the virtual displacements $\delta_j' s$ of all of the degrees of freedom. In this case we only have one, but we could have $\delta_1, \delta_2, \delta_3, \delta_4, \ldots, \delta_j$. Each component of the force in the direction of one of the virtual movements does virtual work.
6. $Q_1\delta_x = -b\dot{x}\delta_x + F(t)\delta_x = \delta W_x^{NC}$

Rearranging the above we arrive at the dynamical equation, which is (Fig. 3.11):

$$m\ddot{x} + kx - mg = -b\dot{x} + F(t) \tag{3.92}$$

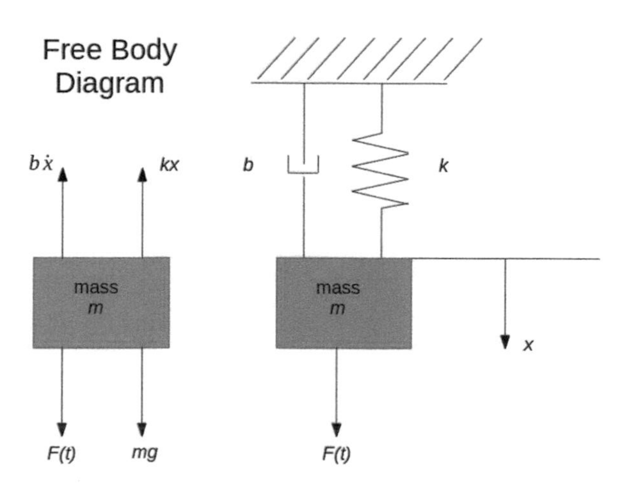

Fig. 3.11 Mass–spring dashpot (damper) single degree of freedom system

Example 7: Pendulum with a Mass and Spring

Lecture 15—Video Times—37:40–1:21:16

This example was presented in the video of Vandiver's Lecture 15 (see Vandiver-2.003SC Engineering Dynamics. Video of Lecture 15: Introduction to Lagrange With Examples) [42]. The system consists of a steel rod with a metal sleeve through which the rod passes, and which is attached to the rod by a spring. The metal sleeve slides along the rod in both the up and down directions without friction. The rod, sleeve, and spring are attached at point A to a hinge and the whole system can swing back and forth as a pendulum. The system contains several sources of kinetic energy and varied forms of potential energy. There is a force $F \cos \omega t$ which is always horizontal, located at the bottom of the metal sleeve, which pushes this system back and forth. The problem is to derive the equations of motion of the system. The motion is planar and there are two rigid bodies involved. Each rigid body has six degrees of freedom, but because of planar motion, each body has a maximum of three degrees of freedom. For planar motion, each rigid body can move in the x and y directions and can rotate about the z axis, where they are not attached together by the spring. The maximum number of degrees of freedom for the two rigid bodies, each acting independently of the other is six. This problem has two degrees of freedom, which are the angular displacement θ of the system from the vertical and x_1 the linear displacement of the sleeve's center of mass along the rod with respect to point A. The coordinate system $X_1 - Y_1$ rotates with the rod and sleeve. The coordinates chosen are independent, since, freezing one of them, the other can still traverse over the full range of values for that coordinate. They are also complete since with the two coordinates, all parts of the system may be located at all times. Furthermore, the generalized coordinates are also holonomic since the number of independent generalized coordinates required to describe the motion of the system equals the number of degrees of freedom. The physical parameters of the system are as follows:

1. Mass of the rod: M_1, moment of inertia in the z direction about the point A: $(I_{zz})_A$, length of the rod: L_1 location of the center of mass of the rod: G_1
2. Mass of the sleeve: M_2, moment of inertia in the z direction about the sleeve's center of mass: $(I_{zz})_G$, length of the sleeve: L_2 location of the center of mass of the sleeve: G_2

In order to proceed, the potential and kinetic energies must be determined. In the calculation of the potential energy, the appropriate reference points must be determined and the un-stretched length of the spring accounted for. Let L_0 be the un-stretched length of the spring. The potential energy of the spring may be written as: $V_{spring} = \frac{1}{2}k\left(x_1 - L_0 - \frac{L_2}{2}\right)^2$ (Fig. 3.12). There are two other sources for potential energy due to gravity, the potential energies of the rod and the sleeve, respectively. In order to calculate the potential energy of the rod, it is recommended to use the equilibrium position of the rod (hanging vertically) as the reference position. The un-stretched length of the spring has no bearing on the calculation

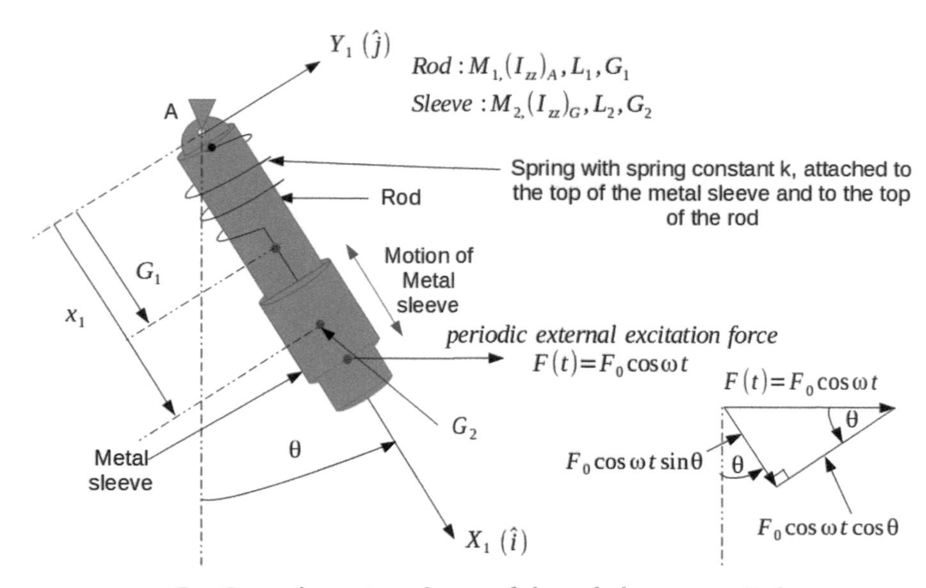

Fig. 3.12 Pendulum with a moveable mass and spring

of the potential energy of the rod. The potential energy is the difference in the heights between the rod in its reference position and the rod which has rotated by an angle θ (see Fig. 3.13).

It should be noted that the change in potential energy from one position to another is path independent (because gravity is a conservative force). In other words, it isn't necessary to carry out the calculation of $-\int mg \cdot dr$. The only thing that must be accounted for is Δh, the change in height (or the change in the vertical position) of the center of gravity of the object between its equilibrium position and its perturbed or rotated position. The potential energy of the rod is: $V_{rod} = \frac{L_1}{2} M_1 g\,(1-\cos\theta)$. Similarly, the potential energy of the sleeve is: $V_{sleeve} = M_2 g\left(L_0 + \frac{L_2}{2} - x_1\cos\theta\right)$. Note that the initial height was $\left(L_0 + \frac{L_2}{2}\right)$ and the final vertical position was $x_1\cos\theta$. The important entity is the difference between the initial and final heights and it is this difference which gives us the potential energy. Adding together the potential energies of the spring, rod, and sleeve results in the overall potential energy, which is (Fig. 3.14):

$$V = V_{spring} + V_{rod} + V_{sleeve} = \frac{1}{2}k\left(x_1 - L_0 - \frac{L_2}{2}\right)^2 + \frac{L_1}{2}M_1 g\,(1-\cos\theta) + M_2 g\left(L_0 + \frac{L_2}{2} - x_1\cos\theta\right) \tag{3.93}$$

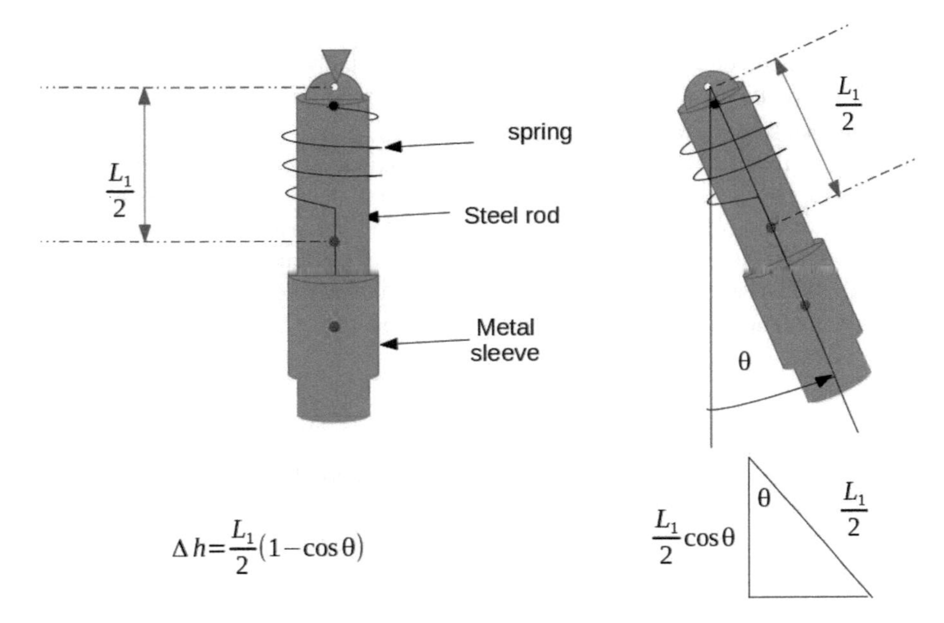

Fig. 3.13 Pendulum with a moveable mass and spring—geometrical definitions I

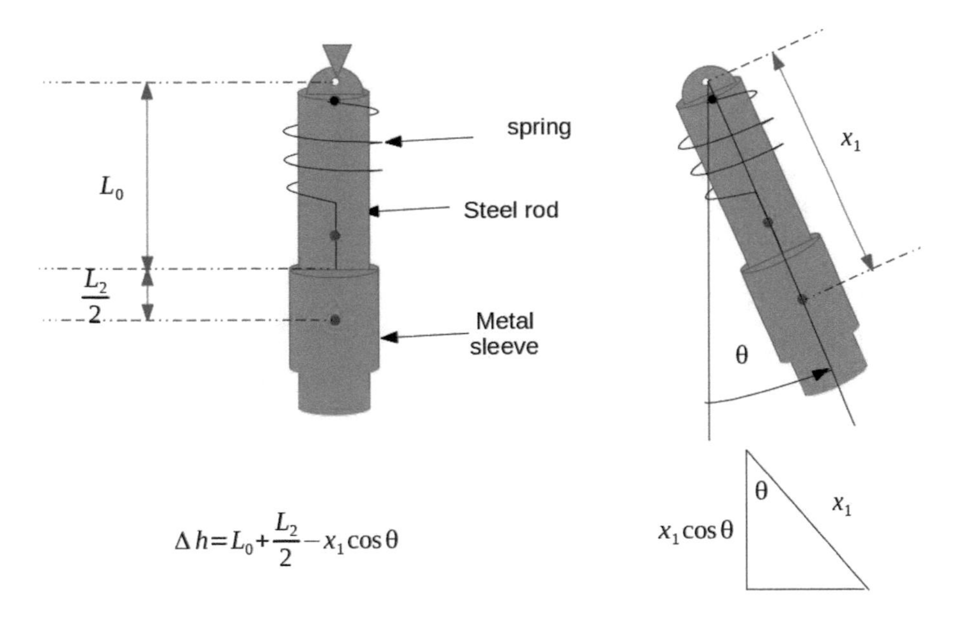

Fig. 3.14 Pendulum with a moveable mass and spring-geometrical definitions II

The kinetic energy T consists of the rotational kinetic energy of the rod, rotational kinetic energy of the sleeve, and the translational kinetic energy of the sleeve. The rotational kinetic energy of the rod is: $T_{rot-rod} = \frac{1}{2}(I_{zz})_A \dot{\theta}^2$. Similarly the rotational kinetic energy of the sleeve is: $T_{rot-sleeve} = \frac{1}{2}(I_{zz})_G \dot{\theta}^2$. The moment of inertia of the rod is calculated by the parallel axis theorem because it is pinned at A. As for the sleeve, since it is in motion both rotationally and in translation, and the center of mass of the sleeve is in motion (due to its translation), then it is more convenient to use the sleeve's moment of inertia about its center of mass. Because of the sleeve's translatory motion, this must also be accounted for in the kinetic energy calculation of the sleeve. The translational kinetic energy of the sleeve is therefore:

$$T_{trans\text{-}sleeve} = \frac{1}{2}M_2 (V_G)_0^T \cdot (V_G)_0 \tag{3.94}$$

The velocity of the center of mass of the sleeve, in inertial coordinates, is: $(V_G)_0$. Note that $(V_G)_0$ is a vector. The velocity $(V_G)_0$ with respect to coordinates $X_1 - Y_1$ is: $(V_G)_0 = \dot{x}_1\hat{i} + \dot{\theta}x_1\hat{j}$. This implies that $(V_G)_0^T \cdot (V_G)_0 = \dot{x}_1^2 + \dot{\theta}^2 x_1^2$. The total kinetic energy of the sleeve is therefore:

$$T_{sleeve} = T_{rot\text{-}sleeve} + T_{trans\text{-}sleeve} = \frac{1}{2}(I_{zz})_G \dot{\theta}^2 + \frac{1}{2}M_2\left(\dot{x}_1^2 + \dot{\theta}^2 x_1^2\right) \tag{3.95}$$

Hence the total system kinetic energy is of the form:

$$T = T_{rod} + T_{sleeve} = \frac{1}{2}(I_{zz})_A \dot{\theta}^2 + \frac{1}{2}(I_{zz})_G \dot{\theta}^2 + \frac{1}{2}M_2\left(\dot{x}_1^2 + \dot{\theta}^2 x_1^2\right) \tag{3.96}$$

Since we have two independent generalized coordinates x_1 and θ, we are required to apply the d'Alembert–Lagrange's equation twice, once for each coordinate. For the x_1 generalized coordinate, the d'Alembert–Lagrange's equation results in:

1. $\frac{d}{dt}\left(\frac{\partial T}{\partial \dot{q}_j}\right) - \frac{\partial T}{\partial q_j} + \frac{\partial V}{\partial q_j} = Q_j$
2. $\frac{d}{dt}\left(\frac{\partial T}{\partial \dot{x}_1}\right) = \frac{d}{dt}\left(2\frac{1}{2}M_2\dot{x}_1\right) = M_2\ddot{x}_1$
3. $-\frac{\partial T}{\partial x_1} = -\frac{1}{2}M_2 2x_1\dot{\theta}_1^2 = -M_2 x_1\dot{\theta}_1^2$. This is the centripetal force acting on the sleeve.
4. $\frac{\partial V}{\partial q_j} = \frac{\partial V}{\partial x_1} = k\left(x_1 - L_0 - \frac{L_2}{2}\right) - M_2 g\cos\theta$
5. Summing the above three terms we have: $M_2\ddot{x}_1 - M_2 x_1\dot{\theta}_1^2 + k\left(x_1 - L_0 - \frac{L_2}{2}\right) - M_2 g\cos\theta = Q_1$.

There should be a centripetal term in the $-X_1$ direction since the system is undergoing circular motion. In addition there is a coriolis force in the Y_1 direction. Note furthermore that the equations are scalar equations and not vector equations. The only thing required to complete the Lagrangian formulation in the generalized x_1 coordinate direction is to find Q_1. The virtual work in the x_1 direction must be determined. The virtual displacement δx_1 is in the x_1 direction. The virtual work is the product of the component in the x_1 direction of the externally applied force $F_0 \cos \omega t$ with the virtual displacement δx_1. The virtual work δW_1 may be written as:

$$\begin{aligned} \delta W_1 &= F(t) \cdot dr = [F_0 \cos \omega t \sin \theta \hat{i} + F_0 \cos \omega t \cos \theta \hat{j}] \cdot \delta x_1 \hat{i} \\ &= F_0 \cos \omega t \sin \theta \delta x_1 = Q_1 \delta x_1 \Rightarrow Q_1 = F_0 \cos \omega t \sin \theta \end{aligned} \tag{3.97}$$

The d'Alembert Lagrangian dynamics in the x_1 direction are therefore:

$$M_2 \ddot{x}_1 - M_2 x_1 \dot{\theta}_1^2 + k \left(x_1 - L_0 - \frac{L_2}{2} \right) - M_2 g \cos \theta = F_0 \cos \omega t \sin \theta \tag{3.98}$$

where the linear acceleration term is $\ddot{x}_1$, the centripetal acceleration is $-x_1 \dot{\theta}_1^2$, the spring force is $k \left(x_1 - L_0 - \frac{L_2}{2} \right)$, and the gravitational acceleration along the rod in the X_1 direction is $-M_2 g \cos \theta$. Once completed, it always helps to check the resulting equations for consistency, based upon classical Newtonian physics. The linear centripetal and coriolis accelerations in classical Newtonian physics are, respectively:

$$\begin{aligned} (\ddot{r})_{Oxy} &= \ddot{x}_1 a_{centripetal} = \Omega \times (\Omega \times r)\,; a_{coriolis} = 2\Omega \times (\dot{r})_{Oxy} \\ r &= x_1 \hat{i};\ (\dot{r})_{Oxy} = \dot{x}_1 \hat{i};\ \Omega = \dot{\theta} \hat{k} \\ \Rightarrow a_{centripetal} &= \dot{\theta} \hat{k} \times \left(\dot{\theta} \hat{k} \times x_1 \hat{i} \right) = -(\dot{\theta})^2 x_1 \hat{i} \\ a_{coriolis} &= 2\dot{\theta} \hat{k} \times \dot{x}_1 \hat{i} = 2\dot{\theta} \dot{x}_1 \hat{j} \end{aligned} \tag{3.99}$$

Questions to be Answered for Consistency Check of $q_1 = x_1$

- Does the equation contain a linear acceleration term along the X_1 axis?
- Is there a centripetal acceleration term only in the x_1 direction?
- Is there no coriolis acceleration term in the x_1 direction?
- Is there a spring force?
- Is there a component of gravity in the direction of motion, up and down the rod?
- Do the sum of the above terms equal any externally applied forces in that direction?

Another test for consistency is to make sure the system satisfies the laws of statics. Assume that all of the time derivatives are zero. In addition, at static equilibrium the system is completely vertical, and so $\theta = 0$, which implies that $\cos \theta = 1$, $\sin \theta = 0$. The dynamic equation along the X_1 direction becomes:

$$k \left(x_1 - L_0 - \frac{L_2}{2} \right) - M_2 g = 0 \Rightarrow k \left(x_1 - L_0 - \frac{L_2}{2} \right) = M_2 g \tag{3.100}$$

This condition states that when static equilibrium is achieved, the spring stretches by the amount $\left(x_1 - L_0 - \frac{L_2}{2}\right)$ in order to compensate for the gravitation force due to the mass of the sleeve $M_2 g$.

As for the d'Alembert–Lagrange equations in the θ direction, we have:

1. $T = T_{rod} + T_{sleeve} = \frac{1}{2}(I_{zz})_A \dot{\theta}^2 + \frac{1}{2}(I_{zz})_G \dot{\theta}^2 + \frac{1}{2}M_2\left(\dot{x}_1^2 + \dot{\theta}^2 x_1^2\right)$
2. $V = V_{spring} + V_{rod} + V_{sleeve} =$
   $\frac{1}{2}k\left(x_1 - L_0 - \frac{L_2}{2}\right)^2 + \frac{L_1}{2}M_1 g(1-\cos\theta) + M_2 g\left(L_0 + \frac{L_2}{2} - x_1\cos\theta\right)$
3. $\frac{d}{dt}\left(\frac{\partial T}{\partial \dot{q}_j}\right) - \frac{\partial T}{\partial q_j} + \frac{\partial V}{\partial q_j} = Q_j$
4. $\frac{d}{dt}\left(\frac{\partial T}{\partial \dot{\theta}}\right) = \frac{d}{dt}\left(\frac{1}{2}[I_{zz}]_A 2\dot{\theta} + \frac{1}{2}[I_{zz}]_G 2\dot{\theta} + \frac{1}{2}M_2 2\dot{\theta}x_1^2\right)$
   $= \frac{d}{dt}\left([I_{zz}]_A\dot{\theta} + [I_{zz}]_G\dot{\theta} + M_2\dot{\theta}x_1^2\right)$
   $= \left([I_{zz}]_A\ddot{\theta} + [I_{zz}]_G\ddot{\theta} + M_2\ddot{\theta}x_1^2 + 2M_2\dot{\theta}x_1\dot{x}_1\right)$
5. $-\frac{\partial T}{\partial \theta} = 0$
6. $\frac{\partial V}{\partial q_j} = \frac{\partial V}{\partial \theta} = \left(\frac{L_1}{2}M_1 g\sin\theta + M_2 g x_1\sin\theta\right)$
7. $\delta W_2 = Q_2\delta\theta = F\cdot dr\ , dr = \left(x_1 + \frac{L_2}{2}\right)\delta\theta\hat{j}\ , F(t) = F_0\cos\omega t\sin\theta\hat{i} + F_0\cos\omega t\cos\theta\hat{j}$
   $\Rightarrow Q_2\delta\theta = F\cdot dr = \left(x_1 + \frac{L_2}{2}\right)F_0\cos\omega t\cos\theta\delta\theta$
   $\Rightarrow Q_2 = \left(x_1 + \frac{L_2}{2}\right)F_0\cos\omega t\cos\theta$
8. Summing the above three terms we have: $\left([I_{zz}]_A\ddot{\theta} + [I_{zz}]_G\ddot{\theta} + M_2\ddot{\theta}x_1^2 + 2M_2\dot{\theta}x_1\dot{x}_1\right) + \left(\frac{L_1}{2}M_1 g\sin\theta + M_2 g x_1\sin\theta\right) = \left(x_1 + \frac{L_2}{2}\right)F_0\cos\omega t\cos\theta$

Remarks

- The dynamics of the second equation describe the swinging motion of the pendulum, as well as the linear motion of the sleeve as it slides up and down along the rod.
- The swinging motion is in the $\hat{j}$ direction. Hence we will end up with an equation in the $\hat{j}$ direction.
- The angle θ speeds up or slows down along its trajectory. As the pendulum reaches its highest angular value at the top of the swing, the angular velocity $\dot{\theta}$ goes to zero, while at the bottom of the swing, when the pendulum is vertical, the angular velocity is at a maximum.
- This implies that we expect to have terms in $\ddot{\theta}$ (Eulerian terms).
- There is also a coriolis term of the form $2M_2\dot{\theta}\dot{x}_1 x_1$ in the $\hat{j}$ direction or along the Y_1 axis. This is to be expected since the sleeve is sliding up and down along the rod and has a non-zero velocity $\dot{x}_1$, while the pendulum is swinging.
- Whenever a body moves radially while it is swinging in a circle, the result will be a coriolis acceleration.
- This implies that the angular momentum of the body is changing and a force is required to change the angular momentum.

- For the generalized force Q_2, the virtual work is $F \cdot dr$. The work is force applied over a distance.
- A virtual deflection in angle times the moment arm gives you a distance. The small displacement dr is the linear trajectory traced out by $\left(x_1 + \frac{L_2}{2}\right)\delta\theta$. The distance x_1 is from the pivot point of the pendulum to the center of gravity of the sleeve and $\frac{L_2}{2}$ is from the sleeve's center of gravity to the end of the sleeve. The sum of the two terms is the effective position or radius of the sleeve or the moment arm with respect to the pivot point and its product with $\delta\theta$, the virtual angle, describes the distance the sleeve has traveled.
- The direction of the angular motion of the sleeve is along the Y_1 axis or in the $\hat{j}$ direction.
- This equation is a torque equation.
- The term $M_2 x_1^2$ looks like the parallel axis theorem for the sleeve. Since we used the $(I_{zz})_G$ which is the sleeve's moment of inertia at its center of mass, then the sum of $(I_{zz})_G + M_2 x_1^2 = (I_{zz})_A$ is the sleeve's moment of inertia around the pivot point of the pendulum.

Questions to be Answered for Consistency Check of $q_2 = \theta$

- Does the equation contain angular acceleration terms?
- Is there a coriolis acceleration term?
- Are there torques due to the masses $M_1 g, M_2 g$, that is, $\left(\frac{L_1}{2} M_1 g \sin\theta + M_2 g x_1 \sin\theta\right)$?
- Do the sum of the above terms equal any externally applied moments?

Example 8: Cart and Pendulum, Lagrange Method

Recitation 8—Video Times—3:57–35:00

This example originates from Vandiver's video of Recitation 8 (see Vandiver-2.003SC Engineering Dynamics. Video of Recitation 8: Cart and Pendulum Lagrange Method [46] and the Recitation 8 Notes [38]).

A cart and pendulum, shown below, consists of a cart of mass, m_1, moving on a horizontal surface, without friction, and acted upon by a spring with spring constant k. From the cart, a pendulum consisting of a uniform rod of length, l, and mass, m_2, is suspended at and pivots about point A (see Fig. 3.15). There are no external forces acting on the system.

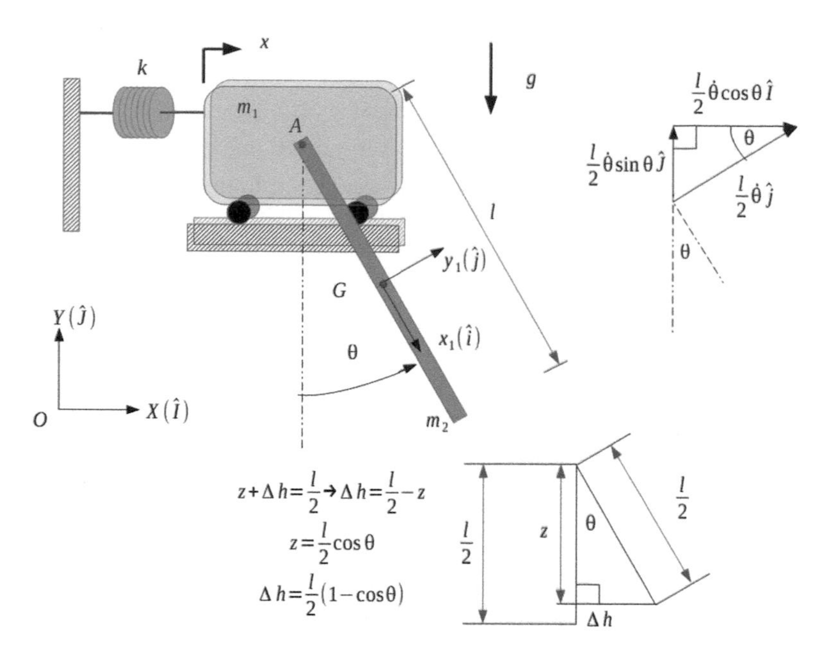

Fig. 3.15 Cart with spring and pendulum

Find the following using Lagrange's method:

1. T, the system's kinetic energy
2. V, the system's potential energy
3. v_G^2, the square of the magnitude of the pendulum's center of gravity

The generalized coordinates are: $q_1 = x; q_2 = \theta$. The linear velocity of the pendulum's center of mass, v_G, is given by:

$$v_G = \dot{x}\hat{I} + \frac{l}{2}\dot{\theta}\hat{j} = \left(\dot{x} + \frac{l}{2}\dot{\theta}\cos\theta\right)\hat{I} + \frac{l}{2}\dot{\theta}\sin\theta\,\hat{J}$$

$$\Rightarrow v_G^T \cdot v_G = v_G^2 = \left(\dot{x}^2 + \frac{l^2\dot{\theta}^2}{4}\cos^2\theta + l\dot{x}\dot{\theta}\cos\theta\right) + \frac{l^2\dot{\theta}^2}{4}\sin^2\theta$$

$$= \left(\dot{x}^2 + \frac{l^2\dot{\theta}^2}{4} + l\dot{x}\dot{\theta}\cos\theta\right) \tag{3.101}$$

The kinetic energy T is:

$$T = \frac{1}{2}m_1\dot{x}^2 + \frac{1}{2}m_2 v_G^2 + \frac{1}{2}I_G\dot{\theta}^2;\ v_G^2 = \left(\dot{x}^2 + \frac{l^2\dot{\theta}^2}{4} + l\dot{x}\dot{\theta}\cos\theta\right)$$

$$\Rightarrow T = \frac{1}{2}m_1\dot{x}^2 + \frac{1}{2}m_2\left(\dot{x}^2 + \frac{l^2\dot{\theta}^2}{4} + l\dot{x}\dot{\theta}\cos\theta\right) + \frac{1}{2}\underbrace{\left(\frac{m_2 l^2}{12}\right)}_{I_G}\dot{\theta}^2 \tag{3.102}$$

where $I_G = \frac{m_2 l^2}{12}$ is the moment of inertia of the pendulum (slender rod) at its center of mass G. Since the point at which the rod is pinned moves, i.e., the rod doesn't rotate about a fixed point, it is more convenient, when calculating the kinetic energy, to separate the motion of the rod into a rotation and translation of its center of mass (see Nielsen et al. [16, p. 4]). Consequently, the parallel axis theorem for the moment of inertia of the rod cannot be used and the moment of inertia about the rod's center of mass must be utilized instead.

The total kinetic energy is made up of the sum of the kinetic energies of all of the rigid bodies in the system. There are two rigid bodies, namely the cart and the pendulum and so each one has its own kinetic energy. The velocity of the cart on wheels is $\dot{x}$. The linear velocity of the pendulum is also $\dot{x}$ at the point at which it is pinned to the cart but it also has an angular velocity of $\frac{1}{2}l\dot{\theta}$ in the $\hat{j}$ direction ($v_G = \dot{\theta}\hat{k} \times \frac{l}{2}\hat{i} = \frac{1}{2}l\dot{\theta}\hat{j}$). Summing the two velocities of the pendulum and transforming the velocity in the $\hat{j}$ direction to components in the $\hat{I}$ and $\hat{J}$ directions gives us the v_G, that is, $v_G = \left(\dot{x} + \frac{l}{2}\dot{\theta}\cos\theta\right)\hat{I} + \frac{l}{2}\dot{\theta}\sin\theta\hat{J}$. This allows us to write the kinetic energy as the sum of the kinetic energies of each individual rigid body (see equation for T above—Eq. (3.102)). In general form, the kinetic energy may be written as:

$$T = \frac{1}{2}m_1\dot{x}^2 + \frac{1}{2}m_2 v_G \cdot v_G + \frac{1}{2}\omega^T I_G \omega \tag{3.103}$$

where $\omega = \dot{\theta}\hat{k}$, $I_G\omega = I_{zz}\omega = \frac{m_2 l^2}{12}\dot{\theta}\hat{k}$, and $\omega^T I_G \omega = \frac{m_2 l^2}{12}\dot{\theta}^2$. The potential energy V may be written as:

$$V = \frac{1}{2}kx^2 + m_2 g\frac{l}{2}(1 - \cos\theta) \tag{3.104}$$

The Lagrangian $\mathcal{L} = T - V$ is:

$$\begin{aligned}\mathcal{L} = &\frac{1}{2}m_1\dot{x}^2 + \frac{1}{2}m_2\left(\dot{x}^2 + \frac{l^2\dot{\theta}^2}{4} + l\dot{x}\dot{\theta}\cos\theta\right)\\ &+\frac{1}{2}\left(\frac{m_2 l^2}{12}\right)\dot{\theta}^2 - \frac{1}{2}kx^2 - m_2 g\frac{l}{2}(1 - \cos\theta)\end{aligned} \tag{3.105}$$

The d'Alembert–Lagrange's equations for $q_1 = x$ are as follows:

1. $\frac{d}{dt}\left(\frac{\partial \mathcal{L}}{\partial \dot{q}_j}\right) - \frac{\partial \mathcal{L}}{\partial q_j} = Q_j$; $\mathcal{L} = T - V$; $\frac{d}{dt}\left(\frac{\partial V}{\partial \dot{q}_j}\right) = 0$
2. $\frac{d}{dt}\left(\frac{\partial T}{\partial \dot{q}_j}\right) - \frac{\partial T}{\partial q_j} + \frac{\partial V}{\partial q_j} = Q_j \Rightarrow \frac{d}{dt}\left(\frac{\partial T}{\partial \dot{x}}\right) - \frac{\partial T}{\partial x} + \frac{\partial V}{\partial x} = Q_1$
3. $\frac{d}{dt}\left(\frac{\partial T}{\partial \dot{x}}\right) = \frac{d}{dt}\left([m_1 + m_2]\dot{x} + \frac{1}{2}m_2 l\dot{\theta}\cos\theta\right)$
   $= [m_1 + m_2]\ddot{x} + \frac{1}{2}m_2 l\ddot{\theta}\cos\theta - \frac{1}{2}m_2 l\dot{\theta}^2\sin\theta$
4. $\frac{\partial T}{\partial x} = 0$
5. $\frac{\partial V}{\partial x} = kx$
6. Summing: $\frac{d}{dt}\left(\frac{\partial T}{\partial \dot{x}}\right) - \frac{\partial T}{\partial x} + \frac{\partial V}{\partial x} = [m_1 + m_2]\ddot{x} + \frac{1}{2}m_2 l\ddot{\theta}\cos\theta - \frac{1}{2}m_2 l\dot{\theta}^2\sin\theta + kx$

Note that $-\frac{1}{2}m_2 l\dot{\theta}^2 \sin\theta$ resembles a centripetal acceleration term. Since there are no non-conservative forces, then $Q_1 = 0$ and the dynamic equation for $q_1 = x$ is:

$$[m_1 + m_2]\ddot{x} + \frac{1}{2}m_2 l\ddot{\theta}\cos\theta - \frac{1}{2}m_2 l\dot{\theta}^2 \sin\theta + kx = 0 \tag{3.106}$$

The d'Alembert–Lagrange's equations for $q_2 = \theta$ are as follows:

1. $\frac{d}{dt}\left(\frac{\partial \mathcal{L}}{\partial \dot{q}_j}\right) - \frac{\partial \mathcal{L}}{\partial q_j} = Q_j;\ \mathcal{L} = T - V;\ \frac{d}{dt}\left(\frac{\partial V}{\partial \dot{q}_j}\right) = 0$
2. $\frac{d}{dt}\left(\frac{\partial \mathcal{L}}{\partial \dot{q}_j}\right) - \frac{\partial \mathcal{L}}{\partial q_j} = Q_j \Rightarrow \frac{d}{dt}\left(\frac{\partial \mathcal{L}}{\partial \dot{\theta}}\right) - \frac{\partial \mathcal{L}}{\partial \theta} = Q_2$
3. $\mathcal{L} = \frac{1}{2}m_1\dot{x}^2 + \frac{1}{2}m_2\left(\dot{x}^2 + \frac{l^2\dot{\theta}^2}{4} + l\dot{x}\dot{\theta}\cos\theta\right) + \frac{1}{2}\left(\frac{m_2 l^2}{12}\right)\dot{\theta}^2 - \frac{1}{2}kx^2 - m_2 g\frac{l}{2}(1 - \cos\theta)$
4. $\frac{d}{dt}\left(\frac{\partial \mathcal{L}}{\partial \dot{\theta}}\right) = \frac{d}{dt}\left(\frac{1}{2}m_2\left[\frac{l^2\dot{\theta}}{2} + l\dot{x}\cos\theta\right] + \frac{m_2 l^2}{12}\dot{\theta}^2\right) =$
   $\frac{1}{2}m_2\left(\ddot{x}l\cos\theta - \dot{x}l\dot{\theta}\sin\theta + \frac{l^2}{2}\ddot{\theta}\right) + \frac{m_2 l^2}{12}\ddot{\theta}$
5. $-\frac{\partial \mathcal{L}}{\partial \theta} = -\left[-\frac{1}{2}m_2\left(l\dot{x}\dot{\theta}\sin\theta\right) - \frac{m_2 gl}{2}\sin\theta\right] = \left[\frac{1}{2}m_2\left(l\dot{x}\dot{\theta}\sin\theta\right) + \frac{m_2 gl}{2}\sin\theta\right]$
6. Summing: $\frac{1}{2}m_2\left(\ddot{x}l\cos\theta - \dot{x}l\dot{\theta}\sin\theta + \frac{l^2}{2}\ddot{\theta}\right) + \frac{m_2 l^2}{12}\ddot{\theta} + \left[\frac{1}{2}m_2\left(l\dot{x}\dot{\theta}\sin\theta\right) + \frac{m_2 gl}{2}\sin\theta\right] = \frac{1}{2}m_2\left(\ddot{x}l\cos\theta + \frac{l^2}{2}\ddot{\theta}\right) + \frac{m_2 l^2}{12}\ddot{\theta} + \frac{m_2 gl}{2}\sin\theta = Q_2 = 0$

Adding a dashpot (or damper) and an external time-varying force $F(t)$ will introduce non-conservative forces into the system which must now be accounted for as generalized forces Q_1 and Q_2, respectively (see Fig. 3.16). The virtual work due to the non-conservative forces is:

$$\delta W^{NC} = \sum_{j=1}^{2} Q_j \delta q_j = \underbrace{(F(t) - b\dot{x})}_{Q_1}\delta x + \underbrace{(0)}_{Q_2}\delta\theta \tag{3.107}$$

Note that the non-conservative force $F(t)$ does not undergo any motion at its point of application, when a virtual displacement of $\delta\theta$ is introduced. The same is true for the non-conservative damper force $-b\dot{x}$, hence $Q_2 = 0$.

The two dynamic equations then become:

$$[m_1 + m_2]\ddot{x} + \frac{1}{2}m_2 l\ddot{\theta}\cos\theta - \frac{1}{2}m_2 l\dot{\theta}^2 \sin\theta + kx + b\dot{x} = F(t)$$

$$\frac{1}{2}m_2\left(\ddot{x}l\cos\theta + \frac{l^2}{2}\ddot{\theta}\right) + \frac{m_2 l^2}{12}\ddot{\theta} + \frac{m_2 gl}{2}\sin\theta = 0 \tag{3.108}$$

Adding another horizontal force $F_2(t)$ at the bottom end of the pendulum will result in (see Fig. 3.17) an additional force in the $\hat{I}$ direction which is: $F_2(t)\hat{I} \cdot \delta x\hat{I} = F_2(t)\delta x$. This implies that $Q_1\delta x = (F(t) - b\dot{x} + F_2(t))\,\delta x \Rightarrow Q_1 = F(t) - b\dot{x} + F_2(t)$. Freezing δx and allowing only $\delta\theta$ motion, some work will

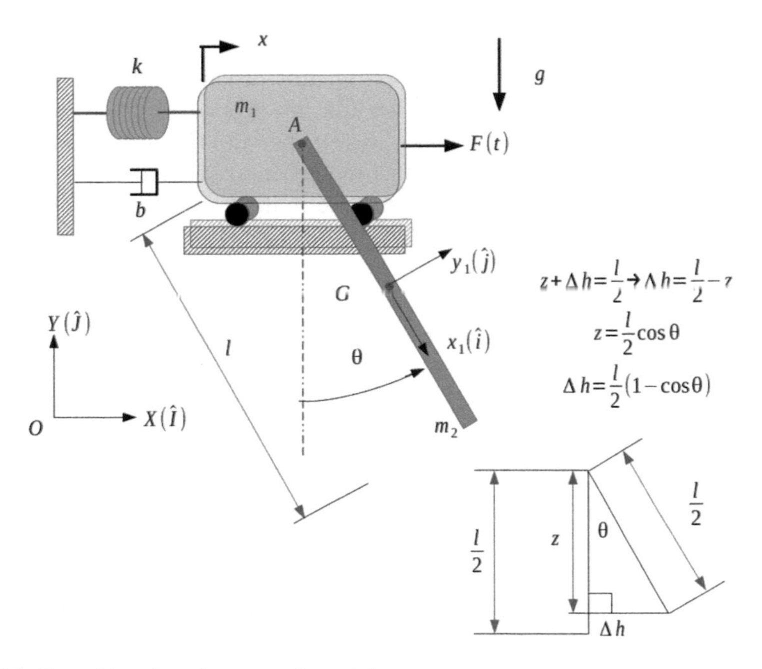

Fig. 3.16 Cart with spring, damper, and pendulum

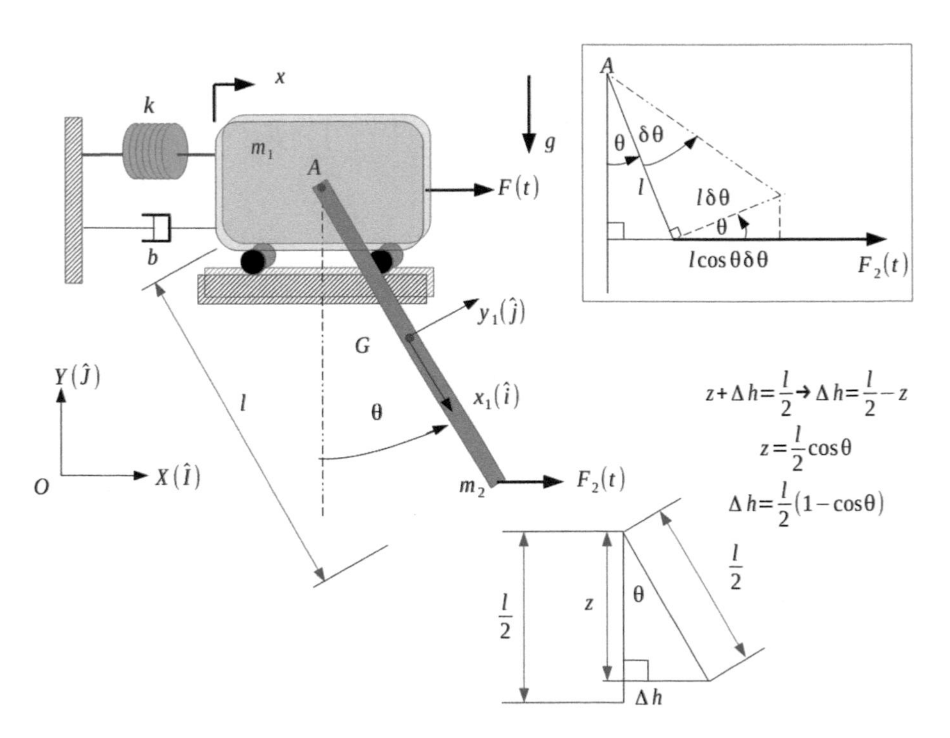

Fig. 3.17 Cart with spring, damper, and pendulum-geometrical considerations

be performed by $F_2(t)$. The variation $\delta\theta$ causes a motion of the slender rod at distance $l\delta\theta$ in the $\hat{j}$ direction. Projecting this distance onto the $\hat{I}$ axis, the distance becomes $l\cos\theta\delta\theta\hat{I}$. This implies that the work performed by $F_2(t)$ is: $\delta W_\theta = Q_2\delta\theta = F_2(t)\hat{I}\cdot l\cos\theta\delta\theta\hat{I} = F_2(t)l\cos\theta\delta\theta$. This further implies that Q_2 becomes: $Q_2 = F_2(t)l\cos\theta$. Note that the generalized force in the x direction has the units of force and the generalized force in the θ direction has the units of torque.

3.9 Lagrange Multipliers for Constrained Systems

The following section is based upon Chap. 2 in the book by L. Meirovitch (L. Meirovitch—Methods of Analytical Dynamics [23, pp. 52–55]) and Chapters 9 and 10 in Fitzpatrick's book [11, pp. 128, 147–148].

Assume without loss of generality that we have a system described by two generalized variables $q_1$and q_2. The two generalized coordinates q_1 and q_2 are not independent and the holonomic (time independent or scleronomous) constraint connecting the two variables is of the form $f(q_1, q_2) = 0$. The Lagrangian $\mathcal{L}$ is a function of both q_1 and q_2 and their derivatives with respect to time, that is: $\mathcal{L} = \mathcal{L}(q_1, q_2, \dot{q}_1, \dot{q}_2)$. From the extended Hamilton's principle (Eq. 3.46), we have:

$$\int_{t_1}^{t_2}(\delta\mathcal{L})dt = 0;\ \mathcal{L} = T - V;\ t = t_1, t_2$$

$$\Rightarrow \int_{t_1}^{t_2}(\delta\mathcal{L})dt = \int_{t_1}^{t_2}\sum_{k=1}^{2}\left[\frac{\partial T}{\partial q_k} - \frac{\partial V}{\partial q_k} - \frac{d}{dt}\left(\frac{\partial T}{\partial \dot{q}_k}\right)\right]\delta q_k dt = 0$$

$$\Rightarrow \int_{t_1}^{t_2}\left(\left[\frac{\partial T}{\partial q_1} - \frac{\partial V}{\partial q_1} - \frac{d}{dt}\left(\frac{\partial T}{\partial \dot{q}_1}\right)\right]\delta q_1 + \left[\frac{\partial T}{\partial q_2} - \frac{\partial V}{\partial q_2} - \frac{d}{dt}\left(\frac{\partial T}{\partial \dot{q}_2}\right)\right]\delta q_2\right)dt = 0$$

$$\Rightarrow \int_{t_1}^{t_2}\left(\left[\frac{\partial \mathcal{L}}{\partial q_1} - \frac{d}{dt}\left(\frac{\partial \mathcal{L}}{\partial \dot{q}_1}\right)\right]\delta q_1 + \left[\frac{\partial \mathcal{L}}{\partial q_2} - \frac{d}{dt}\left(\frac{\partial \mathcal{L}}{\partial \dot{q}_2}\right)\right]\delta q_2\right)dt = 0;\ t = t_1, t_2 \tag{3.109}$$

It was previously noted that the generalized coordinates q_1 and q_2 are not independent and at a fixed instant of time the variation of the constraint equation $f(q_1, q_2) = 0$ becomes:

$$\delta f = \frac{\partial f}{\partial q_1}\delta q_1 + \frac{\partial f}{\partial q_2}\delta q_2 = 0 \Rightarrow \delta q_2 = \frac{\delta f - \frac{\partial f}{\partial q_1}\delta q_1}{\frac{\partial f}{\partial q_2}}$$

$$f = 0 \Rightarrow \delta f = 0 \Rightarrow \delta q_2 = -\frac{\partial f}{\partial q_1}\frac{\partial q_2}{\partial f}\delta q_1 \tag{3.110}$$

Eliminating δq_2 from the d'Alembert–Lagrange equations results in:

$$\int_{t_1}^{t_2}\left(\left[\frac{\partial \mathcal{L}}{\partial q_1}-\frac{d}{dt}\left(\frac{\partial \mathcal{L}}{\partial \dot{q}_1}\right)\right]\delta q_1+\left[\frac{\partial \mathcal{L}}{\partial q_2}-\frac{d}{dt}\left(\frac{\partial \mathcal{L}}{\partial \dot{q}_2}\right)\right]\delta q_2\right)dt=0;\ t=t_1,t_2$$

$$=\int_{t_1}^{t_2}\left(\left[\frac{\partial \mathcal{L}}{\partial q_1}-\frac{d}{dt}\left(\frac{\partial \mathcal{L}}{\partial \dot{q}_1}\right)\right]\delta q_1-\left[\frac{\partial \mathcal{L}}{\partial q_2}-\frac{d}{dt}\left(\frac{\partial \mathcal{L}}{\partial \dot{q}_2}\right)\right]\frac{\partial f}{\partial q_1}\frac{\partial q_2}{\partial f}\delta q_1\right)dt=0;$$

$$t=t_1,t_2$$

$$=\int_{t_1}^{t_2}\left(\left[\frac{\partial \mathcal{L}}{\partial q_1}-\frac{d}{dt}\left(\frac{\partial \mathcal{L}}{\partial \dot{q}_1}\right)\right]\frac{\partial q_1}{\partial f}-\left[\frac{\partial \mathcal{L}}{\partial q_2}-\frac{d}{dt}\left(\frac{\partial \mathcal{L}}{\partial \dot{q}_2}\right)\right]\frac{\partial q_2}{\partial f}\right)\delta q_1 dt=0;$$

$$t=t_1,t_2 \tag{3.111}$$

The above equation must be satisfied for all possible perturbations of δq_1, which implies that the integrand must be identically zero. Letting $\left[\frac{\partial \mathcal{L}}{\partial q_1}-\frac{d}{dt}\left(\frac{\partial \mathcal{L}}{\partial \dot{q}_1}\right)\right]\frac{\partial q_1}{\partial f}=\left[\frac{\partial \mathcal{L}}{\partial q_2}-\frac{d}{dt}\left(\frac{\partial \mathcal{L}}{\partial \dot{q}_2}\right)\right]\frac{\partial q_2}{\partial f}=\lambda(t)$ leads to the following result:

$$\underbrace{\left[\frac{\partial \mathcal{L}}{\partial q_1}-\frac{d}{dt}\left(\frac{\partial \mathcal{L}}{\partial \dot{q}_1}\right)\right]}_{A}\frac{\partial q_1}{\partial f}=\underbrace{\left[\frac{\partial \mathcal{L}}{\partial q_2}-\frac{d}{dt}\left(\frac{\partial \mathcal{L}}{\partial \dot{q}_2}\right)\right]}_{B}\frac{\partial q_2}{\partial f}$$

$$\Rightarrow A\frac{\partial q_1}{\partial f}=B\frac{\partial q_2}{\partial f}=\lambda(t)$$

$$\Rightarrow A-\lambda(t)\frac{\partial f}{\partial q_1}=0;\ B-\lambda(t)\frac{\partial f}{\partial q_2}=0$$

$$\Rightarrow\left[\frac{\partial \mathcal{L}}{\partial q_1}-\frac{d}{dt}\left(\frac{\partial \mathcal{L}}{\partial \dot{q}_1}\right)\right]-\lambda(t)\frac{\partial f}{\partial q_1}=0$$

$$\left[\frac{\partial \mathcal{L}}{\partial q_2}-\frac{d}{dt}\left(\frac{\partial \mathcal{L}}{\partial \dot{q}_2}\right)\right]-\lambda(t)\frac{\partial f}{\partial q_2}=0 \tag{3.112}$$

The Lagrange multiplier, $\lambda(t)$, may be chosen in such a way as to ensure that both of Eq. (3.112) go to zero. Because the virtual work must remain zero, it follows that the generalized force Q_k must act in a direction which is conjugate to the corresponding generalized coordinate. Assuming that the dynamical system in question is conservative, then from Eqs. (3.22), (3.33), and (3.130), the generalized work and the generalized forces may be written as:

$$\delta W_c=\sum_{i=1}^{N}F_i\cdot\delta r_i=\sum_{i=1}^{N}F_i\cdot\sum_{j=1}^{n}\frac{\partial r_i}{\partial q_j}\delta q_j=\sum_{j=1}^{n}\left(\sum_{i=1}^{N}F_i\cdot\frac{\partial r_i}{\partial q_j}\right)\delta q_j=\sum_{j=1}^{n}Q_j\delta q_j$$

$$\Rightarrow Q_j = \sum_{i=1}^{N} F_i \cdot \frac{\partial r_i}{\partial q_j} \quad j = 1, 2, \ldots n$$

$$F_i = -\frac{\partial V}{\partial r_i} \Rightarrow Q_j = -\sum_{i=1}^{N} \frac{\partial V}{\partial r_i} \cdot \frac{\partial r_i}{\partial q_j} = -\frac{\partial V}{\partial q_j} \quad j = 1, 2, \ldots n \tag{3.113}$$

where V is the system's potential energy. By analogy, the generalized constraint force (i.e., the generalized force responsible for maintaining the corresponding constraint) takes the form: $\hat{Q}_k = \lambda(t)\frac{\partial f}{\partial q_k}$. Extending the analysis to n generalized coordinates which are subject to the holonomic constraint $f(q_1, q_2, \ldots, q_n) = 0$, we find that:

$$\begin{gathered}
\left[\frac{\partial \mathcal{L}}{\partial q_1} - \frac{d}{dt}\left(\frac{\partial \mathcal{L}}{\partial \dot{q}_1}\right)\right] - \lambda(t)\frac{\partial f}{\partial q_1} = 0 \\
\left[\frac{\partial \mathcal{L}}{\partial q_2} - \frac{d}{dt}\left(\frac{\partial \mathcal{L}}{\partial \dot{q}_2}\right)\right] - \lambda(t)\frac{\partial f}{\partial q_2} = 0 \\
\vdots \\
\left[\frac{\partial \mathcal{L}}{\partial q_n} - \frac{d}{dt}\left(\frac{\partial \mathcal{L}}{\partial \dot{q}_n}\right)\right] - \lambda(t)\frac{\partial f}{\partial q_n} = 0
\end{gathered} \tag{3.114}$$

The generalization to multiple (m) holonomic constraints of the form $f_j(q_1, q_2, \ldots, q_n)$; $j = 1, 2, \ldots, m$ is straightforward and may be written as:

$$\begin{gathered}
\left[\frac{\partial \mathcal{L}}{\partial q_1} - \frac{d}{dt}\left(\frac{\partial \mathcal{L}}{\partial \dot{q}_1}\right)\right] - \sum_{j=1}^{m} \lambda_j(t)\frac{\partial f_j}{\partial q_1} = 0 \\
\left[\frac{\partial \mathcal{L}}{\partial q_2} - \frac{d}{dt}\left(\frac{\partial \mathcal{L}}{\partial \dot{q}_2}\right)\right] - \sum_{j=1}^{m} \lambda_j(t)\frac{\partial f_j}{\partial q_2} = 0 \\
\vdots \\
\left[\frac{\partial \mathcal{L}}{\partial q_n} - \frac{d}{dt}\left(\frac{\partial \mathcal{L}}{\partial \dot{q}_n}\right)\right] - \sum_{j=1}^{m} \lambda_j(t)\frac{\partial f_j}{\partial q_n} = 0
\end{gathered} \tag{3.115}$$

The term $\frac{\partial \mathcal{L}}{\partial q_k} - \sum_{j=1}^{m} \lambda_j(t)\frac{\partial f_j}{\partial q_k}$ may be viewed as the derivative with respect to the generalized coordinate q_k of an augmented potential function $\mathcal{L} - \sum_{j=1}^{m} \lambda_j(t) f_j$. The idea is to render this augmented potential function stationary by requiring that all the terms of the form $\frac{\partial \mathcal{L}}{\partial q_k} - \sum_{j=1}^{m} \lambda_j(t) \frac{\partial f_j}{\partial q_k}$; $k = 0, 1, \ldots, n$ are identically equal to zero. When the generalized coordinates are not independent but are subject to m non-holonomic constraints having the Pfaffian form:

$$\sum_{k=1}^{n} a_{jk} dq_k + a_{j0} dt = 0;\ j = 1, 2, \ldots, m \tag{3.116}$$

where the coefficients a_{jk}, $k = 1, 2, \ldots, n$ are functions of the generalized coordinates q_i, the virtual displacements are:

$$\sum_{k=1}^{n} a_{jk} \delta q_k = 0;\ j = 1, 2, \ldots, m \tag{3.117}$$

Multiplying the virtual displacements by λ_i, $i = 1, 2, \ldots, m$ and adding all of the results to the d'Alembert–Lagrange equations, we obtain:

$$\frac{d}{dt}\left(\frac{\partial \mathcal{L}}{\partial \dot{q}_k}\right) - \frac{\partial \mathcal{L}}{\partial q_k} = \sum_{l=1}^{m} \lambda_l a_{lk} + Q_k \quad k = 1, 2, \ldots, n \tag{3.118}$$

The terms $\sum_{l=1}^{m} \lambda_l a_{lk}$ $k = 1, 2, \ldots, n$ may be regarded as equivalent forces, which are in fact constraint forces, that is:

$$Q_k^{'} = \sum_{l=1}^{m} \lambda_l a_{lk} \quad k = 1, 2, \ldots, n \tag{3.119}$$

There are $n + m$ equations, n d'Alembert–Lagrange equations of the form: $\frac{d}{dt}\left(\frac{\partial \mathcal{L}}{\partial \dot{q}_k}\right) - \frac{\partial \mathcal{L}}{\partial q_k} = \sum_{l=1}^{m} \lambda_l a_{lk} + Q_k$ $k = 1, 2, \ldots, n$ and m non-holonomic constraint equations $\sum_{k=1}^{n} a_{jk} dq_k + a_{j0} dt = 0;\ j = 1, 2, \ldots, m$. Hence, there are n generalized coordinates and m Lagrange multipliers which must be found. If the constraints are holonomic, that is, $f_l(q_1, q_2, \ldots, q_n, t) = 0$ $l = 1, 2, \ldots, m$, then $a_{lk} = \partial f_l / \partial q_k$.

Example 9: Virtual Work (see Meirovitch [23, pp. 62–63])

The system in Fig. 3.18 is composed of a mass m connected to a weightless link of length L and a spring with a spring coefficient k as shown. In the spring's unstretched position, the link is in the horizontal position. The principle of virtual work may be used to calculate the angle θ corresponding to the equilibrium position of the system for the given configuration. The position of the ends of the link in the equilibrium position is given by:

$$x = L(1 - \cos\theta);\ y = L\sin\theta \tag{3.120}$$

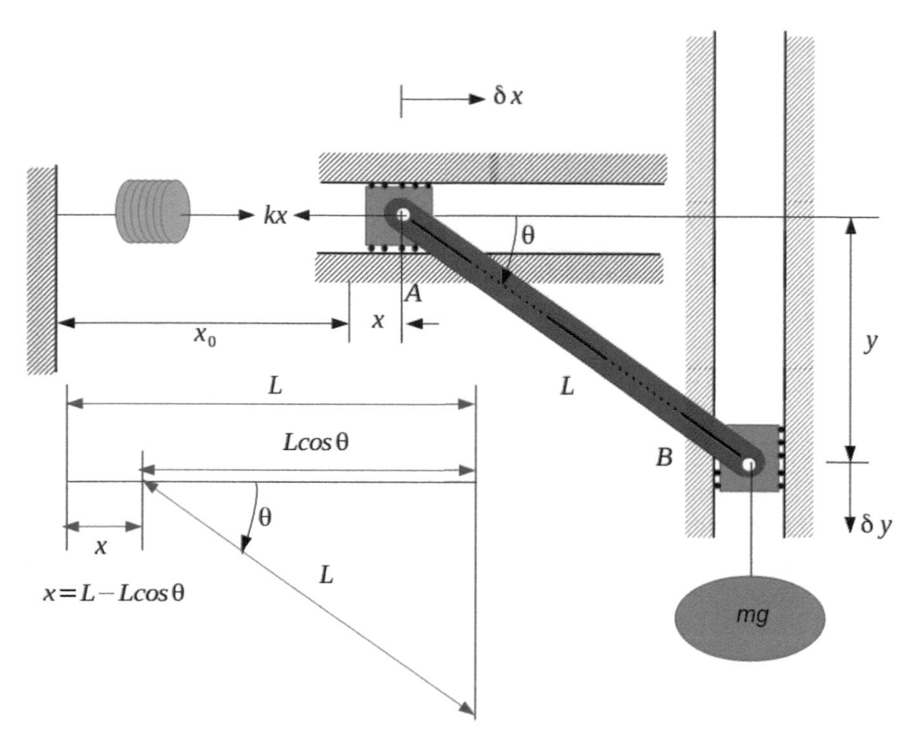

Fig. 3.18 System used to demonstrate the virtual work principle

where x is the amount by which the spring is stretched and y is the distance by which the mass m is lowered below the horizontal. The virtual work is as follows:

$$\overline{\delta W} = -kx\delta x + mg\delta y = 0;\ \delta x = L\sin\theta\delta\theta;\ \delta y = L\cos\theta\delta\theta \tag{3.121}$$

Recall that the variation δ operates on functions in the same way as a derivative. The virtual displacements in the x and y directions are $\delta x = L\sin\theta\delta\theta$ and $\delta y = L\cos\theta\delta\theta$, respectively. Substituting the above values for δx and δy and $x = L(1 - \cos\theta)$ into the equation for virtual work results in:

$$\overline{\delta W} = -kL(1 - \cos\theta)L\sin\theta\delta\theta + mgL\cos\theta\delta\theta = 0 \Rightarrow (1 - \cos\theta)\tan\theta = \frac{mg}{kL} \tag{3.122}$$

The solution to the above transcendental equation results in the equilibrium value of θ. It is also possible to solve this problem with Lagrange multipliers. The constraint equations may be written as:

$$f_1 = x - L(1 - \cos\theta) = 0;\ f_2 = y - L\sin\theta = 0$$
$$\Rightarrow \delta f_1 = \delta x - L\sin\theta\delta\theta = 0;\ \delta f_2 = \delta y - L\cos\theta\delta\theta = 0 \tag{3.123}$$

Multiplying δf_1 by λ_1 and δf_2 by λ_2, respectively, and adding the results to $\overline{\delta W}$ we obtain:

$$\overline{\delta W} = -kx\delta x + mg\delta y + \lambda_1 (\delta x - L\sin\theta\delta\theta) + \lambda_2 (\delta y - L\cos\theta\delta\theta) = 0 \quad (3.124)$$

Replacing x in the above equation by $x = L(1 - \cos\theta)$ and regrouping the terms in the above equation as coefficients of the parameters $\delta x, \delta y, \delta\theta$, we have:

$$[-kL(1 - \cos\theta) + \lambda_1]\,\delta x + [mg + \lambda_2]\,\delta y - [\lambda_1 L\sin\theta + \lambda_2 L\cos\theta]\,\delta\theta \quad (3.125)$$

Equating the coefficients of $\delta x, \delta y$, and $\delta\theta$, respectively, to zero in the above results in:

$$[-kL(1 - \cos\theta) + \lambda_1] = 0;\ [mg + \lambda_2] = 0;\ [\lambda_1 L\sin\theta + \lambda_2 L\cos\theta] = 0 \quad (3.126)$$

The solutions for λ_1, λ_2, and θ then follow, and are:

$$\lambda_1 = mg\cot\theta;\ \lambda_2 = -mg;\ \frac{mg}{kL} = (1 - \cos\theta)\tan\theta \quad (3.127)$$

The Lagrange multiplier λ_2 is simply the reaction force at point A to the mass m, while λ_1 is the horizontal reaction force at point B.

3.10 A Systematic Procedure for Generalized Forces

Recitation 9—Video Times—2:18–14:49, Lecture 16—Video Times—28:16–56:16

The problem of generalized forces and the method by which they may be deduced is presented below and relies in part on the videos and handout (Recitation 9 Notes: Generalized Forces with Double Pendulum Example) of Recitation 9 and Lecture 16 by Prof. Vandiver et al. (Vandiver and Gossard—Video of Lecture 16: Kinematic Approach to Finding Generalized Forces [43] and Vandiver et al. Video of Recitation 9: Generalized Forces [47].)

Consider a three-dimensional rigid body with n generalized coordinates, q_j, upon which N non-conservative forces are brought to bear, and where the distance from the origin O of an inertial coordinate system to the point of application of force F_i on the body is r_i. For the same position vector r_i, the virtual displacement of the point when generalized coordinates are varied is δr_i (see Fig. 3.19). δr_i is the total virtual displacement at the point of application of force F_i and it is the sum of all of the virtual displacements of the generalized coordinates, or:

$$\delta r_i = \frac{\partial r_i}{\partial q_1}\delta q_1 + \frac{\partial r_i}{\partial q_2}\delta q_2 + \cdots + \frac{\partial r_i}{\partial q_n}\delta q_n \quad (3.128)$$

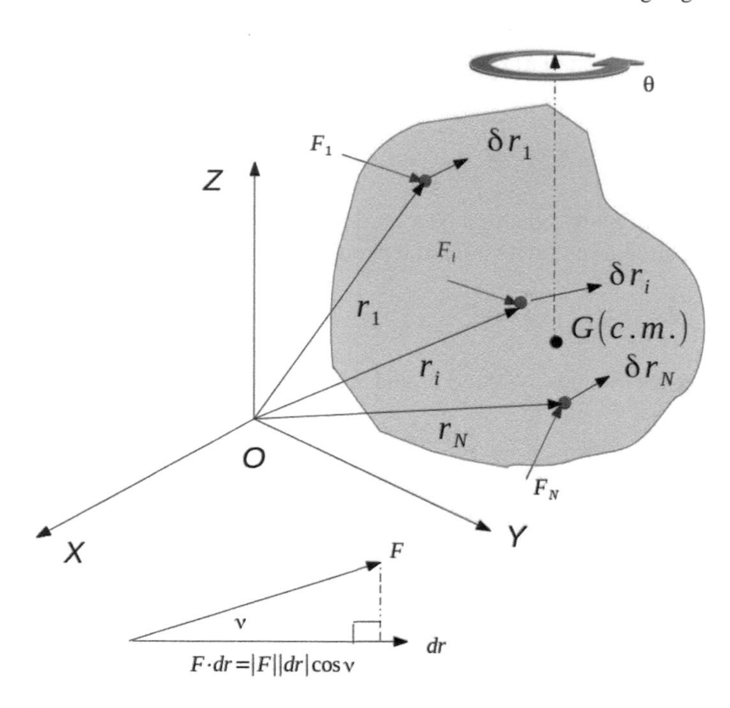

Fig. 3.19 3D rigid body with n generalized coordinates acted upon by N non-conservative forces

The work done by a force is: $dW = F \cdot dr = |F||dr|\cos\nu$, where ν is the angle between the force and the infinitesimal displacement dr. Note that only the component of the force, which is projected in the direction of the motion, is important. The virtual work done by all the non-conservative forces is therefore:

$$\delta W^{NC} = \sum_{i=1}^{N} F_i \cdot \delta r_i = \sum_{i=1}^{N} F_i \cdot \sum_{j=1}^{n} \frac{\partial r_i}{\partial q_j}\delta q_j = \sum_{j=1}^{n}\left(\sum_{i=1}^{N} F_i \cdot \frac{\partial r_i}{\partial q_j}\right)$$
$$\delta q_j = \sum_{j=1}^{n} Q_j \delta q_j \tag{3.129}$$

Hence the generalized forces Q_j are given by:

$$Q_j = \sum_{i=1}^{N} F_i \cdot \frac{\partial r_i}{\partial q_j} \quad j = 1, 2, \ldots n \tag{3.130}$$

If the rigid body executes only planar motion, then it has three degrees of freedom, since it can move in the X direction, the Y direction and rotate with angle θ around its Z axis. The three generalized coordinates can be X, Y, and θ. Accompanying these generalized coordinates are the virtual displacement δX, δY, and $\delta\theta$ and these are used to calculate the virtual work of the rigid body. The generalized forces Q_X, Q_Y, and Q_θ are the entities being sought after.

Example 10: Generalized Forces on a Double Pendulum

Recitation 9—Video Times—2:18–14:49

A double pendulum is shown in Fig. 3.20 with a force F acting at point P. The generalized coordinates for this case are $q_1 = \theta_1$ and $q_2 = \theta_2$ (see Vandiver—2.003SC Engineering Dynamics. Recitation 9 Notes: Generalized Forces with Double Pendulum Example [39]). The location of the force with respect to the origin O is described by the vector r_p which can be written in terms of the $\hat{I}$ and $\hat{J}$ unit vectors as:

$$r_P = (l_1 \sin\theta_1 + l_2 \sin\theta_2)\,\hat{I} + (-l_1 \cos\theta_1 - l_2 \cos\theta_2)\,\hat{J} \tag{3.131}$$

The partial derivatives of r_P with respect to θ_1 and θ_2 are then:

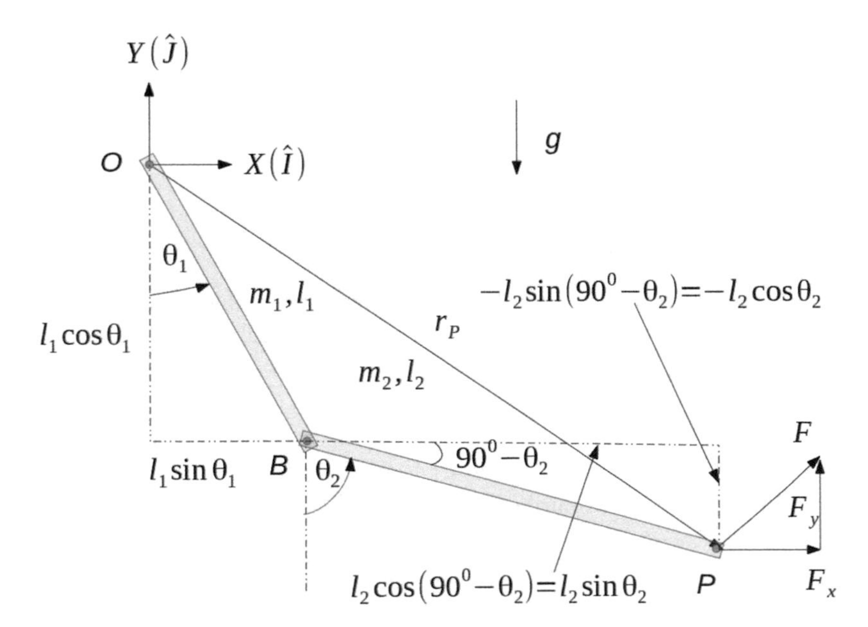

Fig. 3.20 Generalized forces acting on a double pendulum

$$\frac{\partial r_P}{\partial \theta_1} = l_1 \cos\theta_1 \hat{I} + l_1 \sin\theta_1 \hat{J}; \ \frac{\partial r_P}{\partial \theta_2} = l_2 \cos\theta_2 \hat{I} + l_2 \sin\theta_2 \hat{J} \tag{3.132}$$

The applied force F may be decomposed along the two coordinates' directions as: $F = F_x \hat{I} + F_y \hat{J}$. From Eq. (3.130), the generalized forces $Q_{\theta_1}, Q_{\theta_2}$ become:

$$\begin{aligned}
Q_{\theta_1} &= F \cdot \frac{\partial r_P}{\partial \theta_1} = \left(F_x \hat{I} + F_y \hat{J}\right) \cdot \left(l_1 \cos\theta_1 \hat{I} + l_1 \sin\theta_1 \hat{J}\right) \\
&= F_x l_1 \cos\theta_1 + F_y l_1 \sin\theta_1 \\
Q_{\theta_2} &= F \cdot \frac{\partial r_P}{\partial \theta_2} = \left(F_x \hat{I} + F_y \hat{J}\right) \cdot \left(l_2 \cos\theta_2 \hat{I} + l_2 \sin\theta_2 \hat{J}\right) \\
&= F_x l_2 \cos\theta_2 - F_y l_2 \sin\theta_2
\end{aligned} \tag{3.133}$$

The virtual work done on the system by the generalized forces and small displacements $\delta\theta_1, \delta\theta_2$ is therefore:

$$\begin{aligned}
\delta W^{NC} &= Q_{\theta_1}\delta\theta_1 + Q_{\theta_2}\delta\theta_2 = \left(F_x l_1 \cos\theta_1 + F_y l_1 \sin\theta_1\right)\delta\theta_1 \\
&\quad + \left(F_x l_2 \cos\theta_2 + F_y l_2 \sin\theta_2\right)\delta\theta_2
\end{aligned} \tag{3.134}$$

The double pendulum has two natural frequencies and two mode shapes, which may be obtained by writing out the d'Alembert–Lagrange equations.

Additional Remarks on Recitation 9 (See Vandiver-2.003SC Engineering Dynamics. Video of Recitation 9: Generalized Forces [47])

Recitation 9—Video Times—2:18–14:49

1. If a force is added at point B of the double pendulum, with components B_x and B_y, then a vector r_b from the origin to the point of application of the force B would be: $r_b = l_1 \sin\theta_1 \hat{I} - l_1 \cos\theta_1 \hat{J} \Rightarrow \frac{\partial r_b}{\partial \theta_1} = l_1 \cos\theta \hat{I} - l_1 \sin\theta_1 \hat{J}$.
2. The generalized force would then be: $Q_B = \left(B_x \hat{I} + B_y \hat{J}\right) \cdot \frac{\partial r_b}{\partial \theta_1} = \left(B_x \hat{I} + B_y \hat{J}\right) \cdot \left(l_1 \cos\theta \hat{I} - l_1 \sin\theta_1 \hat{J}\right) = B_x l_1 \cos\theta_1 - B_y l_1 \sin\theta_1$.
3. There are therefore two contributions to the total generalized force Q_{θ_1}: $Q_{\theta_1} = \sum_{i=1}^{2} F_i \cdot \frac{\partial r_i}{\partial q_j} \Rightarrow Q_{\theta_1} = F \cdot \frac{\partial r_p}{\partial \theta_1} + B \cdot \frac{\partial r_b}{\partial \theta_1}$.
4. The same procedure as for Q_{θ_1} is used to calculate the generalized force Q_{θ_2}.

Example 11: Torques and Kinetic Energy on a Rotating Cylinder and Disk

Recitation 9—Video Times—15:31–35:40

1. For the system in Fig. 3.21, assume that the coordinate system $x(\hat{i}), y(\hat{j}), z(\hat{k})$ at the instant considered is an inertial coordinate system (we freeze the motion at the instant considered).
2. Since the body B (mass m_2) is axially symmetric, at any instant of time, a set of principal coordinates could be defined for the body B which are parallel to the $x(\hat{i}), y(\hat{j}), z(\hat{k})$ coordinates at the instant considered.
3. The total angular velocity of the body B is: $\omega = \Omega\hat{k} - \omega_1\hat{j}$.
4. The mass m_2 has a linear component $v_B = \Omega\hat{k} \times l\hat{j} = -\Omega l\hat{i}$. This equation describes the instantaneous linear motion of the center of mass of m_2. The mass m_2 is also rotating with angular velocity $\omega = \Omega\hat{k} - \omega_1\hat{j}$ and so the kinetic energy T_2 of mass m_2 is calculated with the following formula (see Eq. 2.16):

$$T = \frac{1}{2}Mv_B^2 + \frac{1}{2}\left[\omega_x^2 I_x + \omega_y^2 I_y + \omega_z^2 I_z - 2\omega_x\omega_y I_{xy} - 2\omega_x\omega_z I_{xz} - 2\omega_y\omega_z I_{yz}\right]$$
$$+M\left[v_{Bx}(\omega_y\overline{z} - \omega_z\overline{y}) + v_{By}(\omega_z\overline{x} - \omega_x\overline{z}) + v_{Bz}(\omega_x\overline{y} - \omega_y\overline{x})\right] \tag{3.135}$$

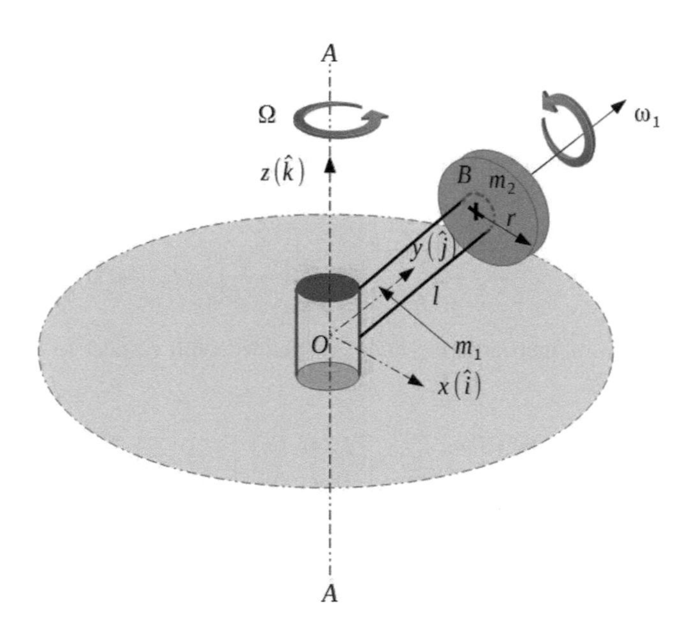

Fig. 3.21 Another example for the calculation of torques and kinetic energy

where $\overline{x}, \overline{y}, \overline{z}$, the centers of gravity of the rigid body are: $\overline{x} = \int xdm/M, \quad \overline{y} = \int ydm/M, \quad \overline{z} = \int zdm/M$, the moments of inertia I_x, I_y, I_z are, respectively: $I_x = \int(y^2 + z^2)dm, \quad I_y = \int(x^2 + z^2)dm, \quad I_z = \int(x^2 + y^2)dm$ and where the cross products of inertia I_{xy}, I_{xz}, I_{yz} are: $I_{xy} = \int xydm, \quad I_{xz} = \int xzdm, \quad I_{yz} = \int yzdm$. It may be shown that:

$$\left[\omega_x^2 I_x + \omega_y^2 I_y + \omega_z^2 I_z - 2\omega_x\omega_y I_{xy} - 2\omega_x\omega_z I_{xz} - 2\omega_y\omega_z I_{yz}\right] = \omega^T I\omega \tag{3.136}$$

5. Since B is axially symmetric, the inertial cross products are zero. In addition, the body B rotates around its center of mass and hence $\overline{x} = \overline{y} = \overline{z} = 0$, and so the general formula for kinetic energy becomes:

$$T_2 = \frac{1}{2}m_2 v_B^2 + \frac{1}{2}\left[\omega_x^2 I_x + \omega_y^2 I_y + \omega_z^2 I_z\right] \tag{3.137}$$

6. Additionally, since $\omega = \Omega\hat{k} - \omega_1\hat{j}$, this implies that $\omega_x = 0, \omega_y = -\omega_1, \omega_z = \Omega$ and so the above formula becomes:

$$T_2 = \frac{1}{2}m_2 v_B^2 + \frac{1}{2}\left[0 + \omega_1^2 I_y + \Omega^2 I_z\right] \tag{3.138}$$

7. Finally $v_B \cdot v_B = \left(-\Omega l\hat{i}\right) \cdot \left(-\Omega l\hat{i}\right) = \Omega^2 l^2$ and so the kinetic energy T_2 becomes:

$$T_2 = \frac{1}{2}m_2\Omega^2 l^2 + \frac{1}{2}\left[\omega_1^2 I_y + \Omega^2 I_z\right] \tag{3.139}$$

8. The kinetic energy of the rod (m_1) is: $T_1 = \frac{1}{2}I_o\Omega^2 = \frac{1}{2}\frac{m_1 l^2}{3}\Omega^2$.
9. The total kinetic energy is:

$$T = T_1 + T_2 = \frac{1}{2}\frac{m_1 l^2}{3}\Omega^2 + \frac{1}{2}m_2\Omega^2 l^2 + \frac{1}{2}\left[\omega_1^2 I_y + \Omega^2 I_z\right] \tag{3.140}$$

10. The moments of inertia of the body B (defined with respect to its center of gravity) are:

$$H_G = \begin{bmatrix} I_x & 0 & 0 \\ 0 & I_y & 0 \\ 0 & 0 & I_z \end{bmatrix} \begin{bmatrix} \omega_x \\ \omega_y \\ \omega_x \end{bmatrix} = \begin{bmatrix} I_x & 0 & 0 \\ 0 & I_y & 0 \\ 0 & 0 & I_z \end{bmatrix} \begin{bmatrix} 0 \\ -\omega_1\hat{j} \\ \Omega\hat{k} \end{bmatrix} = \begin{bmatrix} 0 \\ -I_y\omega_1\hat{j} \\ I_z\Omega\hat{k} \end{bmatrix} \tag{3.141}$$

11. The angular momentum at point O of mass m_2 with velocity $v_B = -\Omega l\hat{i}$ is:

$$H_O = H_G + r_{G/O} \times P_{G/O}$$

$$H_O = H_G + l\hat{j} \times m_2 v_B = -I_y\omega_1\hat{j} + I_z\Omega\hat{k} + \left(l^2 m_2\Omega\right)\hat{k}$$

$$= \left(I_z + l^2 m_2\right)\Omega\hat{k} - I_y\omega_1\hat{j} \tag{3.142}$$

Recall that $\hat{j} \times -\hat{i} = \hat{k}$.

12. What are the torques required to make this system continue to rotate?
13. Torque is the derivative with respect to time of H_O, that is:

$$\sum \tau_{ext} = \frac{dH_O}{dt} + v_{A/O} \times P_{G/O};\ v_{A/O} = v_O = 0$$

$$\frac{dH_O}{dt} = \left(\frac{\partial H_O}{\partial t}\right)_{rotatingframe} + \omega \times H_O$$

$$\Rightarrow \sum \tau_{ext} = \left(\frac{\partial H_O}{\partial t}\right)_{rotatingframe} + \underbrace{v_{A/O}}_{=0} \times P_{G/O} + \omega \times H_O$$

$$= \frac{\partial}{\partial t}\left[\left(I_z + l^2 m_2\right)\Omega\hat{k} - I_y\omega_1\hat{j}\right] + \omega \times H_O$$

$$= 0 + \left(\Omega\hat{k} - \omega_1\hat{j}\right) \times \left[\left(I_z + l^2 m_2\right)\Omega\hat{k} - I_y\omega_1\hat{j}\right]$$

$$= \Omega\hat{k} \times \left(-I_y\omega_1\hat{j}\right) + \left(-\omega_1\hat{j}\right) \times \left(I_z + l^2 m_2\right)\Omega\hat{k}$$

$$= \Omega\omega_1\left(I_y - I_z - l^2 m_2\right)\hat{i} \tag{3.143}$$

where $\sum \tau_{ext}$ is the sum of all externally applied torques. The term $\omega \times H_O$ appears because of the change of direction of two of the unit vectors (both $\hat{i}$ and $\hat{j}$ are continuously changing direction). The $\hat{i}$ and $\hat{j}$ unit vectors are changing direction at the rate of Ω. The explicit time derivative term $\left(\frac{\partial H_O}{\partial t}\right)_{rotatingframe}$ accounts for the remaining time-varying entities in the problem and it equals zero—i.e., there are no time-varying entities except for the unit vectors (both $\hat{i}$ and $\hat{j}$ are continuously changing direction).

Example 12: Spring and Mass on Inclined Face of Moving Cart

Lecture 16—Video Times—9:30–20:45, 37:50–53:29

This problem was taken from: Vandiver 2.003SC Engineering Dynamics. Video of Lecture 16: Kinematic Approach to Finding Generalized Forces [43] and 2.003SC Engineering Dynamics. and Problem Set 6-Solutions [37].

The cart has mass m_o and an inclined surface as shown in Fig. 3.22. A uniform disk of mass m and radius R rolls without slipping on the inclined surface. The disk is restrained by a spring with spring constant K_1, attached at one end to the cart. The other end of the spring attaches to an axle passing through the center of the disk (center of mass of the disk). The cart is also attached to a stationary wall by a spring with spring constant K_2, and a dashpot with constant "b." A horizontal external force $F_1(t)$ is applied at the center of mass of the disk as shown.

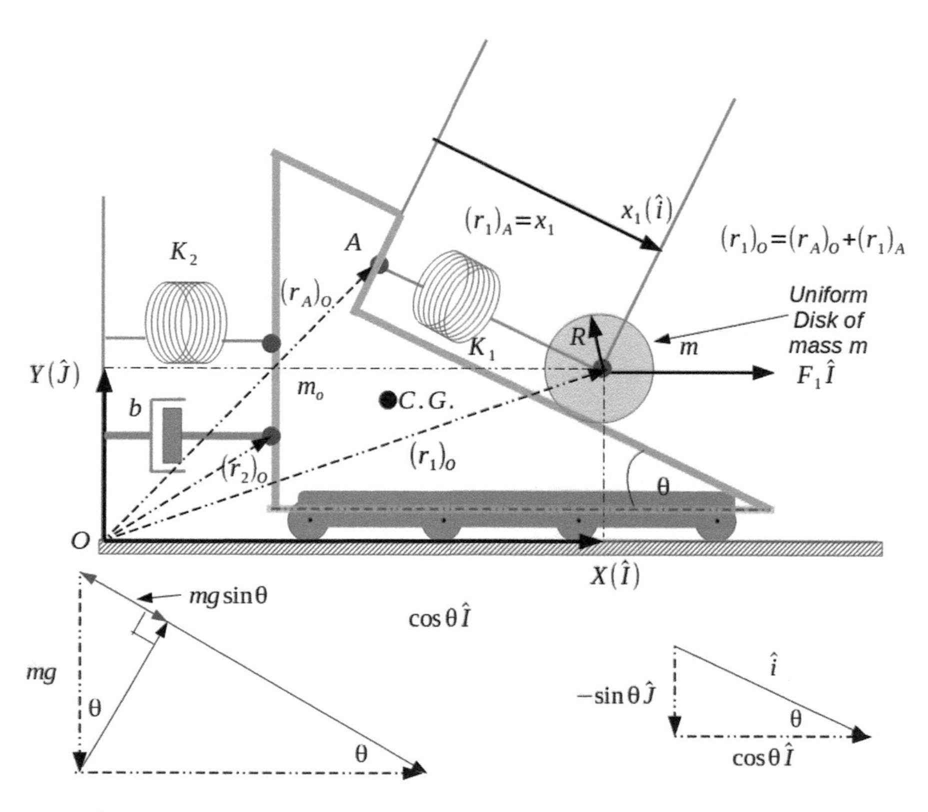

Fig. 3.22 Spring and mass on inclined face of moving cart

1. **Kinetic Energy T and Potential Energy V:**
   (a) There are two generalized coordinates which completely describe the motion of this system $X(\hat{I})$ and $x_1(\hat{i})$.
   (b) Assume that there are no external non-conservative forces being applied to the system, that is, $F_1\hat{I} = 0$ and $b\dot{X} = 0$.
   (c) Assume further that the static position of spring K_1 exactly balances out the gravitational force on the disk in the generalized $x_1(\hat{i})$ direction, that is: $K_1 x_{1_{static}} = mg\sin\theta$.
   (d) The only remaining term in the potential energy in the $x_1(\hat{i})$ direction is due to the stretching of the K_1 spring beyond its equilibrium position, that is: $V_{K_1} = \frac{1}{2}K_1 x_1^2$.
   (e) There is no gravitational potential due to the cart in the $X(\hat{I})$ direction since the cart doesn't undergo any change in height along its trajectory (its trajectory is horizontal only). The only potential energy in the $X(\hat{I})$ direction is due to the spring K_2 and it is: $V_{K_2} = \frac{1}{2}K_2 X^2$.
   (f) Adding the two potential energy terms results in the total system potential energy, which turns out to be: $V = V_{K_1} + V_{K_2} = \frac{1}{2}K_1 x_1^2 + \frac{1}{2}K_2 X^2$.
   (g) This is a planar motion problem and the axis of rotation of the disk is a principal axis, perpendicular to the plane of translational motion of the system or in the $z\hat{k}$ direction.
   (h) The disk rotates with an angular velocity of $\omega_z = \dot{\theta} = -\frac{\dot{x}_1}{R}$.
   (i) The velocity of the center of mass of the disk, with respect to the origin O of the inertial coordinate system, equals the velocity of the cart as a whole with respect to O and the velocity of the center of mass of the disk with respect to the cart.
   (j) In mathematical terms, we have: $(v_G)_O = (v_C)_O + (v_G)_C = \dot{X}\hat{I} + \dot{x}_1\hat{i}$, where $(v_G)_O$ is the velocity of the center of mass of the disk with respect to O, $(v_C)_O$ is the velocity of the cart wrt O, and $(v_G)_C$ is the velocity of the center of mass of the disk wrt the cart.
   (k) The unit vector in the $\hat{i}$th direction may be rewritten in the $\hat{I}$, $\hat{J}$ coordinate system as: $\hat{i} = \cos\theta\hat{I} - \sin\theta\hat{J}$.
   (l) The $(v_G)_C$ term then becomes: $(v_G)_C = \dot{x}_1\left(\cos\theta\hat{I} - \sin\theta\hat{J}\right)$.
   (m) The kinetic energy of the disk with respect to the origin O is therefore: $T_{disk} = \frac{1}{2}m(v_G)_O\cdot(v_G)_O + \frac{1}{2}I_z\omega_z^2 = \frac{1}{2}m\left(\dot{X}^2 + 2\dot{X}\dot{x}_1\cos\theta + \dot{x}_1^2\cos^2\theta + \dot{x}_1^2\sin^2\theta\right) + \frac{1}{2}I_z\omega_z^2 = \frac{1}{2}m\left(\dot{X}^2 + 2\dot{X}\dot{x}_1\cos\theta + \dot{x}_1^2\right) + \frac{1}{2}I_z\omega_z^2$.
   (n) The kinetic energy of the cart, as it moves in the $X(\hat{I})$ direction is: $T_{cart} = \frac{1}{2}m_o\dot{X}^2$.
   (o) The total kinetic energy of the system is the sum of the kinetic energies of the cart and the disk and it is equal to: $T = T_{cart} + T_{disk} = \frac{1}{2}m_o\dot{X}^2 + \frac{1}{2}m\left(\dot{X}^2 + 2\dot{X}\dot{x}_1\cos\theta + \dot{x}_1^2\right) + \frac{1}{2}I_z\omega_z^2$.
   (p) The distance traveled in the $x_1\hat{i}$ direction by the disk is: $x_1 = R\theta$. This implies that $\dot{x}_1 = R\dot{\theta}$.

(q) For the disk, its mass moment of inertia is: $I_z = \frac{1}{2}mR^2$. In addition, $\omega_z = \dot{\theta} \Rightarrow \omega_z^2 =$, and so the total kinetic energy becomes: $T = \frac{1}{2}m_o\dot{X}^2 + \frac{1}{2}m\left(\dot{X}^2 + 2\dot{X}\dot{x}_1\cos\theta + \dot{x}_1^2\right) + \frac{1}{2}\frac{mR^2\dot{\theta}^2}{2} = \frac{1}{2}m_o\dot{X}^2 + \frac{1}{2}m\left(\dot{X}^2 + 2\dot{X}\dot{x}_1\cos\theta + \dot{x}_1^2\right) + \frac{1}{2}\frac{m\dot{x}_1^2}{2}$.

2. **Equations of Motion Derived by Means of Lagrange's Equation:**
   (a) The Lagrangian $\mathcal{L} = T - V$ is: $\mathcal{L} = \frac{1}{2}m_o\dot{X}^2 + \frac{1}{2}m\left(\dot{X}^2 + 2\dot{X}\dot{x}_1\cos\theta + \dot{x}_1^2\right) + \frac{1}{2}\frac{m\dot{x}_1^2}{2} - \left(\frac{1}{2}K_1x_1^2 + \frac{1}{2}K_2X^2\right)$.
   (b) Lagrange's equation for the generalized variable X is: $\frac{d}{dt}(\frac{\partial\mathcal{L}}{\partial\dot{X}}) - \frac{\partial\mathcal{L}}{\partial X} = Q_X = 0$.
   (c) Lagrange's equation for the generalized variable x_1 is similarly: $\frac{d}{dt}(\frac{\partial\mathcal{L}}{\partial\dot{x}_1}) - \frac{\partial\mathcal{L}}{\partial x_1} = Q_{x_1} = 0$.
   (d) Since there are no external non-conservative forces acting on the system, the generalized forces Q_X and Q_{x_1} are both equal to zero.
   (e) $\frac{\partial\mathcal{L}}{\partial\dot{X}} = (m_o + m)\dot{X} + m\dot{x}_1\cos\theta$
   (f) $\frac{d}{dt}(\frac{\partial\mathcal{L}}{\partial\dot{X}}) = (m_o + m)\ddot{X} + m\ddot{x}_1\cos\theta$
   (g) $-\frac{\partial\mathcal{L}}{\partial X} = K_2X$
   (h) $\frac{d}{dt}(\frac{\partial\mathcal{L}}{\partial\dot{X}}) - \frac{\partial\mathcal{L}}{\partial X} = (m_o + m)\ddot{X} + m\ddot{x}_1\cos\theta + K_2X = 0$
   (i) $\frac{\partial\mathcal{L}}{\partial\dot{x}_1} = m\dot{X}\cos\theta + m\dot{x}_1 + \frac{1}{2} + m\dot{x}_1$
   (j) $\frac{d}{dt}(\frac{\partial\mathcal{L}}{\partial\dot{x}_1}) = m\ddot{X}\cos\theta + \frac{3}{2}m\ddot{x}_1$
   (k) $-\frac{\partial\mathcal{L}}{\partial x_1} = K_1x_1$
   (l) $\frac{d}{dt}(\frac{\partial\mathcal{L}}{\partial\dot{x}_1}) - \frac{\partial\mathcal{L}}{\partial x_1} = m\ddot{X}\cos\theta + \frac{3}{2}m\ddot{x}_1 + K_1x_1 = 0$

   Summarizing, the two equations of motion are:

$$(m_o + m)\ddot{X} + m\ddot{x}_1\cos\theta + K_2X = 0$$

$$m\ddot{X}\cos\theta + \frac{3}{2}m\ddot{x}_1 + K_1x_1 = 0 \tag{3.144}$$

3. **Generalized Forces:**
   Assume now that there are two external non-conservative forces being applied to the system, that is, $F_1\hat{I}$ and $-b\dot{X}\hat{I}$ (the force from the dashpot or damper). This implies that $Q_X \neq 0$ and $Q_{x_1} \neq 0$. The position vector from the origin O of the inertial coordinate system to the point of application of the external non-conservative force F_1 is $(r_1)_O$. Since the total motion of this point (the point of application of the external non-conservative force) is made up of the motion of the main cart plus the motion of the disk relative to the main cart, the position vector $(r_1)_O$ may be written as: $(r_1)_O = (r_A)_O + (r_1)_A$, where $(r_A)_O$ is the position vector of point A with respect to the origin O and $(r_1)_A$ is the point of application of the external non-conservative force F_1 with respect to the point A on the cart. The vector from the origin to point A may be written in inertial

coordinates as $(r_A)_O = X\hat{I} + Y\hat{J}$. The unit vector $\hat{i}$ may be expressed in terms of unit vectors $\hat{I}$ and $\hat{J}$ as follows: $\hat{i} = \cos\theta\hat{I} - \sin\theta\hat{J}$. Hence, $x_1\hat{i}$ becomes: $x_1(\cos\theta\hat{I} - \sin\theta\hat{J})$. The position vector $(r_1)_O$ may therefore be written as:

$$(r_1)_O = \underbrace{X\hat{I} + Y\hat{J}}_{(r_A)_O} + \underbrace{x_1(\cos\theta\hat{I} - \sin\theta\hat{J})}_{(r_1)_A} \tag{3.145}$$

Two forces in the $X\hat{I}$ direction contribute to the generalized force Q_X, where the contribution of F_1 to Q_X is $(Q_X)_{F_1}$. The virtual work due to F_1 is shown to be:

$$(Q_x)_{F_1}\delta X = F_1\hat{I} \cdot \frac{\partial(r_1)_O}{\partial X}\delta X\hat{I} \tag{3.146}$$

However,

$$\frac{\partial(r_1)_O}{\partial X} = \hat{I} \Rightarrow (Q_X)_{F_1}\delta X = F_1\hat{I} \cdot \hat{I}\delta X = F_1\delta X \tag{3.147}$$

The virtual work in the x_1 direction due to F_1 is: $(Q_1)_{x_1}\delta x_1$ and it turns out to be:

$$(Q_1)_{x_1}\delta x_1 = F_1\hat{I} \cdot \frac{\partial r_1}{\partial x_1}\delta x_1\hat{i} \tag{3.148}$$

As was demonstrated above r_1 in terms of the inertial coordinates is: $r_1 = x_1(\cos\theta\hat{I} - \sin\theta\hat{J})$, and so the partial derivative of r_1 with respect to x_1 becomes:

$$r_1 = x_1(\cos\theta\hat{I} - \sin\theta\hat{J}); \quad \frac{\partial r_1}{\partial x_1} = (\cos\theta\hat{I} - \sin\theta\hat{J}) \tag{3.149}$$

Hence the virtual work in the x_1 direction due to F_1 is:

$$(Q_1)_{x_1}\delta x_1 = F_1\hat{I} \cdot (\cos\theta\hat{I} - \sin\theta\hat{J})\delta x_1 = F_1\cos\theta\delta x_1 \Rightarrow (Q_1)_{x_1} = F_1\cos\theta \tag{3.150}$$

Given that the wheels of the cart don't slip, then the friction at the point of contact with the wheels of the cart doesn't do any work. In addition, a small virtual deflection, δx_1, doesn't make the cart move so that there are no other forces in the problem that move when the system undergoes a small movement in x_1. Hence, the total $(Q_1)_{x_1}$ is: $(Q_1)_{x_1} = F_1\cos\theta$. We found, previously that Q_X, the generalized force due to the motion of the cart (due to F_1) was: $(Q_X)_{F_1} = F_1$. The vector from the origin to the point of application of the force due to the dashpot is $(r_2)_O$ and the magnitude of the dashpot force is $-b\dot{X}$. The position vector $(r_2)_O$ may be written in the form:

$$(r_2)_O = (X - a_1)\hat{I} + (Y - a_2)\hat{J} \tag{3.151}$$

The virtual work due to the dashpot is:

$$(Q_X)_{b\dot{X}}\delta X = (Q_X)_{F_2}\delta X = -b\dot{X}\hat{I}\cdot\frac{\partial(r_2)_O}{\partial X}\delta X\hat{I} = -b\dot{X}\hat{I}\cdot\delta X\hat{I} = -b\dot{X}\delta X \tag{3.152}$$

Hence the total Q_X is:

$$Q_X = (Q_X)_{F_1} + (Q_X)_{F_2} = F_1 - b\dot{X} \tag{3.153}$$

So whenever a position vector r_i to the point of application of an external non-conservative force F_i can be specified, then it can be used in the following formula:

$$Q_j\delta q_j = \sum_{i=1}^{N} F_i\cdot\frac{\partial r_i}{\partial q_j}\delta q_j \quad j = 1, 2, \ldots n \tag{3.154}$$

The foregoing procedure should be performed for each force that is applied. It involves taking the derivative with respect to the coordinate q_j and multiplying it by δq_j, which is the virtual work done by each of these forces. All of the virtual work terms for each of the applied non-conservative forces are added together to get the total virtual work done due to a deflection at the particular generalized coordinate, q_j. For example, in the case of Q_X, there are two contributions because two forces were exerted on the main cart, F_1 and $-b\dot{X}$. And so the summation in this problem, when the generalized variable is X, is for the two contributions, F_1 and F_2.

Remarks

- In this problem, the two generalized coordinates are X in the inertial system which describes the motion of the cart, and x_1 which describes the motion of the disk relative to the cart and it enables the distance vector $(r_1)_O$ to be written as follows: $(r_1)_O = X\hat{I} + Y\hat{J} + x_1(\cos\theta\hat{I} - \sin\theta\hat{J})$.
- The term $x_1(\cos\theta\hat{I} - \sin\theta\hat{J})$ is only relative to the cart.
- The motion of the cart plus the motion of the point relative to the cart yields the total motion. The generalized coordinates were chosen in order to allow for a description of those two motions.
- Application of a force to the disk may cause the cart to move; however, it should be remembered that the generalized forces are what is desired, i.e., the application of a force to the disk which may cause the cart to move is not the problem to be solved when looking for the generalized forces.
- In the search for the generalized forces, motion of only one generalized coordinate at a time, with all other generalized coordinates "frozen," should be performed. For example, while the disk is allowed to undergo a small deviation in its location, the main cart should not move.

- The consequence of that motion must be determined. The motion of the disk does a little virtual work because there's a generalized force operating on it over an infinitesimal distance. If the cart's position is perturbed, and the disk's position remains unchanged, the whole cart including the disk moves. But the amount that the disk moves is exactly equal to the amount that the cart moves because the disk's position with respect to the cart is frozen.
- The motion of the cart, with the disk frozen (i.e., the relative position between the disk and the cart is frozen or fixed) allows for the calculation of Q_X.
- Even though the force F_1 is applied to the disk, the disk moves when the cart moves. But the cart doesn't move when the relative position between the cart and the disk changes. The disk is free to move when the position of the cart is frozen.
- Recalling the discussion about **complete** and **independent** coordinates, x_1 is independent of X.
- Freezing x_1 and making a change in X implies that the cart as a whole moves. However freezing X still allows x_1 to move. These coordinates are therefore independent.

Example 13: Two Carts Connected By a Spring (Video of Lecture 16—Vandiver [43])

Lecture 16—Video Times—56:40–1:03:35

Question: Given two masses connected by a spring, if the first mass is "pulled," will the second mass be "pulled" along by the first mass? Will the two masses "pull each other along"? Why will there not be the "pull-along" effect on the second mass from the first mass? Two carts both on wheels, with a spring in between them, are depicted in Fig. 3.23.

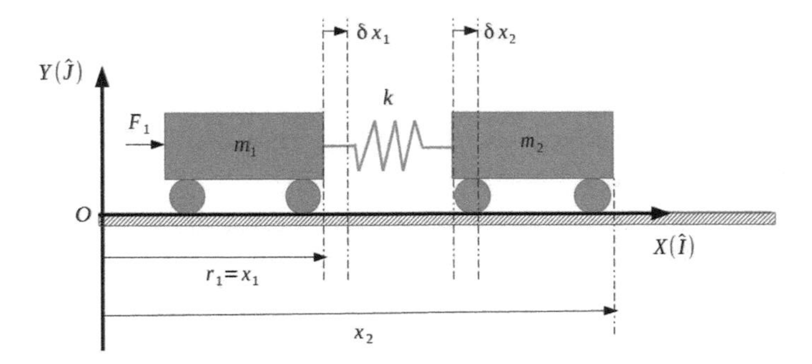

Fig. 3.23 Two carts connected by a spring

This is a planar motion problem. Each of these bodies is capable of planar motion in the x and y directions and each can have a rotation. Rotation and motion in the y direction are not however allowed (the cart has two wheels and can only move in the X direction) so that the number of degrees of freedom of the system is 3*2-c = 6-2-2 = 2 degrees of freedom. How many generalized coordinates are needed? The generalized coordinate for the first mass will be x_1 and for the second mass it will be x_2.

The kinetic and potential energies of the system are:

$$T = \frac{1}{2}m_1\dot{x}_1^2 + \frac{1}{2}m_2\dot{x}_2^2;\ V = \frac{1}{2}k\left(x_1 - (x_2 - l_0)\right)^2 \tag{3.155}$$

where l_0 is the un-stretched length of the spring. The virtual work due to x_1 is therefore:

$$Q_{x_1}\delta x_1 = F_1 \cdot \frac{\partial r_1}{\partial x_1}\delta x_1 = F_1\delta x_1 \Rightarrow Q_{x_1} = F_1;\ r_1 = x_1 \tag{3.156}$$

where Q_{x_1} is the generalized force due to F_1 (it's equal to F_1) and it's the generalized force exerted on the mass m_1 (on the first cart). The virtual work in the x_2 direction is:

$$Q_{x_2}\delta x_2 = 0 \tag{3.157}$$

There are no external non-conservative forces on the second cart, so that $Q_{x_2} = 0$. When computing the generalized forces, all of the movements except one are frozen and the work done due to the unfrozen motion must be determined. In reality, a force F_1 on the system will result in this whole system moving to the right.

Applying a steady force F_1 on cart 1, the whole system will move to the right-hand side. But for the purpose of computing the generalized force on each mass, fix the positions of the masses at some instant of time, and then for one coordinate at a time, cause a little virtual deflection and determine how much work gets done.

Example 14: (Example 7—Continued)—Pendulum with a Mass and Spring (see Vandiver et al. [42])

Lecture 15—Video Times—1:05:00–1:21:16

The system consists of a steel rod with a metal sleeve on the outside of the rod and attached to the rod by a spring. The metal sleeve slides up and down along the rod. The rod, sleeve, and spring are attached at point A to a hinge and the whole system can swing back and forth as a pendulum. The system has multiple sources of kinetic energy and multiple forms of potential energy. For purposes of the problem, there is a horizontal force $F \cos \omega t$ located at the metal sleeve which pushes this system back and forth. The problem is to derive the equations of motion of the system. The motion is planar and there are two rigid bodies involved. Each rigid body has six degrees of freedom, but because of planar motion, each body has a maximum of three degrees of freedom. For planar motion, each rigid body can move in the x and y directions and can rotate about the z axis, where they are not attached together by the spring. The maximum number of degrees of freedom for the two rigid bodies, each acting independently of the other is six. This problem has two degrees of freedom, which are the angular displacement θ of the system from the vertical and x_1 the linear displacement of the sleeve's center of mass along the rod with respect to point A. The coordinate system $X_1 - Y_1$ rotates with the rod and sleeve. The coordinates chosen are independent, since, freezing one of them, the other can still traverse over the full range of values for that coordinate. They are also complete since with the two coordinates, all parts of the system may be located at all times. Furthermore, the generalized coordinates are also holonomic since the number of independent generalized coordinates required to describe the motion of the system equals the number of degrees of freedom (Fig. 3.24).

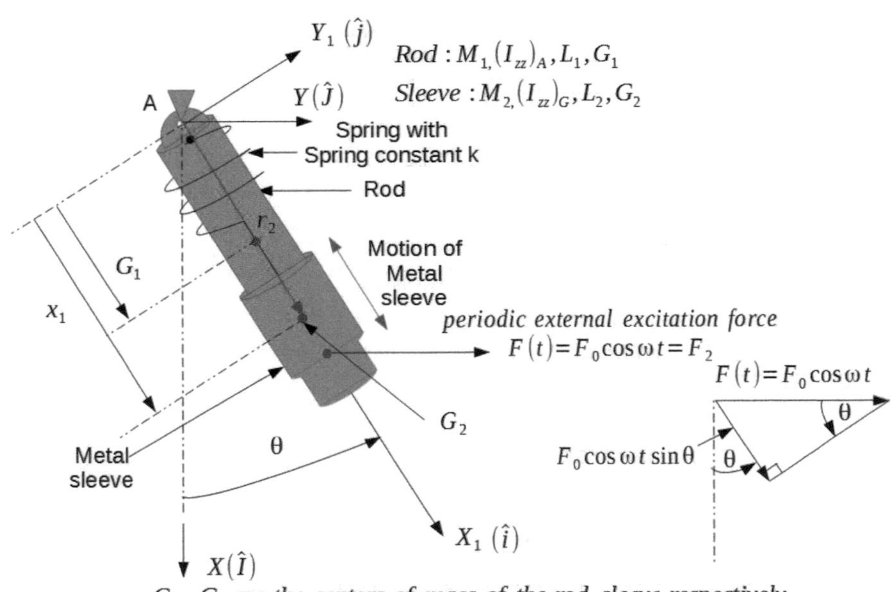

Fig. 3.24 Pendulum with a mass and spring—continued

The equations of motion of the system were:

$$M_2\ddot{x}_1 - M_2x_1\dot{\theta}_1^2 + k\left(x_1 - L_0 - \frac{L_2}{2}\right) - M_2g\cos\theta = Q_{x_1}$$

$$\left([I_{zz}]_A\,\ddot{\theta} + [I_{zz}]_G\,\ddot{\theta} + M_2\ddot{\theta}x_1^2 + 2M_2\dot{\theta}x_1\dot{x}_1\right) + \left(\frac{L_1}{2}M_1g\sin\theta + M_2gx_1\sin\theta\right)$$
$$= \left(x_1 + \frac{L_2}{2}\right)F_0\cos\omega t\cos\theta = Q_\theta \tag{3.158}$$

where $[I_{zz}]_A$ is the moment of inertia around the Z axis of the rod at point A and $[I_{zz}]_G$ is the moment of inertia of the sleeve at its center of mass. In addition, due to the parallel axis theorem, the term $M_2x_1^2$ is added to $[I_{zz}]_G$ in order to compensate for the fact that rotation is at the point A and not at the mass center of the sleeve. The terms $\left(\frac{L_1}{2}M_1g\sin\theta + M_2gx_1\sin\theta\right)$ are the moments due to gravity acting on the rod and sleeve, respectively. The virtual work along x_1 may be written as:

$$Q_{x_1}\delta x_1 = F_2\hat{J}\cdot\frac{\partial r_2}{\partial x_1}\delta x_1\hat{i};\; r_2\hat{i} = r_2\cos\theta\hat{I} + r_2\sin\theta\hat{J};\; \hat{i} = \cos\theta\hat{I} + \sin\theta\hat{J}$$
$$r_2 = x_1;\; \frac{\partial r_2}{\partial x_1} = 1 \Rightarrow Q_{x_1}\delta x_1 = F_2\hat{J}\cdot(\cos\theta\hat{I} + \sin\theta\hat{J})\delta x_1 = F_2\sin\theta\delta x_1 \tag{3.159}$$

The virtual work in the θ direction may be written in a similar manner as follows:

$$F_2\hat{J}\cdot\frac{\partial r_2}{\partial \theta}\delta\theta r_2\hat{i} = r_2\cos\theta\hat{I} + r_2\sin\theta\hat{J};\; \frac{\partial r_2}{\partial \theta} = -r_2\sin\theta\hat{I} + r_2\cos\theta\hat{J}$$
$$\Rightarrow F_2\hat{J}\cdot\frac{\partial r_2}{\partial \theta}\delta\theta = F_2\hat{J}\cdot\left(r_2\cos\theta\hat{J} - r_2\sin\theta\hat{I}\right) = F_2r_2\cos\theta \tag{3.160}$$

At static equilibrium, all of the derivatives and second derivatives with respect to time tend to zero and the angle $\theta = 0$, and so the dynamic Eq. (3.158) becomes:

$$k\left(x_1 - L_0 - \frac{L_2}{2}\right) = M_2g$$

where $\left(x_1 - L_0 - \frac{L_2}{2}\right)$ is the amount by which the spring is stretched due to M_2g.

3.11 Practice Finding Equations of Motion—D'Alembert–Lagrange

Lecture 17 [44]—Video Times—0:00–9:00

This material was adapted from Vandiver and Gossard-2.003SC Engineering Dynamics. Video of Lecture 17 [44].

The relevant equations for the direct method of finding the equations of motion for a rigid body are the following:

1. Newton's 2nd law: $\sum F_{ext} = \frac{dP_{G/O}}{dt} = ma_{G/O}$.
2. Torque about an arbitrary point A: $\tau_A = \left(\frac{dH_A}{dt}\right)_O + v_{A/O} \times P_{G/O} = \left(\frac{dH_G}{dt}\right)_O + r_{G/A} \times ma_{G/O}$. If the point A is at the center of mass, i.e., at point G, then simplifications can be made. If the point A is moving, then the term $v_{A/O} \times P_{G/O}$ must be included in the equation for torque. The new equation is: $\left(\frac{dH_G}{dt}\right)_O + r_{G/A} \times ma_{G/O}$.
3. Angular momentum about point A: $H_A = H_G + r_{A/G} \times P_{G/O}$.

Problems in mechanics may be analyzed from either the direct (Newton Euler) or the d'Alembert–Lagrangian points of view. Several problems will be treated by both the direct and d'Alembert–Lagrangian methods. The question of which method to choose and the ease of use of either method will be discussed.

Hockey Puck Problem

Lecture 17 [44]—Video Times—9:10–40:56

We have a mass (puck) sliding on a horizontal frictionless plane (ice) with a moment of inertia about the Z axis of $(I_Z)_G = M\kappa^2$, where κ is the radius of gyration of the puck and M is its mass (Fig. 3.25).

The puck has a cord wrapped around it and it is being pulled with a force of F_1. The puck's radius is R. The puck can both slide on the ice and rotate around its Z_p axis. In order to write the equations of motion, the number of degrees of freedom must be determined. Which method is the simplest, Lagrange or the direct method? How many independent coordinates are required? Is this problem one of planar motion? It is a planar motion problem with one rotational degree of freedom (around the Z axis) and two translation degrees of freedom (in the $X - Y$ plane). With three degrees of freedom, three equations of motion are required. One way of solving this problem is by the direct method with the torque equation around the center of mass. Another way would be to use the Lagrange method. The direct method will be used to solve this problem with the X, axis aligned with the force F_1, which implies that $\theta = 0°$. The three independent coordinates are X, Y, θ. There are no external forces in the Y direction which implies that $\ddot{y} = 0$. The only external

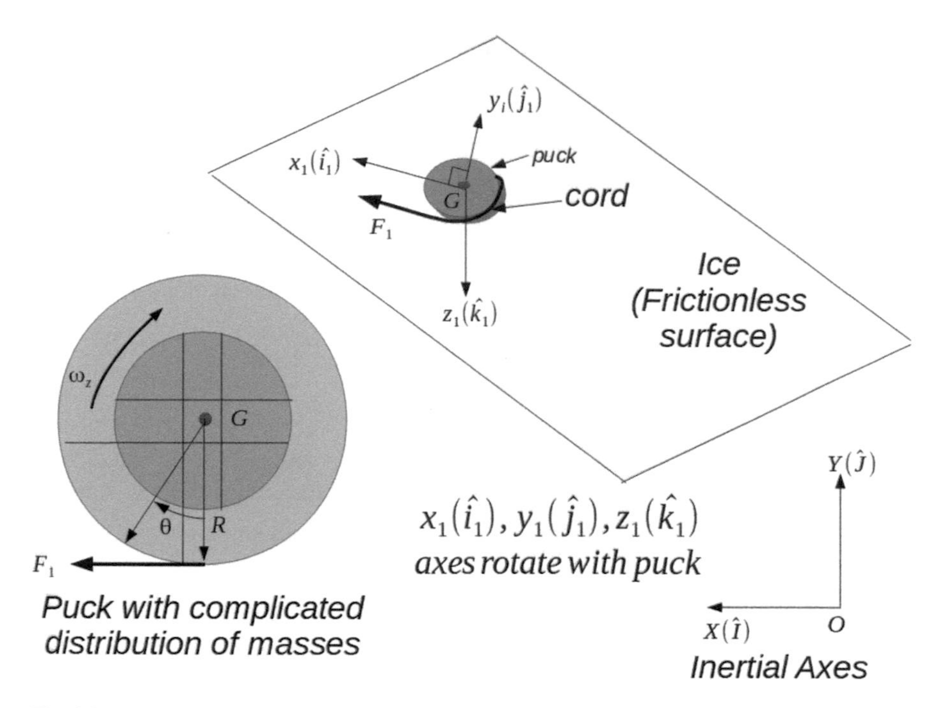

Fig. 3.25 Puck sliding on a horizontal frictionless plane

force in X direction is F_1 and it equals the mass times the acceleration, that is, $F_1 = Ma$. The force equations are:

$$\sum F_y = m\ddot{y} = 0 \Rightarrow \ddot{y} = 0; \sum F_x = F_1\hat{I} = M\ddot{x} \tag{3.161}$$

Rotation of the inertial frame X, Y, Z to be aligned with F_1 has eliminated the need for breaking the force F_1 into components $F_1 \cos\theta$ in the X direction and $F_1 \sin\theta$ in the Y direction. It has also resulted in $\ddot{y} = 0$. The third equation is the equation for the torque and it is:

$$\sum \tau_{/G} = \frac{dH_{/G}}{dt} = \frac{d\left((I_z)_G\omega_z\right)}{dt} = (I_z)_G\dot{\omega}_z = (I_z)_G\ddot{\theta} = F_1 R\hat{K} \tag{3.162}$$

where $-R\hat{J} \times F_1\hat{I} = F_1 R\hat{k}$ is the external applied torque.

Remark Although the equation $\sum F_y = m\ddot{y} = 0$ seems trivial, it shouldn't be discarded. It is a valid unconstrained equation of motion with a degree of freedom in that direction. It was achieved by a judicious alignment of the inertial axes and the fact that θ was nullified as a result. Otherwise three non-zero accelerations would have resulted.

Attention will now be turned towards solving the problem by means of the d'Alembert–Lagrangian equations. It is equally simple, except for the derivation of the generalized forces. The kinetic energy T is:

$$T = \underbrace{\frac{1}{2}(I_Z)_G\,\dot{\theta}^2}_{rotational\ K.E.} + \underbrace{\frac{1}{2}M\left(\dot{x}^2 + \dot{y}^2\right)}_{translational\ K.E} \tag{3.163}$$

Since there are no springs and no differences in heights in the X, Y plane, the potential energy is simply:

$$V = 0 \tag{3.164}$$

d'Alembert–Lagrange's equation is:

$$\frac{d}{dt}\frac{\partial T}{\partial \dot{q}_j} - \frac{\partial T}{\partial q_j} + \frac{\partial V}{\partial q_j} = Q_j \tag{3.165}$$

Letting the generalized coordinates be: $q_1 = \theta, q_2 = x, q_3 = y$ and carrying out the operations indicated in the above equation (d'Alembert–Lagrange's equation) results in:

1. $\frac{\partial T}{\partial \dot{q}_1} = \frac{\partial T}{\partial \dot{\theta}} = (I_Z)_G\,\dot{\theta}$
2. $\frac{d}{dt}\frac{\partial T}{\partial \dot{q}_1} = \frac{d}{dt}(I_Z)_G\,\dot{\theta} = (I_Z)_G\,\ddot{\theta}$
3. $\frac{\partial T}{\partial q_1} = \frac{\partial T}{\partial \theta} = 0$
4. $\frac{\partial V}{\partial q_1} = 0$
5. $\frac{d}{dt}\frac{\partial T}{\partial \dot{\theta}} - \frac{\partial T}{\partial \theta} + \frac{\partial V}{\partial \theta} = Q_\theta \Rightarrow (I_Z)_G\,\ddot{\theta} = Q_\theta$
6. $\frac{\partial T}{\partial \dot{q}_2} = \frac{\partial T}{\partial \dot{x}} = M\dot{x}$
7. $\frac{d}{dt}\frac{\partial T}{\partial \dot{q}_2} = \frac{d}{dt}\frac{\partial T}{\partial \dot{x}} = M\ddot{x}$
8. $\frac{\partial T}{\partial q_2} = \frac{\partial T}{\partial x} = 0$
9. $\frac{\partial V}{\partial q_1} = \frac{\partial V}{\partial q_1} = 0$
10. $\frac{d}{dt}\frac{\partial T}{\partial \dot{x}} - \frac{\partial T}{\partial x} + \frac{\partial V}{\partial x} = Q_x \Rightarrow M\ddot{x} = Q_x$
11. $M\ddot{y} = Q_y$

Now, what remains is to find the three generalized forces Q_θ, Q_x, Q_y. Assuming that our X axis is parallel to the force F_1, and there is a virtual displacement δy, there would not be any virtual work since δy is perpendicular to F_1 and virtual work $\delta W = F_1 \cdot \delta y = 0$. This implies that $Q_y = 0$. The approach to solving for the remaining two generalized forces will be to use the rigorous kinematic theory. The steps are as follows:

1. Return to the situation where the force F_1 and the inertial $X(\hat{I})$ axis are not aligned (see Fig. 3.26).
2. Draw a position vector from the origin O of the coordinate frame to the point of application of the force, point D on Fig. 3.26—position vector $r_{D/O}$.

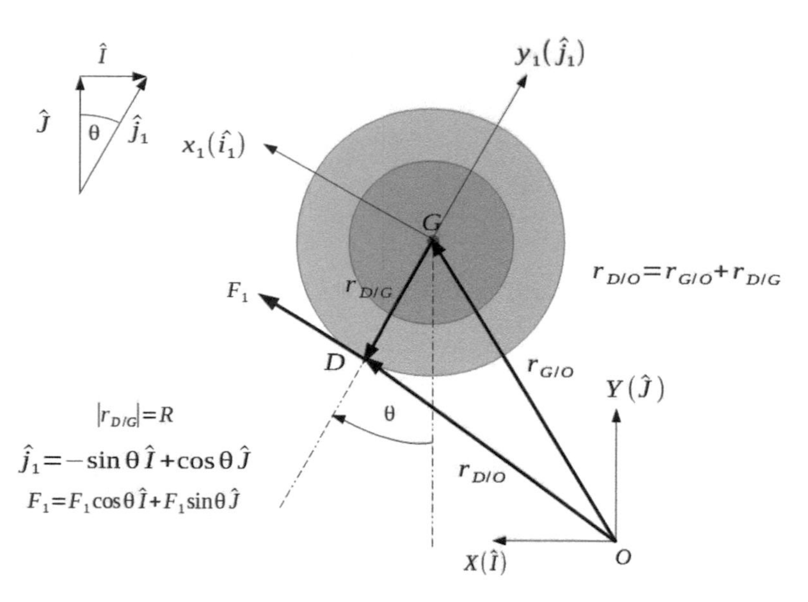

Fig. 3.26 Generalized forces for sliding puck problem—I

3. Draw position vectors $r_{G/O}$ from O to the center of mass G and $r_{G/D}$ from G to D.
4. Using vector algebra, it can easily be shown that $r_{D/O} = r_{G/O} + r_{D/G}$ (see Fig. 3.26).
5. $r_{G/O} = X\hat{I} + Y\hat{J}$
6. Define a coordinate system $x_1(\hat{i}_1)$, $y_1(\hat{j}_1)$ which rotates with the rigid body.
7. $r_{D/G} = -R\hat{j}_1$
8. $r_{D/O} = r_{G/O} + r_{D/G} = X\hat{I} + Y\hat{J} - R\hat{j}_1$
9. $\hat{j}_1$ expressed in an inertial frame is: $\hat{j}_1 = -\sin\theta\hat{I} + \cos\theta\hat{J}$.
10. Combining terms results in: $r_{D/O} = X\hat{I} + Y\hat{J} - R\hat{j}_1 = X\hat{I} + Y\hat{J} + R\sin\theta\hat{I} - R\cos\theta\hat{J} = (X + R\sin\theta)\,\hat{I} + (Y - R\cos\theta)\,\hat{J}$.
11. F_1 in the inertial coordinate system becomes: $F_1\hat{i}_1 = F_1\cos\theta\hat{I} + F_1\sin\theta\hat{J}$.
12. The virtual work in the $X\hat{I}$ direction is:

$$Q_2 \cdot \delta r = Q_x\delta x = F_1\cos\theta\hat{I} \cdot \frac{\partial r_{D/O}}{\partial x}\delta x$$

From item 8, the partial derivative with respect to X of $r_{D/O}$ is 1 and so

$$Q_2 \cdot \delta r = F_1\cos\theta\hat{I} \cdot \frac{\partial r_{D/O}}{\partial x}\delta x = F_1\cos\theta\delta x \Rightarrow Q_x = F_1\cos\theta$$

13. The virtual work in the $Y\hat{J}$ direction is:

$$Q_3 \cdot \delta r = Q_y \delta y = F_1 \sin\theta \hat{J} \cdot \frac{\partial r_{D/O}}{\partial y} \delta y$$

Once again, from item 8, the partial derivative with respect to Y of $r_{D/O}$ is 1 and so:

$$Q_3 \cdot \delta r = F_1 \sin\theta \hat{J} \cdot \frac{\partial r_{D/O}}{\partial y} \delta y = F_1 \sin\theta \delta y \Rightarrow Q_y = F_1 \sin\theta$$

14. The virtual work in the θ direction is:

$$Q_\theta \delta\theta = F_1 \cdot \frac{\partial r_{D/O}}{\partial \theta} \delta\theta = F_1 \left(\cos\theta \hat{I} + \sin\theta \hat{J}\right) \cdot \frac{\partial r_{D/O}}{\partial \theta} \delta\theta$$

$$r_{D/O} = (X + R\sin\theta)\,\hat{I} + (Y - R\cos\theta)\,\hat{J}$$

$$\frac{\partial r_{D/O}}{\partial \theta} = R\cos\theta \hat{I} + R\sin\theta \hat{J}$$

$$\Rightarrow Q_\theta \delta\theta = F_1 \left(\cos\theta \hat{I} + \sin\theta \hat{J}\right) \cdot R\left(\cos\theta \hat{I} + \sin\theta \hat{J}\right) \delta\theta$$

$$= F_1 R \left(\cos^2\theta + \sin^2\theta\right) \delta\theta = F_1 R \delta\theta\, Q_\theta \Rightarrow Q_\theta = F_1 R \tag{3.166}$$

The torque Q_θ is not dependent on the angle θ, which is logical. Had the axes been oriented initially in such a way that the force F_1 was aligned with the $X(\hat{I})$ axis and perpendicular to the $Y(\hat{J})$ axis, then the angle θ would have been zero. The sine and cosine terms then do not appear in the equation for $r_{D/O}$ and it is impossible to calculate the partial derivative $\frac{\partial r_{D/O}}{\partial \theta}$ required in order to obtain the generalized force Q_θ. *The conclusion is that in order to carry out this procedure, the configuration and orientation of the system should be in its most general form. Once the above procedure has been completed, the problem can be simplified by letting* $\theta = 0$. For this problem the intuitive method would have yielded results much more quickly. For the intuitive method we could do the following:

1. For the system with the axes $X(\hat{I}), Y(\hat{J}), Z(\hat{K})$, allow a small deflection δx, then the virtual work would be $Q_x \delta x = F_1 \cos\theta \delta x$, which would mean that the generalized force $Q_x = F_1 \cos\theta$
2. Similarly for δy, the virtual work would be $Q_y \delta y = F_1 \sin\theta \delta y$ and so the generalized force Q_y would be $Q_y = F_1 \sin\theta$
3. For a small displacement $\delta\theta$, the puck would move a distance $R\delta\theta$ and so the virtual work would be $Q_\theta \delta\theta \cdot F_1 = R\delta\theta \cdot F_1 = RF_1 \delta\theta$, which implies that $Q_\theta = RF_1$. Note that the motion $R\delta\theta$ is in the same direction as F_1.

3.12 A Note on Equivalent Forces and Torques

Lecture 12 [41]—Video Times—46:23–50:02, Lecture 17 [44]—Video Times—40:56–45:29

This material was adapted from Vandiver and Gossard-2.003SC Engineering Dynamics. Video of Lecture 12 (see [41]) and the Video of Lecture 17 (see [44]). A rigid body has a center of mass G and an applied force F (see Fig. 3.27). The force has a line of action and the perpendicular to that line of action is of length d from the force's line of action to the center of mass. Since the line of action of the force doesn't pass through the center of mass, the force exerts a torque (or moment) on the body of $\tau = r \times F = Fd$. Figure 3.27a and b depict two entirely equivalent situations, in that the two upper forces in Fig. 3.27b cancel each other leaving only the force F on the lower left. The force F on the lower left side of Fig. 3.27b and the force F on the upper right-hand side of Fig. 3.27b form a couple which is equivalent to the couple in Fig. 3.27a of $\tau = Fd$, both in magnitude and direction. This implies that the system in Fig. 3.27b is equivalent to Fig. 3.27c which has a force F whose line of action passes through the center of mass G, along with a torque (or moment) $\tau = Fd$ of magnitude and direction the same as Fig. 3.27a. In the general case (Fig. 3.27d), where many forces are being applied to the body, an equivalent system consists of a force F_{Total} whose line of action passes through the center of mass, and a torque τ_{Total} (see Fig. 3.27e). The force F_{Total} is: $F_{Total} = \sum_{i=1}^{N} F_i$. Similarly the total torque is: $\tau_{Total} = \sum_{i=1}^{N} r_{i/G} \times F_i$.

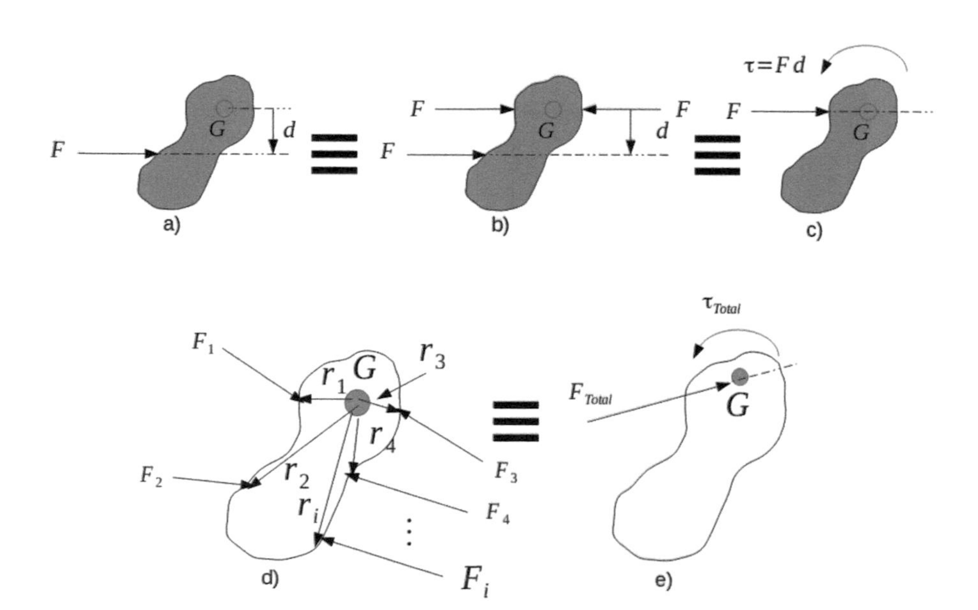

Fig. 3.27 Geometry for equivalent forces and torques

Hockey Puck Problem—Continued

Lecture 17 [44]—Video Times—45:29–49:56

Returning to the hockey puck problem, the force F_1 is in the direction as shown in Fig. 3.28a. Carrying out the same procedure as discussed above, the system is entirely equivalent to Fig. 3.28c, with a net force F_1 in the positive $\hat{I}$direction acting at the mass center and a torque of $\tau = RF_1\hat{K}$ also acting at the center of the mass.

The generalized force Q_x is obtained from the virtual work due to the motion of a generalized coordinate. The virtual work in the X direction is (the kinematic approach is used):

$$\begin{aligned} Q_x\delta_x &= F_1\hat{I}\cdot\frac{\partial r_1}{\partial x}\delta_x;\, r_1 = X\hat{I}+Y\hat{J};\, \frac{\partial r_1}{\partial x} = \hat{I}\ \frac{\partial r_1}{\partial y} = \hat{J} \\ \Rightarrow Q_x\delta_x &= F_1\hat{I}\cdot\hat{I}\delta_x \\ Q_y\delta_y &= F_1\hat{I}\cdot\hat{J}\delta_y = 0 \Rightarrow Q_x = F_1;\, Q_y = 0 \end{aligned} \tag{3.167}$$

As for Q_θ, the total torque $\tau_T = RF_1\hat{K}$ and the corresponding virtual work becomes:

$$Q_\theta\delta\theta = (\tau_T)_G\cdot\delta\theta = RF_1\hat{K}\cdot\hat{K}\delta\theta \Rightarrow Q_\theta = RF_1 \tag{3.168}$$

Note that no derivatives were required in order to obtain the Q_θ term. The lesson to be learned is that if the problem is shifted to the center of mass of the rigid body, the r and its derivatives become much easier to compute and the application of the torque becomes very obvious.

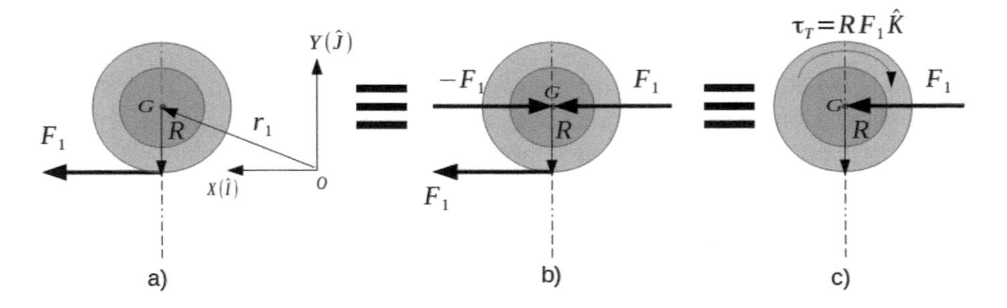

Fig. 3.28 Generalized forces for hockey puck problem—II

Example 15: Pendulum with a Plane of Symmetry (Video of Lecture 17—Vandiver [44])

Lecture 17—Video Times—50:28–1:00:04

The following figure contains an odd shaped pendulum with one plane of symmetry. The plane of symmetry is parallel to the plane of the paper or perpendicular to the pin about which the body rotates. The axis of rotation at point A is perpendicular to the plane of symmetry, while the center of mass is at G. Any rigid body with a plane of symmetry can have an axis perpendicular to its plane of symmetry about which it can rotate as a pendulum. The pendulum in planar motion has three degrees of freedom (translatory motion in the plane and one angular motion degree of freedom) and two constraints (no translatory motion because it is pinned at one point), which leaves one degree of freedom. The axis about which the pendulum rotates is a principal axis of the body with mass moment of inertia $(I_z)_A = (I_z)_G + ML^2$ (from the parallel axis theorem—see Beer et al. [4, pp 514]). Applying the general equation of the direct method (see page 139) about point A, the equation of motion is:

$$\tau_A = \frac{dH_A}{dt} = \frac{d[(I_z)_A\dot{\theta}\hat{K}]}{dt} = (I_z)_A\ddot{\theta}\hat{K} = \sum \tau_{ext} \tag{3.169}$$

The free body diagram of this pendulum is depicted in Fig. 3.29. The reaction forces at the pin at point A are R_x and R_y and Mg is the gravitational force. The torque which results from Mg with respect to point A is $R \times F = L\sin\theta\hat{I} \times Mg(-\hat{J}) = -MgL\sin\theta\hat{K}$. The direction of the torque is negative and points into the plane of the paper. The torque equation then becomes:

$$(I_z)_A\ddot{\theta}\hat{K} = \sum \tau_{ext} = -MgL\sin\theta\hat{K} \Rightarrow (I_z)_A\ddot{\theta} + MgL\sin\theta = 0 \tag{3.170}$$

The following equation is the generic undamped equation of motion for any single degree of freedom pendulum made out of a rigid body, and rotating about an axis that is a principal axis, no matter what the shape of the body. The equation is:

$$\boxed{(I_z)_A\ddot{\theta} + MgL\sin\theta = 0} \tag{3.171}$$

If it can be rotated about one of its principal axes and not through G, the center of mass of the body, any single degree of freedom system will oscillate. If the axis of rotation passes through G, the torque due to gravitational forces is zero and no oscillation will take place. If there is a damping force, then it will appear on the right-hand side of the above equation. If a given body has a single plane of symmetry, then the above equation applies when the axis of rotation is perpendicular to that plane of symmetry and the body in question becomes a pendulum. The

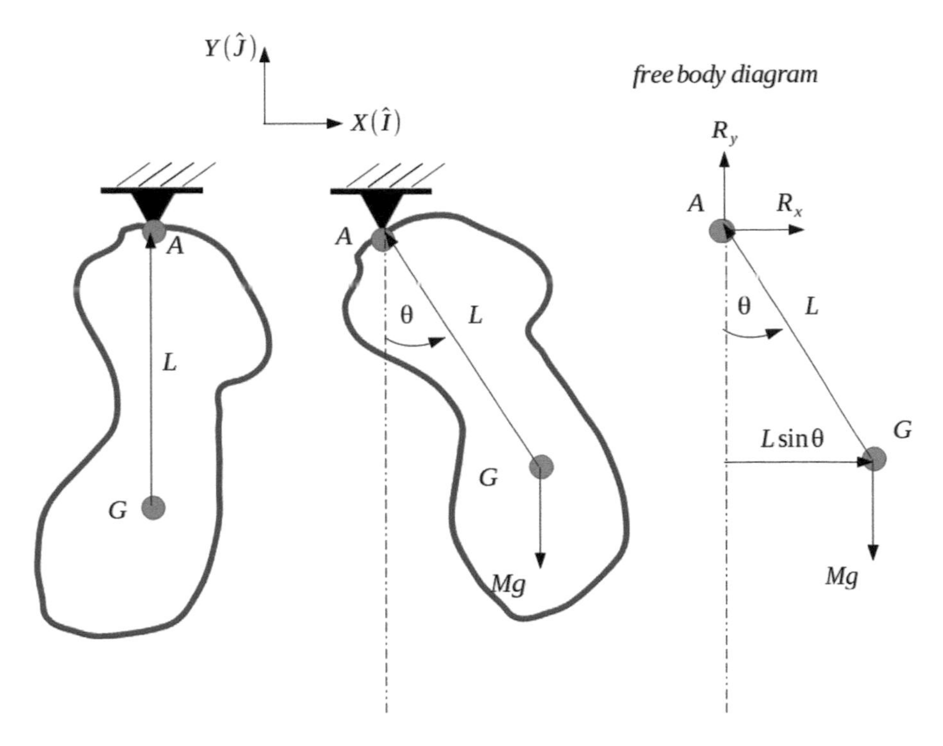

Fig. 3.29 Pendulum with a plane of symmetry

moment of inertia of I_z with respect to A or $(I_z)_A$ can be easily obtained from the parallel axis theorem if $(I_z)_G$ is given (Fig. 3.29).

Example 16: Atwood's Machine (Video of Lecture 17—Vandiver [44]+ Video of Lecture 10—Vandiver [40])

Lecture 17—Video Times—1:01:00–1:05:00, Lecture 10—Video Times—17:52–31:50

The Atwood's machine is depicted in Fig. 3.30. There are three rigid bodies involved in planar motion, which implies that there are potentially nine degrees of freedom. However, due to constraints, the two masses can neither rotate nor move in the $\hat{I}$ direction; therefore, each mass has one degree of freedom. In addition the pulley can only rotate, leaving it with only one degree of freedom. The rope connecting the two masses doesn't slip, which implies that $R\theta = y$, thus imposing an additional constraint on the system. The masses are connected, so that when one moves upwards by a distance of say y, the other mass moves down by the same amount. So,

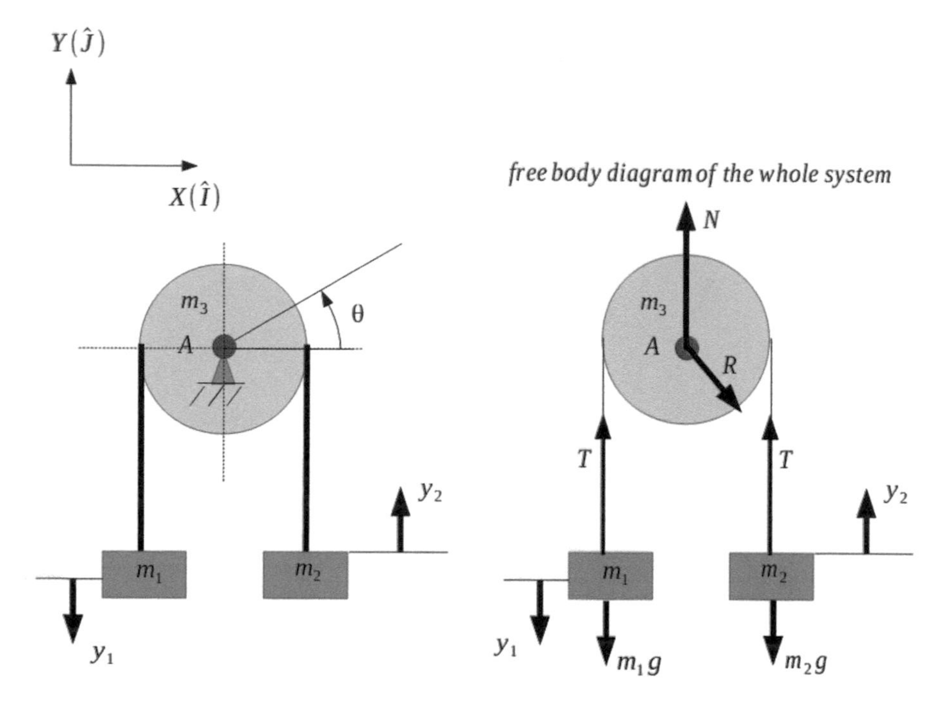

Fig. 3.30 Atwood's machine

finally, we end up with a one degree of freedom system. Both masses are initially at the same starting positions with respect to the $\hat{J}$ coordinate, when they are released. If the pulley is not massless, then the moment of inertia of the pulley system is: $I_A = m_3\kappa^2$, where m_3 is the pulley's mass and κ is its radius of gyration. From the free body diagram of the whole system it is apparent that the reaction force N on the pivot point at A doesn't introduce any moments. The moments are created by $m_1 g$ and $m_2 g$ and are: $(m_1 g R - m_2 g R)\hat{K}$. The tensions in the cord T are internal to the system and are therefore not of concern. Another way of looking at it is to note that the T which supports mass m_1 creates an equal and opposite moment to the T which supports mass m_2, thus resulting in **zero moments produced by T**. The angular momentum with respect to point A that is h_A may be written as:

$$r_{m_1/A} \times P_{m_1/O} + r_{m_2/A} \times P_{m_2/O} = (m_1 + m_2)\, R\dot{y}\hat{k}$$

since $r_{m_1/A} = r_{m_2/A} = R$ and $P_{m_1/O} = m_1\dot{y}\hat{k}$, $P_{m_2/O} = m_2\dot{y}\hat{k}$. The time derivative of h_A turns out to be:

$$\frac{dh_A}{dt} = (m_1 + m_2)\, R\ddot{y}\hat{k}$$

Note that lowercase h was used since the masses m_1 and m_2 are considered to be mass particles. The sum of the torques with respect to point A is therefore:

$$\sum \tau_A = \frac{dh_A}{dt} + \underbrace{v_{A/O}}_{=0} \times P_O$$

$$\sum \tau_A = (m_1 g R - m_2 g R)\hat{K} = \frac{dh_A}{dt} = (m_1 + m_2)\, R\ddot{y}\hat{k}$$

$$\ddot{y} = \frac{m_1 - m_2}{m_1 + m_2} g \tag{3.172}$$

The term $v_{A/O} = 0$ since the point A doesn't move with respect to the point O. In this problem, it was assumed that the pulley was massless. If that isn't the case, then a term would have to be added to h_A and to the time derivative of h_A to account for the mass of the pulley such that: $\frac{dh_A}{dt} = (m_1 + m_2)\, R\ddot{y}\hat{k} + I_A\ddot{\theta}$. It's possible to relate $\ddot{\theta}$ to $\ddot{y}$, since $y = R\theta \Rightarrow \ddot{y} = R\ddot{\theta}$.

Example 17: Falling Stick Problem (Lecture 10-2.003J/1.053J—Dynamics and Control I—Peacock [27])

A stick with uniformly distributed mass m and length L slides without (or with) friction as it falls. For the moment, assume that there is a friction force F. Its center of mass is located at $G(x_c, y_c)$ as in Fig. 3.31. The position vector r_c of the center of mass is:

$$r_c = x_c\hat{I} + y_c\hat{J} = x_c\hat{I} + \frac{L}{2}\sin\theta\,\hat{J} \tag{3.173}$$

The linear momentum P_G is:

$$P_G = m\dot{r}_c = m\dot{x}_c\hat{I} + \frac{mL}{2}\dot{\theta}\cos\theta\,\hat{J}$$

The force equation which is the time derivative of the linear momentum is:

$$\sum F_{ext} = \frac{dP_G}{dt} = m\ddot{x}_c\hat{I} + \frac{mL}{2}\left(\ddot{\theta}\cos\theta - \dot{\theta}^2\sin\theta\right)\hat{J} \tag{3.174}$$

Assuming for the moment that there are no forces in the $X(\hat{I})$ direction, then $m\ddot{x}_c = 0$. In the $Y(\hat{J})$ direction, the sum of forces leads to: $N - mg = \frac{mL}{2}\left(\ddot{\theta}\cos\theta - \dot{\theta}^2\sin\theta\right) \Rightarrow N = mg + \frac{mL}{2}\left(\ddot{\theta}\cos\theta - \dot{\theta}^2\sin\theta\right)$. The moment equation at the center of mass may be written as:

$$\tau_{ext} = \frac{dH_G}{dt}\hat{K} = \frac{d(I_z)_G\dot{\theta}}{dt}\hat{K} = (I_z)_G\ddot{\theta}\hat{K} = -N\frac{L}{2}\cos\theta\,\hat{K} \tag{3.175}$$

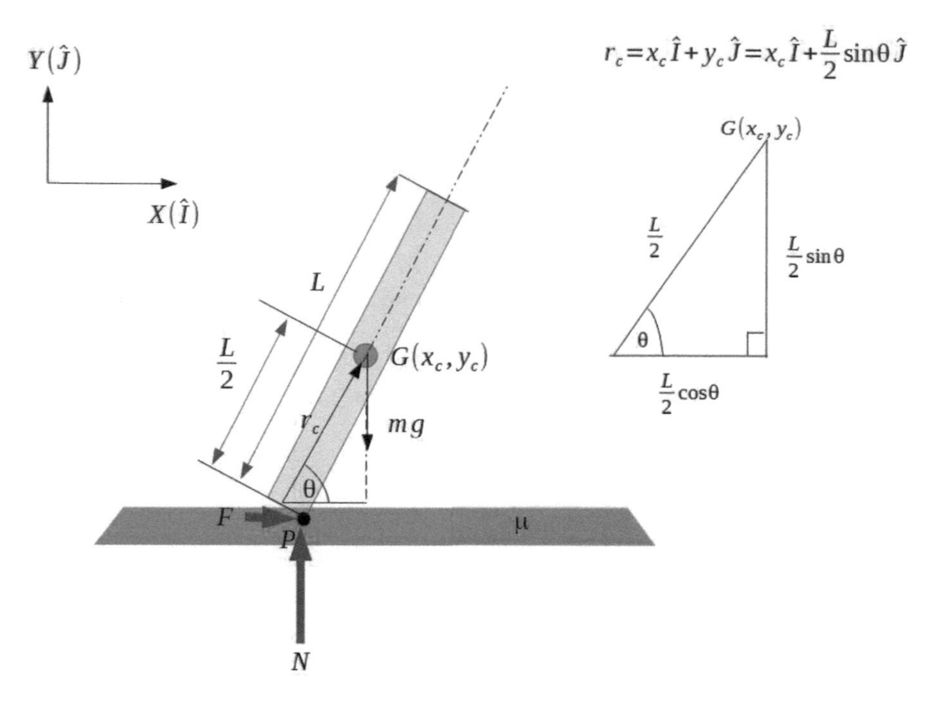

Fig. 3.31 Falling stick problem

where $(I_z)_G$ is the moment of inertia of the stick in the $\hat{K}$ direction (directed out of the page) with respect to its center of mass. Combining the result previously derived for N with the moment equation, we have:

$$(I_z)_G\ddot{\theta} = -\left[mg + \frac{mL}{2}\left(\ddot{\theta}\cos\theta - \dot{\theta}^2\sin\theta\right)\right]\frac{L}{2}\cos\theta$$

$$\Rightarrow \boxed{\left((I_z)_G + \frac{mL^2}{4}\cos^2\theta\right)\ddot{\theta} - \frac{mL^2}{4}\dot{\theta}^2\sin\theta\cos\theta + \frac{mgL}{2}\cos\theta = 0} \tag{3.176}$$

If there is a friction force F, then the acceleration in the $X(\hat{I})$ direction becomes (Fig. 3.31):

$$\boxed{m\ddot{x}_c = \mu N \Rightarrow \ddot{x}_c = \frac{\mu}{m}N = \frac{\mu}{m}\left[mg + \frac{mL}{2}\left(\ddot{\theta}\cos\theta - \dot{\theta}^2\sin\theta\right)\right]} \tag{3.177}$$

3.13 Lagrangians, Hamiltonians, and the Legendre Transformation

In classical mechanics, the Lagrangian is used to derive a system of n second-order differential equations (for a system with n degrees of freedom), while the Hamiltonian formulation results in a system of $2n$ first-order differential equations called Hamilton's equations with some special properties. In particular, the derivatives (states) in this system are uncoupled, and the differential equations can be derived from a single (scalar) function called the Hamiltonian (see Van Brunt [36, pp. 159–170]). Hence, from a computational point of view, the Hamiltonian formulation is advantageous. Note that the n second-order differential equations which are the outcome of the d'Alembert–Lagrange procedure may be converted into a system of $2n$ first-order differential equations by designating the generalized velocities $\dot{q}_i = v_i$ as additional states. However the resulting set of first-order differential equations may or may not have coupled states or may be non-symmetric with respect to q_i and v_i. The Hamiltonian, on the other hand, uses generalized velocities q_i and generalized momenta p_i defined as:

$$p_i = \frac{\partial \mathcal{L}}{\partial \dot{q}_i}, \quad i = 1, 2, \ldots, n \tag{3.178}$$

The resulting $2n$ Hamiltonian equations of motion for q_i and p_i have an elegant symmetric uncoupled form which are referred to as *canonical equations*. The Hamiltonian, as will be shown in the sequel, may be obtained from a Legendre transformation of the d'Alembert–Lagrangian equations. It should be noted that both formalisms lead to equations of motion describing the same trajectory.

Legendre Transformations

Contact transformations play an important role in differential equations and geometry. A contact transformation depends on the derivatives of the dependent variables and one of the simplest and most effective of the contact transformations is known as the Legendre transformation. This transformation provides the link between the d'Alembert–Lagrange and Hamilton's equations. Let $f(x)$ be a smooth convex function such that $f''(x) > 0$ and let p be the derivative (or the slope) of $y = f(x)$ at x (with respect to the independent variable x), that is, $p = f'(x)$, where $(\cdot)'$ denotes the derivative with respect to the independent variable. The derivative p is a strictly monotonic function of x. This implies that there exists a single value of the slope p for each given value of x and vice versa. The positive definiteness of the second derivative of f, that is, $f''(x) > 0$, suggests that $f(x)$ is strictly convex in shape. Geometrically, under these conditions, any point on $f(x)$ is determined uniquely by the slope of its tangent line. Regarding $x(p)$ to be a function of p, it is possible to transform from the coordinates $(x, f(x))$ to $(p, H(p))$ by introduction of the following function which is a simple example of a Legendre transformation:

$$H(p) = -y(x) + px \tag{3.179}$$

A notable feature of the above transformation is that it is an *involution*, i.e., the transformation is its own inverse. To see this, note that

$$\frac{dH(p)}{dp} = -\frac{d}{dp}y(x) + \frac{dpx}{dp} = -\frac{dy}{dx}\frac{dx}{dp} + p\frac{dx}{dp} + x\frac{dp}{dp}$$
$$= \underbrace{(-y'(x) + p)}_{=0}\frac{dx}{dp} + x = x$$
$$-H(p) + px = -(-y(x) + px) + px = y(x) \tag{3.180}$$

Note that $(-y'(x) + p) = 0$ since $y'(x) = p$ by definition. The above calculations show that if we apply the transformation to the pair $(p, H(p))$ we are able to recover $(x, y(x))$. Summarizing the above, we have:

$$\begin{bmatrix} p = \frac{df}{dx} & f(x) = -H(p) + px \\ \Updownarrow & \Updownarrow \\ x = \frac{dH}{dp} & H(p) = -f(x) + px \end{bmatrix} \tag{3.181}$$

Consider the following function of n variables $u_i,\ i = 1, \ldots, n$ (see Meirovitch [23, pp. 91–97]), $F = F(u_1, u_2, \ldots, u_n)$ and define a new set of n variables $v_i,\ i = 1, \ldots, n$ such that:

$$v_i = \frac{\partial F}{\partial u_i}\ , i = 1, 2, \ldots, n \tag{3.182}$$

In order to ensure the independence of the new set of variables v_i, the determinant of the Jacobian of the gradient of F must not equal zero or the Hessian matrix must not be singular, that is:

$$\begin{bmatrix} \frac{\partial v_1}{\partial u_1} & \frac{\partial v_1}{\partial u_2} & \cdots & \frac{\partial v_1}{\partial u_n} \\ \vdots & \vdots & \cdots & \vdots \\ \frac{\partial v_n}{\partial u_1} & \frac{\partial v_n}{\partial u_2} & \cdots & \frac{\partial v_n}{\partial u_n} \end{bmatrix} = \begin{bmatrix} \frac{\partial^2 F}{\partial u_1 u_1} & \frac{\partial^2 F}{\partial u_1 u_2} & \cdots & \frac{\partial^2 F}{\partial u_1 u_n} \\ \vdots & \vdots & \cdots & \vdots \\ \frac{\partial^2 F}{\partial u_n u_1} & \frac{\partial^2 F}{\partial u_n u_2} & \cdots & \frac{\partial^2 F}{\partial u_n u_n} \end{bmatrix} \neq 0 \tag{3.183}$$

This will allow us to solve for the $u_i's$ in terms of the $v_i's$. A new function $G = \sum_{i=1}^{n} u_i v_i - F$, which is an n-dimensional **Legendre transformation**, is defined and the variables u_i are expressed in such a manner so that G may be written only in terms of the $v_i's$, that is, $G = G(v_1, v_2, \ldots, v_n)$. Since the sets of variables u_i and v_i are independent, they may each be assigned infinitesimal arbitrary variations $\delta v_i,\ \delta u_i$ allowing the variation of G to become:

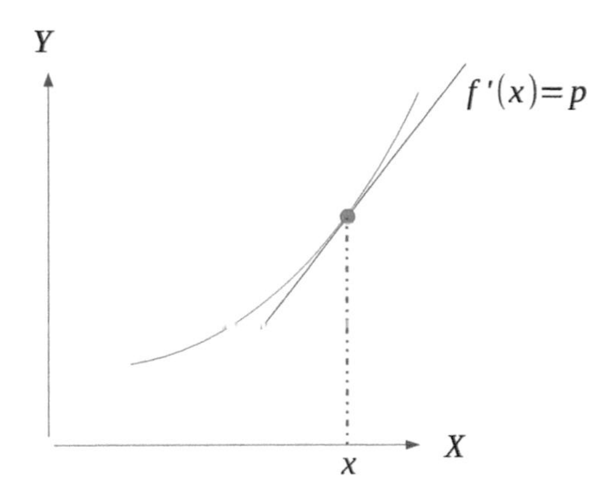

Fig. 3.32 Strictly convex function used in the derivation of a simple legendre transformation

$$\delta G = \sum_{i=1}^{n} (u_i \delta v_i + v_i \delta u_i) - \sum_{i=1}^{n} \frac{\partial F}{\partial u_i} \delta u_i$$

$$\sum_{i=1}^{n} \left[u_i \delta v_i + \left(v_i - \frac{\partial F}{\partial u_i} \right) \delta u_i \right] \quad (3.184)$$

Seeing that $G = G(v_1, v_2, \ldots, v_n)$, it follows that:

$$\delta G = \sum_{i=1}^{n} \frac{\partial G}{\partial v_i} \delta v_i \quad (3.185)$$

Based on the fact that $v_i = \frac{\partial F}{\partial u_i}$ (see Eq. 3.182), Eqs. (3.184) and (3.185) become (Fig. 3.32):

$$\delta G = \sum_{i=1}^{n} [u_i \delta v_i] = \sum_{i=1}^{n} \frac{\partial G}{\partial v_i} \delta v_i$$

$$\Rightarrow u_i = \frac{\partial G}{\partial v_i}, \quad i = 1, \ldots, n \quad (3.186)$$

Hamiltonian Function

Using the foregoing results, the *Hamiltonian function* is defined to be:

$$\mathcal{H} = \sum_{i=1}^{n} \frac{\partial \mathcal{L}}{\partial \dot{q}_i} \dot{q}_i - \mathcal{L} = \sum_{i=1}^{n} p_i \dot{q}_i - \mathcal{L} \quad (3.187)$$

The Hamiltonian function may be written in vector shorthand as:

$$\mathcal{H}(q_1, q_2, \ldots, q_n, p_1, p_2, \ldots, p_n, t) = \mathcal{H}(\mathbf{q}, \mathbf{p}, t) \tag{3.188}$$

where $\mathbf{q}, \mathbf{p}$ are both vectors of length n and $\mathbf{q}$ is the vector of generalized coordinates, while $\mathbf{p}$ is the vector of generalized momenta. Notice that the Hamiltonian is a Legendre transformation of the Lagrangian. The variation of the Hamiltonian function $\mathcal{H}$, which appears in Eq. (3.187), may be written in the form:

$$\delta\mathcal{H} = \sum_{i=1}^{n}\left(\dot{q}_i\delta p_i + p_i\delta\dot{q}_i - \frac{\partial\mathcal{L}}{\partial q_i}\delta q_i - \frac{\partial\mathcal{L}}{\partial\dot{q}_i}\delta\dot{q}_i\right) = \sum_{i=1}^{n}\left(\dot{q}_i\delta p_i - \frac{\partial\mathcal{L}}{\partial q_i}\delta q_i\right) \tag{3.189}$$

since $p_i = \frac{\partial\mathcal{L}}{\partial\dot{q}_i}$ from Eq. (3.178). The variation of $\mathcal{H}$ is also calculated to be:

$$\delta\mathcal{H}(\mathbf{q}, \mathbf{p}, t) = \sum_{i=1}^{n}\left(\frac{\partial\mathcal{H}}{\partial q_i}\delta q_i + \frac{\partial\mathcal{H}}{\partial p_i}\delta p_i\right) \tag{3.190}$$

and thus we have from Eqs. (3.189) and (3.190) that:

$$\dot{q}_i = \frac{\partial\mathcal{H}}{\partial p_i}, \quad -\frac{\partial\mathcal{L}}{\partial q_i} = \frac{\partial\mathcal{H}}{\partial q_i}, \quad i = 1, 2, \ldots, n \tag{3.191}$$

Noticing that $G = \sum_{i=1}^{n} u_i v_i - F$ has the same form as the Hamiltonian function $\mathcal{H} = \sum_{i=1}^{n} p_i\dot{q}_i - \mathcal{L}$, the terms in Eq. (3.191) are the result of the Legendre transformation.

Assuming that we have a holonomic system, that is, the generalized coordinates are independent of each other, and there are no generalized non-conservative forces acting on our system ($Q_i = 0; i = 1, 2, \ldots, n$), the d'Alembert–Lagrange equations are (see Eq. (3.56)):

$$\begin{aligned}\frac{d}{dt}\left(\frac{\partial T}{\partial\dot{q}_i}\right) - \frac{\partial T}{\partial q_i} + \frac{\partial V}{\partial q_i} &= 0; i = 1, 2, \ldots, n \\ \Rightarrow \frac{d}{dt}\left(\frac{\partial\mathcal{L}}{\partial\dot{q}_i}\right) - \frac{\partial\mathcal{L}}{\partial q_i} &= 0\end{aligned} \tag{3.192}$$

From Eqs. (3.187), (3.191), and (3.192), the following relations result:

$$p_i = \frac{\partial\mathcal{L}}{\partial\dot{q}_i}; \; \dot{p}_i = \frac{d}{dt}\left(\frac{\partial\mathcal{L}}{\partial\dot{q}_i}\right); \; \frac{d}{dt}\left(\frac{\partial\mathcal{L}}{\partial\dot{q}_i}\right) = \frac{\partial\mathcal{L}}{\partial q_i} = -\frac{\partial\mathcal{H}}{\partial q_i}$$

$$\Rightarrow \dot{p}_i = -\frac{\partial \mathcal{H}}{\partial q_i}; \dot{q}_i = \frac{\partial \mathcal{H}}{\partial p_i}, \ i = 1, 2, \ldots, n \tag{3.193}$$

The $2n$ set of first-order differential equations (n for $\dot{q}_i$ and n for $\dot{p}_i$) in Eq. (3.193) are generally referred to as the *Hamiltonian canonical equations*. The derivative with respect to time of the Hamiltonian $\mathcal{H}$ may be written as:

$$\dot{p}_i = -\frac{\partial \mathcal{H}}{\partial q_i}; \dot{q}_i = \frac{\partial \mathcal{H}}{\partial p_i}, \ i = 1, 2, \ldots, n$$

$$\frac{d}{dt}\{\mathcal{H}(q_1, q_2, \ldots, q_n, p_1, p_2, \ldots, p_n, t)\} = \sum_{i=1}^{n}\left(\frac{\partial \mathcal{H}}{\partial q_i}\dot{q}_i + \frac{\partial \mathcal{H}}{\partial p_i}\dot{p}_i\right) + \frac{\partial \mathcal{H}}{\partial t}$$

$$= \sum_{i=1}^{n}(-\dot{p}_i\dot{q}_i + \dot{q}_i\dot{p}_i) + \frac{\partial \mathcal{H}}{\partial t} = \frac{\partial \mathcal{H}}{\partial t} \Rightarrow \frac{d\mathcal{H}}{dt} = \frac{\partial \mathcal{H}}{\partial t} \tag{3.194}$$

Equation (3.194) was obtained by substitution of the Hamiltonian canonical equations (see 3.193) into the time derivative of $\mathcal{H}$. If the system's constraints are holonomic while the system itself is subjected to generalized non-conservative forces, that is, forces which are not derivable from a potential function, then the Q_i terms which appear in 3.56 must be taken into account. Equation (3.56) is repeated below for the reader's convenience.

$$\frac{d}{dt}\left(\frac{\partial T}{\partial \dot{q}_i}\right) - \frac{\partial T}{\partial q_i} + \frac{\partial V}{\partial q_i} = Q_i; i = 1, 2, \ldots, n$$

$$\Rightarrow \frac{d}{dt}\left(\frac{\partial \mathcal{L}}{\partial \dot{q}_i}\right) - \frac{\partial \mathcal{L}}{\partial q_i} = Q_i; i = 1, 2, \ldots, n \tag{3.195}$$

Since the generalized momenta p_i were defined to be (see Eq. 3.178): $p_i = \frac{\partial \mathcal{L}}{\partial \dot{q}_i}$, $i = 1, 2, \ldots, n$, then, based upon Eq. (3.195), the generalized momenta take the following form:

$$\dot{p}_i = Q_i + \frac{\partial \mathcal{L}}{\partial q_i}, \quad i = 1, 2, \ldots, n \tag{3.196}$$

and the Hamiltonian canonical equations become:

$$\dot{q}_i = \frac{\partial \mathcal{H}}{\partial p_i}; \dot{p}_i = Q_i - \frac{\partial \mathcal{H}}{\partial q_i}, \quad i = 1, 2, \ldots, n \tag{3.197}$$

In the event that m constraints are non-holonomic (the generalized coordinates are not independent), that is (see Eq. 3.116):

$$\sum_{i=1}^{n} a_{ji}\dot{q}_i + a_{j0} = 0;\ j = 1, 2, \ldots, m$$

and the generalized forces are derivable from a potential function (conservative forces), which means that $Q_i = 0\ ,\ i = 1, 2, \ldots, n$, then with the aid of Lagrange multipliers (see Eq. 3.118), the d'Alembert–Lagrange equations become:

$$\frac{d}{dt}\left(\frac{\partial \mathcal{L}}{\partial \dot{q}_i}\right) - \frac{\partial \mathcal{L}}{\partial q_i} = \sum_{l=1}^{m} \lambda_l a_{li}\ \ , i = 1, 2, \ldots, n \tag{3.198}$$

Once again the Hamiltonian canonical equations become:

$$\dot{p}_i = -\frac{\partial \mathcal{H}}{\partial q_i} + \sum_{l=1}^{m} \lambda_l a_{li};\ \dot{q}_i = \frac{\partial \mathcal{H}}{\partial p_i}\ \ , i = 1, 2, \ldots, n \tag{3.199}$$

Finally, for non-conservative forces which are derivable from Rayleigh's dissipation function F, the d'Alembert–Lagrange equations and the corresponding Hamiltonian canonical equations are:

$$\frac{d}{dt}\left(\frac{\partial \mathcal{L}}{\partial \dot{q}_i}\right) - \frac{\partial \mathcal{L}}{\partial q_i} + \frac{\partial F}{\partial \dot{q}_i} = 0;\ i = 1, 2, \ldots, n$$

$$\Rightarrow \dot{p}_i = -\frac{\partial \mathcal{H}}{\partial q_i} - \frac{\partial F}{\partial \dot{q}_i};\ \dot{q}_i = \frac{\partial \mathcal{H}}{\partial p_i}\ \ , i = 1, 2, \ldots, n \tag{3.200}$$

Outline of the Procedure to Obtain the Hamiltonian (see Greiner [15, pp. 327–331])

1. Form the Lagrangian $\mathcal{L}$
2. Calculate the vector of generalized momenta (see Eq. 3.178) $p_i = \frac{\partial \mathcal{L}}{\partial \dot{q}_i},\ i = 1, 2, \ldots, n$
3. The Hamiltonian (Eq. 3.187) is written as:

$$\mathcal{H} = \sum_{i=1}^{n} \frac{\partial \mathcal{L}(q_i, \dot{q}_i, t)}{\partial \dot{q}_i}\dot{q}_i - \mathcal{L}(q_i, \dot{q}_i, t) = \sum_{i=1}^{n} p_i\dot{q}_i - \mathcal{L}(q_i, \dot{q}_i, t)$$

4. The total differential of the Lagrangian $\mathcal{L}(q_i, \dot{q}_i, t)$ is:

$$d\mathcal{L} = \sum_{i=1}^{n} \frac{\partial \mathcal{L}}{\partial q_i}dq_i + \sum_{i=1}^{n} \frac{\partial \mathcal{L}}{\partial \dot{q}_i}d\dot{q}_i + \frac{\partial \mathcal{L}}{\partial t} \tag{3.201}$$

5. The d'Alembert–Lagrange equations for a system with conservative forces and holonomic constraints are (see Eq. (3.192)):

$$\frac{d}{dt}\left(\frac{\partial \mathcal{L}}{\partial \dot{q}_i}\right) - \frac{\partial \mathcal{L}}{\partial q_i} = 0 \Rightarrow \dot{p}_i = \frac{\partial \mathcal{L}}{\partial q_i}$$

6. The total differential of $\mathcal{H}(\mathbf{p}, \mathbf{q}, t)$, utilizing the calculated total differential of $\mathcal{L}$ in Eq. (3.201), is of the form:

$$\begin{aligned} d\mathcal{H}(\mathbf{p}, \mathbf{q}, t) &= \sum_{i=1}^{n} (p_i d\dot{q}_i + \dot{q}_i dp_i) - \sum_{i=1}^{n} \frac{\partial \mathcal{L}}{\partial q_i} dq_i - \sum_{i=1}^{n} \frac{\partial \mathcal{L}}{\partial \dot{q}_i} d\dot{q}_i - \frac{\partial \mathcal{L}}{\partial t} dt \\ &= \sum_{i=1}^{n} (p_i d\dot{q}_i + \dot{q}_i dp_i - \dot{p}_i dq_i - p_i d\dot{q}_i) - \frac{\partial \mathcal{L}}{\partial t} dt \\ &= \sum_{i=1}^{n} (\dot{q}_i dp_i - \dot{p}_i dq_i) - \frac{\partial \mathcal{L}}{\partial t} dt \end{aligned} \tag{3.202}$$

7. The canonical Hamiltonian equations are easily derived from Eq. (3.202) as follows:

$$\dot{q}_i = \frac{\partial \mathcal{H}}{\partial p_i}; \; \dot{p}_i = \frac{\partial \mathcal{L}}{\partial q_i} = -\frac{\partial \mathcal{H}}{\partial q_i}; \; \frac{\partial \mathcal{H}}{\partial t} = -\frac{\partial \mathcal{L}}{\partial t}$$

Remarks

1. For a system consisting of N particles with holonomic, scleronomic constraints, subjected to conservative internal forces, the kinetic energy is: $T = \frac{1}{2}\sum_{i=1}^{N} m_i \dot{\mathbf{r}}_i \cdot \dot{\mathbf{r}}_i \;\; , i = 1, 2, \ldots, N$.
2. Since the constraints are time independent, there exist transformations between the vector of Cartesian distances and the generalized coordinates, that is, $\mathbf{r}_l = \mathbf{r}_l(q_l), l = 1, 2, \ldots, n$. It is therefore possible to recast $\dot{\mathbf{r}}_l$ in the following form:

$$\dot{\mathbf{r}}_l = \sum_{l=1}^{n} \frac{\partial \mathbf{r}_l}{\partial q_l} \dot{q}_l \tag{3.203}$$

3. Hence, the kinetic energy becomes:

$$\begin{aligned} T &= \frac{1}{2}\sum_{i=1}^{N} m_i \left(\sum_{j=1}^{n} \frac{\partial \mathbf{r}_i}{\partial q_j} \dot{q}_j\right) \cdot \left(\sum_{k=1}^{n} \frac{\partial \mathbf{r}_i}{\partial q_k} \dot{q}_k\right) \\ &= \frac{1}{2}\left\{ m_1 \left[\frac{\partial \mathbf{r}_1}{\partial q_1} \dot{q}_1 + \frac{\partial \mathbf{r}_1}{\partial q_2} \dot{q}_1 + \cdots + \frac{\partial \mathbf{r}_1}{\partial q_n} \dot{q}_n\right]\right. \end{aligned}$$

$$\left[\frac{\partial \mathbf{r}_1}{\partial q_1}\dot{q}_1 + \frac{\partial \mathbf{r}_1}{\partial q_2}\dot{q}_1 + \cdots + \frac{\partial \mathbf{r}_1}{\partial q_n}\dot{q}_n\right]\Bigg\}$$

$$+\frac{1}{2}\left\{m_2\left[\frac{\partial \mathbf{r}_2}{\partial q_1}\dot{q}_1 + \frac{\partial \mathbf{r}_2}{\partial q_2}\dot{q}_2 + \cdots + \frac{\partial \mathbf{r}_2}{\partial q_n}\dot{q}_n\right]\right.$$

$$\left.\left[\frac{\partial \mathbf{r}_2}{\partial q_1}\dot{q}_1 + \frac{\partial \mathbf{r}_2}{\partial q_2}\dot{q}_2 + \cdots + \frac{\partial \mathbf{r}_2}{\partial q_n}\dot{q}_n\right]\right\}$$

$$\vdots$$

$$+\frac{1}{2}\left\{m_N\left[\frac{\partial \mathbf{r}_N}{\partial q_1}\dot{q}_1 + \frac{\partial \mathbf{r}_N}{\partial q_2}\dot{q}_2 + \cdots + \frac{\partial \mathbf{r}_N}{\partial q_n}\dot{q}_n\right]\right.$$

$$\left.\left[\frac{\partial \mathbf{r}_N}{\partial q_1}\dot{q}_1 + \frac{\partial \mathbf{r}_N}{\partial q_2}\dot{q}_2 + \cdots + \frac{\partial \mathbf{r}_N}{\partial q_n}\dot{q}_n\right]\right\}$$

$$= \frac{1}{2}\sum_{j=1}^{n}\sum_{k=1}^{n} m_1\left[\frac{\partial \mathbf{r}_1}{\partial q_j}\frac{\partial \mathbf{r}_1}{\partial q_k}\right]\dot{q}_j\dot{q}_k + \frac{1}{2}\sum_{j=1}^{n}\sum_{k=1}^{n} m_2\left[\frac{\partial \mathbf{r}_2}{\partial q_j}\frac{\partial \mathbf{r}_2}{\partial q_k}\right]\dot{q}_j\dot{q}_k$$

$$+\cdots+\frac{1}{2}\sum_{j=1}^{n}\sum_{k=1}^{n} m_N\left[\frac{\partial \mathbf{r}_N}{\partial q_j}\frac{\partial \mathbf{r}_N}{\partial q_k}\right]\dot{q}_j\dot{q}_k \tag{3.204}$$

4. Designating α_{jk} to be: $\alpha_{jk} = \sum_{i=1}^{N} m_i \left[\frac{\partial \mathbf{r}_i}{\partial q_j}\frac{\partial \mathbf{r}_i}{\partial q_k}\right]$ in Eq. (3.204), we arrive at:

$$T = \frac{1}{2}\sum_{j=1}^{n}\sum_{k=1}^{n}\alpha_{jk}\dot{q}_j\dot{q}_k \tag{3.205}$$

5. Since the kinetic energy term in Eq. (3.205) is a homogeneous function of order **two** with respect to the generalized velocities $\dot{q}_j$, application of Euler's theorem for homogeneous functions (see below: A Short Note on Euler's Theorem for Homogeneous Functions) to the kinetic energy term T results in:

$$\sum_{i=1}^{N}\frac{\partial T}{\partial \dot{q}_i}\cdot\dot{q}_i = 2T \tag{3.206}$$

6. However, from the definition of the Hamiltonian for a holonomic, scleronomic system with conservative forces and the definition of the Lagrangian (see Eq. 3.187), we have:

$$\mathcal{H} = \sum_{i=1}^{n}\frac{\partial \mathcal{L}}{\partial \dot{q}_i}\dot{q}_i - \mathcal{L} = \sum_{i=1}^{n}\frac{\partial T}{\partial \dot{q}_i}\cdot\dot{q}_i - (T-V) = 2T - T + V = T + V \tag{3.207}$$

For a holonomic, scleronomic system with conservative forces, the Hamiltonian turns out to be the total energy of the system. The implication of this result is that the work required in order to find the Hamiltonian need not be carried out by beginning with a determination of the Lagrangian.

A Short Note on Euler's Theorem for Homogeneous Functions

Any function $f(x)$ that possesses the characteristic mapping: $x \to \lambda x;\ f(x) \to \lambda f(x)$ is said to be homogeneous, with respect to x, to degree 1. By the same token, if $f(x)$ obeys the mapping: $x \to \lambda x;\ f(x) \to \lambda^k f(x)$, then $f(x)$ is homogeneous with respect to x to degree k. In general, a multivariable function $f(x_1, x_2, x_3, \ldots, x_k)$ is said to be homogeneous to degree n in variables $(x_1, x_2, x_3, \ldots, x_k)$ if **for any value of** $\boldsymbol{\lambda}$, $f(\lambda x_1, \lambda x_2, \ldots, \lambda x_k) = \lambda^n f(x_1, x_2, \ldots, x_k)$. It shall now be demonstrated that for a homogeneous function to degree n,

$$\sum_{i=1}^{k} x_i \frac{\partial f(x_1, x_2, \ldots, x_k)}{\partial x_i} = nf(x_1, x_2, \ldots, x_k)$$

Proof

$$f(\lambda x_1, \lambda x_2, \ldots, \lambda x_k) = \lambda^n f(x_1, x_2, \ldots, x_k)$$

$$\frac{\partial}{\partial \lambda}\left\{\lambda^n f(x_1, x_2, \ldots, x_k)\right\} = n\lambda^{n-1} f(x_1, x_2, \ldots, x_k)$$

$$\frac{\partial}{\partial \lambda}\{f(\lambda x_1, \lambda x_2, \ldots, \lambda x_k)\} =$$

$$\frac{\partial\{f(\lambda x_1, \lambda x_2, \ldots, \lambda x_k)\}}{\partial \lambda x_1}\frac{\partial \lambda x_1}{\partial \lambda} + \frac{\partial\{f(\lambda x_1, \lambda x_2, \ldots, \lambda x_k)\}}{\partial \lambda x_2}\frac{\partial \lambda x_2}{\partial \lambda} + \ldots$$

$$+ \frac{\partial\{f(\lambda x_1, \lambda x_2, \ldots, \lambda x_k)\}}{\partial \lambda x_k}\frac{\partial \lambda x_k}{\partial \lambda}$$

$$= \frac{\partial\{f(\lambda x_1, \lambda x_2, \ldots, \lambda x_k)\}}{\partial \lambda x_1} x_1 + \frac{\partial\{f(\lambda x_1, \lambda x_2, \ldots, \lambda x_k)\}}{\partial \lambda x_2} x_2 + \cdots$$

$$+ \frac{\partial\{f(\lambda x_1, \lambda x_2, \ldots, \lambda x_k)\}}{\partial \lambda x_k} x_k = n\lambda^{n-1} f(x_1, x_2, \ldots, x_k)$$

$$\lambda = 1;\ \frac{\partial\{f(x_1, x_2, \ldots, x_k)\}}{\partial x_1} x_1 + \frac{\partial\{f(x_1, x_2, \ldots, x_k)\}}{\partial x_2} x_2 + \ldots$$

$$+ \frac{\partial\{f(x_1, x_2, \ldots, x_k)\}}{\partial x_k} x_k = nf(x_1, x_2, \ldots, x_k) \tag{3.208}$$

Example 18: Simple Pendulum (see Greiner [15, pp. 332–334])

The solution of the simple pendulum problem was obtained in **Example 3** by means of the d'Alembert–Lagrange formula (see Example 3.7 Eq. (3.64)) and the derived dynamical equation was:

$$m_1 h_1^2 \ddot{\theta}_1 + m_1 g h_1 \sin\theta_1 = 0 \Rightarrow \ddot{\theta}_1 + \frac{g}{h_1}\sin\theta_1 = 0 \tag{3.209}$$

The Lagrangian $\mathcal{L}$ was shown to be:

$$\mathcal{L} = T - V = \frac{1}{2} m_1 h_1^2 (\dot{\theta}_1)^2 + m_1 g h_1 \cos\theta_1 \tag{3.210}$$

From the Lagrangian, with generalized coordinate $q_i = \theta_1$, the generalized momentum is:

$$p_{\theta_1} = \frac{\partial \mathcal{L}}{\partial \dot{\theta}_1} = m_1 h_1^2 \dot{\theta}_1 \Rightarrow \dot{p}_{\theta_1} = m_1 h_1^2 \ddot{\theta}_1 \tag{3.211}$$

The procedure for deriving the dynamics by means of the Hamiltonian is now presented in outline form as follows:

1. The Hamiltonian $\mathcal{H}$ is:

$$\mathcal{H} = \sum_{i=1}^{n} \frac{\partial \mathcal{L}}{\partial \dot{q}_i}\dot{q}_i - \mathcal{L} = \dot{\theta}_1 \frac{\partial}{\partial \dot{\theta}_1}\left\{\frac{1}{2} m_1 h_1^2 (\dot{\theta}_1)^2 + m_1 g h_1 \cos\theta_1\right\}$$
$$- \frac{1}{2} m_1 h_1^2 (\dot{\theta}_1)^2 - m_1 g h_1 \cos\theta_1$$
$$= m_1 h_1^2 (\dot{\theta}_1)^2 - \frac{1}{2} m_1 h_1^2 (\dot{\theta}_1)^2 - m_1 g h_1 \cos\theta_1 = \frac{1}{2} m_1 h_1^2 (\dot{\theta}_1)^2 - m_1 g h_1 \cos\theta_1$$

2. The canonical Hamiltonian equations were presented above in the form:

$$\dot{q}_i = \frac{\partial \mathcal{H}}{\partial p_i};\ \dot{p}_i = \frac{\partial \mathcal{L}}{\partial q_i} = -\frac{\partial \mathcal{H}}{\partial q_i};\ p_i = \frac{\partial \mathcal{L}}{\partial \dot{q}_i}$$

and so the canonical equations become:

$$\dot{q}_i = \dot{\theta}_1;\ \dot{p}_i = \frac{\partial \mathcal{L}}{\partial q_i} = -\frac{\partial \mathcal{H}}{\partial q_i}$$
$$-\frac{\partial \mathcal{H}}{\partial \theta_1} = \frac{\partial}{\partial \theta_1}\left\{-\frac{1}{2} m_1 h_1^2 (\dot{\theta}_1)^2 + m_1 g h_1 \cos\theta_1\right\} = -m_1 g h_1 \sin\theta_1$$
$$\Rightarrow m_1 h_1^2 \ddot{\theta}_1 = -m_1 g h_1 \sin\theta_1 \Rightarrow m_1 h_1^2 \ddot{\theta}_1 + m_1 g h_1 \sin\theta_1 = 0$$

which is identical to Eq. (3.209).

Example 19: Central Motion (see Greiner [15, pp. 331–332])

Assume that a particle with mass m is subjected to the action of a single central force. A central force is a force that points from the particle directly towards a fixed point in space, the center of rotation, and whose magnitude only depends on the distance of the object to the center, that is, $V(r)$, where r is the distance of the particle to the center or origin of the force. The particle is undergoing planar motion while the force is radial and the motion can therefore be described by the two polar coordinates r, ϕ (see the following figure). The solution procedure is the following:

1. Calculate the Lagrangian $\mathcal{L}$:

$$\begin{aligned}
\mathbf{r} &= r\cos\phi\hat{i} + r\sin\phi\hat{j} \\
\mathbf{v} &= \dot{r}\cos\phi\hat{i} - r\dot{\phi}\sin\phi\hat{i} + \dot{r}\sin\phi\hat{j} + r\dot{\phi}\cos\phi\hat{j} \\
&= \hat{i}\left(\dot{r}\cos\phi - r\dot{\phi}\sin\phi\right) + \hat{j}\left(\dot{r}\sin\phi + r\dot{\phi}\cos\phi\right) \\
|\mathbf{v}|^2 &= \mathbf{v}^T\cdot\mathbf{v} = \dot{r}^2(\cos^2\phi + \sin^2\phi) - 2r\dot{r}\dot{\phi}\cos\phi\sin\phi \\
&\quad +2r\dot{r}\dot{\phi}\cos\phi\sin\phi + r^2\dot{\phi}^2\sin^2\phi + r^2\dot{\phi}^2\cos^2\phi \\
&= \dot{r}^2 + r^2\dot{\phi}^2
\end{aligned}$$

2. The Lagrangian is:

$$\mathcal{L} = T - V(r) = \frac{1}{2}m|\mathbf{v}|^2 - V(r) = \frac{1}{2}m\left(\dot{r}^2 + r^2\dot{\phi}^2\right) - V(r)$$

3. The two generalized momenta (corresponding to the two generalized velocities $\dot{r}, \dot{\phi}$, respectively) are:

$$p_r = \frac{\partial\mathcal{L}}{\partial\dot{r}} = m\dot{r};\ p_\phi = \frac{\partial\mathcal{L}}{\partial\dot{\phi}} = mr^2\dot{\phi}$$

4. The Hamiltonian $\mathcal{H}$ becomes:

$$\begin{aligned}
\mathcal{H} &= \sum_{i=1}^{2}\frac{\partial\mathcal{L}}{\partial\dot{q}_i}\dot{q}_i - \mathcal{L} = \frac{\partial\mathcal{L}}{\partial\dot{r}}\dot{r} + \frac{\partial\mathcal{L}}{\partial\dot{\phi}}\dot{\phi} - \frac{1}{2}m\left(\dot{r}^2 + r^2\dot{\phi}^2\right) + V(r) \\
&= m\dot{r}^2 + mr^2\dot{\phi}^2 - \frac{1}{2}m\left(\dot{r}^2 + r^2\dot{\phi}^2\right) + V(r) = \frac{1}{2}m\left(\dot{r}^2 + r^2\dot{\phi}^2\right) + V(r) \\
&= \frac{p_r^2}{2m} + \frac{p_\phi^2}{2mr^2} + V(r)
\end{aligned}$$

5. The components $\dot{r}, \dot{\phi}$ are calculated from:

$$\dot{r} = \frac{\partial\mathcal{H}}{\partial p_r} = \frac{p_r}{m};\ \dot{\phi} = \frac{\partial\mathcal{H}}{\partial p_\phi} = \frac{p_\phi}{mr^2}$$

6. In similar fashion, the components $\dot{p}_r$, $\dot{p}_\phi$ are calculated from:

$$\dot{p}_r = -\frac{\partial \mathcal{H}}{\partial r} = \frac{p_\phi^2}{2mr^3} - \frac{\partial V(r)}{\partial r}; \; \dot{p}_\phi = -\frac{\partial \mathcal{H}}{\partial \phi} = 0$$

The result above ($\frac{\partial \mathcal{H}}{\partial \phi} = 0$) indicates that for motion subjected to a central force, the angular momentum is preserved This statement may be understood by transformation of the coordinates from Cartesian to polar and evaluating the acceleration in the transformed coordinate frame as follows:

(a) The velocity in radial coordinates is: $\mathbf{v} = \dot{r}\hat{i}_r + r\dot{\phi}\hat{i}_\phi$
(b) The acceleration may be derived as follows:

$$\frac{d\hat{i}_r}{dt} = \dot{\phi}\hat{i}_\phi; \; \frac{d\hat{i}_\phi}{dt} = -\dot{\phi}\hat{i}_r$$

$$\begin{aligned}\mathbf{a} = \frac{d\mathbf{v}}{dt} &= \ddot{r}\hat{i}_r + \dot{r}\frac{d\hat{i}_r}{dt} + \dot{r}\dot{\phi}\hat{i}_\phi + r\ddot{\phi}\hat{i}_\phi + r\dot{\phi}\frac{d\hat{i}_\phi}{dt} \\ &= \hat{i}_r\left(\ddot{r} - r\dot{\phi}^2\right) + \hat{i}_\phi\left(r\ddot{\phi} + 2\dot{\phi}\dot{r}\right)\end{aligned}$$

(c) Since the problem involves a central force, there is no acceleration directed along the $\hat{i}_\phi$ axis, only along the $\hat{i}_r$ or the radial axis. This implies that

$$m\left(r\ddot{\phi} + 2\dot{r}\dot{\phi}\right) = 0 \Rightarrow \frac{d}{dt}\left(r^2\dot{\phi}\right) = r\left(r\ddot{\phi} + 2\dot{r}\dot{\phi}\right) = 0$$

(d) Note that the term $\left(mr^2\dot{\phi}\right)$ is the angular momentum of the system and it remains constant throughout the trajectory of the mass m.

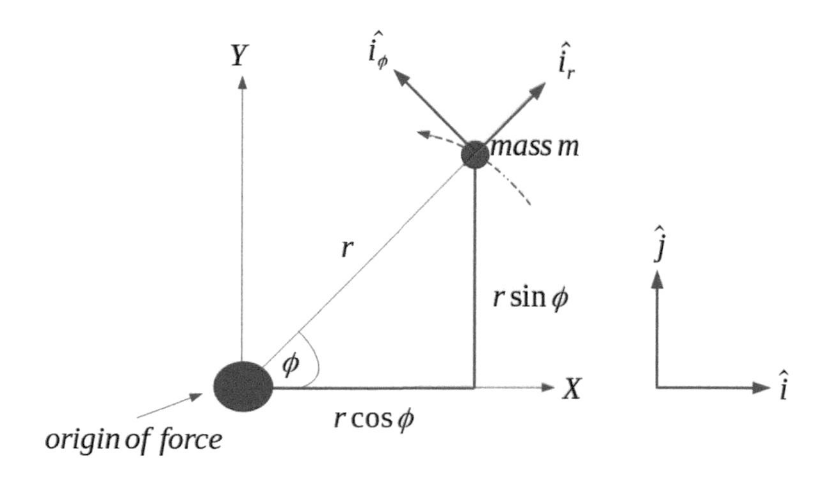

Chapter 4
Quasi-Coordinates and Quasi-Velocities

Lagrangian dynamical systems with non-holonomic constraints is the main topic of this chapter. This category of systems is first transformed into a "standard" d'Alembert–Lagrangian form by means of a set of Lagrangian multipliers applied to the constraints. The concepts of quasi-coordinates and quasi-velocities are then introduced. While quasi-coordinates are not necessarily related to physical generalized coordinates, quasi-velocities do have some physical significance. The d'Alembert–Lagrange method for the derivation of dynamical equations is appropriately modified in order to incorporate the non-holonomic constraints, in the form of relations on quasi-velocity constraints. Two examples which serve to illustrate the procedure are provided, the second example being more significant as it is the prototype of a wide variety of three-wheeled mobile robots. The main drawback of the aforementioned procedure, as portrayed by the given examples, is the effort involved in performing the required transformations and matrix operations. A simpler and more straightforward approach involves the transformation of the Lagrangian and its calculation in body-centered non-inertial coordinates, where the non-holonomic constraints in the form of quasi-velocities are incorporated directly into the modified d'Alembert–Lagrange equations. Two non-trivial examples which use the transformed d'Alembert–Lagrange equations are calculated, the first involving the derivation of a six degree of freedom set of motion equations for a rigid aircraft, while in the second, the equations of motion of a quadcopter vehicle are obtained.

A. W. Pila, *Introduction To Lagrangian Dynamics*,
https://doi.org/10.1007/978-3-030-22378-6_4

4.1 Definitions and Recapitulation

The transformation from one coordinate frame to another, namely from inertial to body axes (and vice versa) was presented and discussed in Chap. 1, Eq. 1.18 and is repeated here for velocities as follows:

$$\begin{bmatrix} u \\ v \\ w \end{bmatrix} = \underbrace{\begin{bmatrix} \alpha_{11} & \alpha_{12} & \alpha_{13} \\ \alpha_{21} & \alpha_{22} & \alpha_{23} \\ \alpha_{31} & \alpha_{32} & \alpha_{33} \end{bmatrix}}_{T_{B/I}} \begin{bmatrix} u_I \\ v_I \\ w_I \end{bmatrix}; \quad \begin{bmatrix} u_I \\ v_I \\ w_I \end{bmatrix} = \underbrace{\begin{bmatrix} \alpha_{11} & \alpha_{21} & \alpha_{31} \\ \alpha_{12} & \alpha_{22} & \alpha_{33} \\ \alpha_{13} & \alpha_{23} & \alpha_{33} \end{bmatrix}}_{T_{I/B}} \begin{bmatrix} u \\ v \\ w \end{bmatrix} \tag{4.1}$$

where u_I, v_I, and w_I are the inertial velocities of a given body and u, v, w are the body's velocities in body-centered coordinates. The Euler transformation from body to inertial components is simply the matrix transpose of the Euler matrix in Eq. 1.19 and turns out to be:

$$\begin{bmatrix} u_I \\ v_I \\ w_I \end{bmatrix} = \begin{bmatrix} \cos\psi\cos\theta & \cos\psi\sin\phi\sin\theta - \cos\phi\sin\psi & \sin\phi\sin\psi + \cos\phi\cos\psi\sin\theta \\ \cos\theta\sin\psi & \cos\phi\cos\psi + \sin\phi\sin\psi\sin\theta & \cos\phi\sin\psi\sin\theta - \cos\psi\sin\phi \\ -\sin\theta & \cos\theta\sin\phi & \cos\phi\cos\theta \end{bmatrix} \begin{bmatrix} u \\ v \\ w \end{bmatrix}$$

$$\Rightarrow \begin{bmatrix} u_I \\ v_I \\ w_I \end{bmatrix} = \underbrace{\begin{bmatrix} \alpha_{11} & \alpha_{21} & \alpha_{31} \\ \alpha_{12} & \alpha_{22} & \alpha_{33} \\ \alpha_{13} & \alpha_{23} & \alpha_{33} \end{bmatrix}}_{T_{I/B}} \begin{bmatrix} u \\ v \\ w \end{bmatrix}$$

$$\Rightarrow \begin{bmatrix} u \\ v \\ w \end{bmatrix} = \underbrace{\begin{bmatrix} \alpha_{11} & \alpha_{12} & \alpha_{13} \\ \alpha_{21} & \alpha_{22} & \alpha_{23} \\ \alpha_{31} & \alpha_{32} & \alpha_{33} \end{bmatrix}}_{T_{B/I}} \begin{bmatrix} u_I \\ v_I \\ w_I \end{bmatrix} \tag{4.2}$$

It then follows that:

$$\begin{bmatrix} \frac{\partial u}{\partial u_I} & \frac{\partial v}{\partial u_I} & \frac{\partial w}{\partial u_I} \\ \frac{\partial u}{\partial v_I} & \frac{\partial v}{\partial v_I} & \frac{\partial v}{\partial v_I} \\ \frac{\partial u}{\partial w_I} & \frac{\partial v}{\partial w_I} & \frac{\partial w}{\partial w_I} \end{bmatrix} = \begin{bmatrix} \alpha_{11} & \alpha_{21} & \alpha_{31} \\ \alpha_{12} & \alpha_{22} & \alpha_{33} \\ \alpha_{13} & \alpha_{23} & \alpha_{33} \end{bmatrix} = T_{I/B} \tag{4.3}$$

The d'Alembert–Lagrange equation expresses in summary form the principle of virtual work and may be written as:

$$\sum_{e=1}^{n} \left[Q_e - \left\{ \frac{d}{dt}\left(\frac{\partial T}{\partial \dot{q}_e}\right) - \frac{\partial T}{\partial q_e} \right\} \right] \delta q_e = 0 \tag{4.4}$$

where T is the kinetic energy of a given system. The above equation is valid for both holonomic and non-holonomic constraints. If the variations δq_e were independent of each other, the terms within the square brackets would be zero for each e and hence the equation would revert to the standard equation for Lagrangian dynamics. Since this is not the case for non-holonomic constraints, no simplification can occur because some of the δq_e terms are dependent on others.

In the above equation it was assumed that all forces which were derived from a potential function were accounted for in the Lagrangian (kinetic energy terms) and all nonpotential forces were designated as Q_e. The generalized coordinates are related by non-holonomic (and scleronomic or time independent) constraints of the type:

$$\sum_{e=1}^{n} a_{le} dq_e = 0; \quad l = 1, 2, \ldots, m \tag{4.5}$$

where a_{le} $(e = 0, 1, 2, \ldots, n)$ are functions of the coordinates δq_e. It then follows from the above differential that the virtual displacements are not independent, but related and of the form:

$$\sum_{e=1}^{n} a_{le} \delta q_e = 0; \quad l = 1, 2, \ldots, m \tag{4.6}$$

Multiplying each sum $\sum_{e=1}^{n} a_{le}\delta q_e = 0; \quad l = 1, 2, \ldots, m$ by λ_l and summing over $e = 1, 2, \ldots, n$ results in (see Sect. 3.9, pp. 118–121):

$$\begin{aligned}
&\{\lambda_1 (a_{11}\delta q_1 + a_{12}\delta q_2 + \cdots + a_{1n}\delta q_n) + \lambda_2 (a_{21}\delta q_1 + a_{22}\delta q_2 + \cdots + a_{2n}\delta q_n) \\
&\quad + \cdots + \lambda_m (a_{m1}\delta q_1 + a_{m2}\delta q_2 + \cdots + a_{mn}\delta q_n)\}\delta q_e = \sum_{e=1}^{n} \left(\sum_{l=1}^{m} \lambda_l a_{le} \right) \delta q_e
\end{aligned} \tag{4.7}$$

While the virtual displacements δ_e are still not independent, a judicious choice of the multipliers $\lambda_l; \quad l = 1, 2, \ldots, m$ will render the bracketed coefficients of $\delta q_e, \ e = n - m + 1, \ldots, n$ to be 0. The remaining δq_e are independent so that they can be chosen at will, and thus, the coefficients of $\delta q_e, \ e = 1, 2, \ldots, n - m$ are zero. Substituting the results of Eq. 4.7 into the d'Alembert–Lagrange equation 4.4, we have:

$$\begin{aligned}
&\sum_{e=1}^{n} \left[Q_e - \left\{ \frac{d}{dt} \left(\frac{\partial T}{\partial \dot{q}_e} \right) - \frac{\partial T}{\partial q_e} - \left(\sum_{l=1}^{m} \lambda_l a_{le} \right) \right\} \right] \delta q_e = 0 \\
&\Rightarrow \left[Q_1 - \left\{ \frac{d}{dt} \left(\frac{\partial T}{\partial \dot{q}_1} \right) - \frac{\partial T}{\partial q_1} - \left(\sum_{l=1}^{m} \lambda_l a_{l1} \right) \right\} \right] \delta q_1
\end{aligned}$$

$$+\left[Q_2-\left\{\frac{d}{dt}\left(\frac{\partial T}{\partial \dot{q}_2}\right)-\frac{\partial T}{\partial q_2}-\left(\sum_{l=1}^{m}\lambda_2 a_{l2}\right)\right\}\right]\delta q_2$$
$$+\cdots+\left[Q_n-\left\{\frac{d}{dt}\left(\frac{\partial T}{\partial \dot{q}_n}\right)-\frac{\partial T}{\partial q_n}-\left(\sum_{l=1}^{m}\lambda_l a_{ln}\right)\right\}\right]\delta q_n=0 \tag{4.8}$$

Letting $Q'_e=\sum_{l=1}^{m}\lambda_l a_{le},\quad e=1,2,\ldots,n$, the above set of d'Alembert–Lagrange equations may be written as:

$$\sum_{e=1}^{n}\left[Q_e-\left\{\frac{d}{dt}\left(\frac{\partial T}{\partial \dot{q}_e}\right)-\frac{\partial T}{\partial q_e}-Q'_e\right\}\right]\delta q_e=0$$
$$\Rightarrow\left[Q_1-\left\{\frac{d}{dt}\left(\frac{\partial T}{\partial \dot{q}_1}\right)-\frac{\partial T}{\partial q_1}-Q'_1\right\}\right]\delta q_1$$
$$+\left[Q_2-\left\{\frac{d}{dt}\left(\frac{\partial T}{\partial \dot{q}_2}\right)-\frac{\partial T}{\partial q_2}-Q'_2\right\}\right]\delta q_2$$
$$+\ldots\left[Q_n-\left\{\frac{d}{dt}\left(\frac{\partial T}{\partial \dot{q}_n}\right)-\frac{\partial T}{\partial q_n}-Q'_n\right\}\right]\delta q_n=0 \tag{4.9}$$

Setting the sum $Q_e+Q'_e=\mathcal{Q}_e$, the collection of d'Alembert–Lagrange equations becomes:

$$\sum_{e=1}^{n}\left[\mathcal{Q}_e-\left\{\frac{d}{dt}\left(\frac{\partial T}{\partial \dot{q}_e}\right)-\frac{\partial T}{\partial q_e}\right\}\right]\delta q_e=0$$
$$\Rightarrow\left[\mathcal{Q}_1-\left\{\frac{d}{dt}\left(\frac{\partial T}{\partial \dot{q}_1}\right)-\frac{\partial T}{\partial q_1}\right\}\right]\delta q_1+\left[\mathcal{Q}_2-\left\{\frac{d}{dt}\left(\frac{\partial T}{\partial \dot{q}_2}\right)-\frac{\partial T}{\partial q_2}\right\}\right]\delta q_2$$
$$+\ldots\left[\mathcal{Q}_n-\left\{\frac{d}{dt}\left(\frac{\partial T}{\partial \dot{q}_n}\right)-\frac{\partial T}{\partial q_n}\right\}\right]\delta q_n=0 \tag{4.10}$$

It then follows that:

$$\left[\frac{d}{dt}\left(\frac{\partial T}{\partial \dot{q}_1}\right)-\frac{\partial T}{\partial q_1}-\mathcal{Q}_1\right]\delta q_1=0$$
$$\left[\frac{d}{dt}\left(\frac{\partial T}{\partial \dot{q}_2}\right)-\frac{\partial T}{\partial q_2}-\mathcal{Q}_2\right]\delta q_2=0$$

$$\vdots$$

$$\left[\frac{d}{dt}\left(\frac{\partial T}{\partial \dot{q}_n}\right) - \frac{\partial T}{\partial q_n} - Q_n\right]\delta q_n = 0 \tag{4.11}$$

The above set of d'Alembert–Lagrange equations may be represented in matrix form as follows:

$$\{\delta q\}^T\left(\frac{d}{dt}\left\{\frac{\partial T}{\partial \dot{q}}\right\} - \left\{\frac{\partial T}{\partial q}\right\} - \{Q\}\right) = 0 \tag{4.12}$$

where

$$\left\{\frac{\partial T}{\partial \dot{q}}\right\} = \begin{bmatrix} \left(\frac{\partial T}{\partial \dot{q}_1}\right) \\ \left(\frac{\partial T}{\partial \dot{q}_2}\right) \\ \vdots \\ \left(\frac{\partial T}{\partial \dot{q}_n}\right) \end{bmatrix}; \quad \left\{\frac{\partial T}{\partial q}\right\} = \begin{bmatrix} \frac{\partial T}{\partial q_1} \\ \frac{\partial T}{\partial q_2} \\ \vdots \\ \frac{\partial T}{\partial q_n} \end{bmatrix}; \quad \{Q\} = \begin{bmatrix} Q_1 \\ Q_2 \\ \vdots \\ Q_n \end{bmatrix}$$

and where furthermore:

$$\{q\} = \begin{bmatrix} q_1 \\ q_2 \\ \vdots \\ q_n \end{bmatrix}; \quad \{\dot{q}\} = \begin{bmatrix} \dot{q}_1 \\ \dot{q}_2 \\ \vdots \\ \dot{q}_n \end{bmatrix}; \quad \{\delta q\} = \begin{bmatrix} \delta q_1 \\ \delta q_2 \\ \vdots \\ \delta q_n \end{bmatrix}$$

4.2 Quasi-Coordinates and Quasi-Velocities

The following two sections are based upon the paper by Cameron and Book [8].

By definition, the instantaneous position of a system may be unambiguously described in terms of its generalized coordinates q_j. We can accordingly describe the instantaneous velocity of a system in terms of its generalized velocities $\dot{q}_j$. However, it turns out that this characterization of velocity is not unique. Consider, for the moment, angular motion. The three Euler angles ψ, θ, ϕ could be the designated generalized coordinates and the angular velocity components $[p_b, q_b, r_b]^T$ could be used to describe the system's angular velocity. However, with the exception of planar motion, an angular velocity component is seldom the rate at which an Euler angle varies. The relationship between the body's angular rates $[p_b, q_b, r_b]^T$ and the Euler angular rates $\dot{\psi}, \dot{\theta}, \dot{\phi}$ is the following:

$$\begin{bmatrix} p_b \\ q_b \\ r_b \end{bmatrix} = \begin{bmatrix} \dot{\phi} - \dot{\psi}\sin\theta \\ \dot{\theta}\cos\phi + \dot{\psi}\cos\theta\sin\phi \\ -\dot{\theta}\sin\phi + \dot{\psi}\cos\theta\cos\phi \end{bmatrix} \tag{4.13}$$

Rate variables such as p_b are called *quasi-velocities* and are denoted by $\dot{\gamma}_j$. Accordingly the symbol γ_j is termed a *quasi-coordinate*. The prefix "quasi" signifies that γ_j need not necessarily have any relation to a physical generalized coordinate, whereas the $\dot{\gamma}_j$ parameters will have some physical significance (see Ginsburg [13, pp. 589]).

Quasi-velocities $\dot{\gamma}_s \quad s = 1, 2, \ldots, N$ may be modeled as linear functions of the time derivatives of the generalized coordinates and are of the form:

$$\dot{\gamma}_s = \sum_{j=1}^{N} \Theta_{js}\dot{q}_j = \Theta_{1s}\dot{q}_1 + \Theta_{2s}\dot{q}_2 + \cdots + \Theta_{Ns}\dot{q}_N \quad s = 1, 2, \ldots, N \tag{4.14}$$

which implies that:

$$\begin{aligned} \dot{\gamma}_1 &= \Theta_{11}\dot{q}_1 + \Theta_{21}\dot{q}_2 + \Theta_{31}\dot{q}_3 + \cdots + \Theta_{N1}\dot{q}_N \\ \dot{\gamma}_2 &= \Theta_{12}\dot{q}_1 + \Theta_{22}\dot{q}_2 + \Theta_{32}\dot{q}_3 + \cdots + \Theta_{N2}\dot{q}_N \\ \dot{\gamma}_3 &= \Theta_{13}\dot{q}_1 + \Theta_{23}\dot{q}_2 + \Theta_{33}\dot{q}_3 + \cdots + \Theta_{N3}\dot{q}_N \\ &\vdots \\ \dot{\gamma}_N &= \Theta_{1N}\dot{q}_1 + \Theta_{2N}\dot{q}_2 + \Theta_{3N}\dot{q}_3 + \cdots + \Theta_{NN}\dot{q}_N \end{aligned} \tag{4.15}$$

or in matrix form:

$$\begin{bmatrix} \dot{\gamma}_1 \\ \dot{\gamma}_2 \\ \dot{\gamma}_3 \\ \vdots \\ \dot{\gamma}_N \end{bmatrix} = \underbrace{\begin{bmatrix} \Theta_{11} & \Theta_{21} & \Theta_{31} & \cdots & \Theta_{N1} \\ \Theta_{12} & \Theta_{22} & \Theta_{32} & \cdots & \Theta_{N2} \\ \Theta_{13} & \Theta_{23} & \Theta_{33} & \cdots & \Theta_{N3} \\ \vdots & \vdots & \vdots & \cdots & \vdots \\ \Theta_{1N} & \Theta_{2N} & \Theta_{3N} & \cdots & \Theta_{NN} \end{bmatrix}}_{[\vartheta]^T} \begin{bmatrix} \dot{q}_1 \\ \dot{q}_2 \\ \dot{q}_3 \\ \vdots \\ \dot{q}_N \end{bmatrix}; \quad \begin{bmatrix} \dot{\gamma}_1 \\ \dot{\gamma}_2 \\ \dot{\gamma}_3 \\ \vdots \\ \dot{\gamma}_N \end{bmatrix} = \{\dot{\Gamma}\} \tag{4.16}$$

where the coefficients Θ_{js} are known functions of the generalized coordinates $q_k,\ k = 1, 2, \ldots, N$. The above matrix equation may be rewritten in the following abbreviated form:

$$\{\dot{\Gamma}\} = [\vartheta]^T \{\dot{q}\} \tag{4.17}$$

Assuming that $[\vartheta]$ is invertible, that is, $\left[\vartheta^T\right]^{-1} = [\beta]$, will allow for a solution to be obtained for $\{\dot{q}\}$ by writing:

$$\{\dot{q}\} = [\beta] \{\dot{\Gamma}\} \tag{4.18}$$

It may easily be shown that when $[\beta][\vartheta]^T = [I]$, then $[\vartheta]^T [\beta] = [I]$ and $[\beta]^T [\vartheta] = [I]$ as follows:

$$\begin{aligned} &[\beta][\vartheta]^T = [I] \Rightarrow [\beta][\vartheta]^T [\beta] = [\beta] \Rightarrow [\vartheta]^T [\beta] = [I] \\ &[\vartheta]^T [\beta] = [I] \Rightarrow \left([\vartheta]^T [\beta]\right)^T = [\beta]^T [\vartheta] = [I] \end{aligned} \tag{4.19}$$

The relationship between $[\beta]$ and $[\vartheta]$, as noted in Eq. 4.19, will allow the kinetic energy T to be expressed as a function of the generalized coordinates $q_k,\ k = 1, 2, \ldots, N$ and the quasi-velocities $\dot{\gamma}_s,\ s = 1, 2, \ldots, N$. The unmodified kinetic energy is a function of the generalized coordinates and their time derivatives of the form $T = T(q; \dot{q})$ and it is transformed to $\bar{\boldsymbol{T}}(q; \dot{\Gamma})$, by replacing the generalized velocity vector $\dot{q}$ by the vector of quasi-velocities $\dot{\Gamma}$, that is: $T(q; \dot{q}) \rightarrow \bar{\boldsymbol{T}}(q; \dot{\Gamma})$.

The d'Alembert–Lagrange equations repeated below for convenience are:

$$\{\delta q\}^T \left(\frac{d}{dt} \left\{ \frac{\partial T}{\partial \dot{q}} \right\} - \left\{ \frac{\partial T}{\partial q} \right\} - \{Q\} \right) = 0$$

The component parts of the above equation include the following vector terms:

$$\{\partial T / \partial \dot{q}\} \text{ and } \{\partial T / \partial q\}$$

Recalling that $\bar{\boldsymbol{T}}$ was formed from T by replacing the generalized velocities $\dot{q}_k$ in T, with the quasi-velocities $\dot{\gamma}_k$, implies that T and $\bar{\boldsymbol{T}}$ are intimately related, that is, $T(q; \dot{q}) \rightarrow \bar{\boldsymbol{T}}(q; \dot{\Gamma})$. Hence the derivative $\{\partial T/\partial \dot{q}\}$ for any index k may be written as:

$$\frac{\partial T}{\partial \dot{q}_k} = \sum_{i=1}^{N} \frac{\partial \bar{\boldsymbol{T}}}{\partial \dot{\gamma}_i} \frac{\partial \dot{\gamma}_i}{\partial \dot{q}_k} = \sum_{i=1}^{N} \Theta_{ki} \frac{\partial \bar{\boldsymbol{T}}}{\partial \dot{\gamma}_i}, \quad k = 1, 2, \ldots, N \tag{4.20}$$

The foregoing is a valid operation since:

$$\frac{\partial T}{\partial \dot{q}_k} = \frac{\partial \bar{\boldsymbol{T}}}{\partial \dot{\gamma}_1} \frac{\partial \dot{\gamma}_1}{\partial \dot{q}_k} + \frac{\partial \bar{\boldsymbol{T}}}{\partial \dot{\gamma}_2} \frac{\partial \dot{\gamma}_2}{\partial \dot{q}_k} + \cdots + \frac{\partial \bar{\boldsymbol{T}}}{\partial \dot{\gamma}_N} \frac{\partial \dot{\gamma}_N}{\partial \dot{q}_k} \tag{4.21}$$

and

$$\dot{\gamma}_1 = \Theta_{11}\dot{q}_1 + \Theta_{21}\dot{q}_2 + \cdots + \Theta_{N1}\dot{q}_N$$

$$\dot{\gamma}_2 = \Theta_{12}\dot{q}_1 + \Theta_{22}\dot{q}_2 + \cdots + \Theta_{N2}\dot{q}_N$$

$$\dot{\gamma}_3 = \Theta_{13}\dot{q}_1 + \Theta_{23}\dot{q}_2 + \cdots + \Theta_{N3}\dot{q}_N$$

$$\vdots$$

$$\dot{\gamma}_N = \Theta_{1N}\dot{q}_1 + \Theta_{2N}\dot{q}_2 + \cdots + \Theta_{NN}\dot{q}_N \tag{4.22}$$

It then follows that:

$$\frac{\partial \dot{\gamma}_1}{\partial \dot{q}_k} = \Theta_{k1},\ \frac{\partial \dot{\gamma}_2}{\partial \dot{q}_k} = \Theta_{k2},\ \frac{\partial \dot{\gamma}_3}{\partial \dot{q}_k} = \Theta_{k3}, \ldots, \frac{\partial \dot{\gamma}_i}{\partial \dot{q}_k} = \Theta_{ki}, \ldots, \frac{\partial \dot{\gamma}_N}{\partial \dot{q}_k} = \Theta_{kN} \tag{4.23}$$

and hence $\frac{\partial T}{\partial \dot{q}_k} = \frac{\partial \bar{\boldsymbol{T}}}{\partial \dot{\gamma}_1}\frac{\partial \dot{\gamma}_1}{\partial \dot{q}_k} + \frac{\partial \bar{\boldsymbol{T}}}{\partial \dot{\gamma}_2}\frac{\partial \dot{\gamma}_2}{\partial \dot{q}_k} + \cdots + \frac{\partial \bar{\boldsymbol{T}}}{\partial \dot{\gamma}_N}\frac{\partial \dot{\gamma}_N}{\partial \dot{q}_k}$ becomes:

$$\frac{\partial T}{\partial \dot{q}_k} = \frac{\partial \bar{\boldsymbol{T}}}{\partial \dot{\gamma}_1}\Theta_{k1} + \frac{\partial \bar{\boldsymbol{T}}}{\partial \dot{\gamma}_2}\Theta_{k2} + \cdots + \frac{\partial \bar{\boldsymbol{T}}}{\partial \dot{\gamma}_N}\Theta_{kN} = \sum_{i=1}^{N} \frac{\partial \bar{\boldsymbol{T}}}{\partial \dot{\gamma}_i}\Theta_{ki} \tag{4.24}$$

Writing the $\partial T/\partial \dot{q}_k,\ k = 1, 2, \ldots, N$ terms as a column vector:

$$\begin{bmatrix} \frac{\partial T}{\partial \dot{q}_1} \\ \frac{\partial T}{\partial \dot{q}_2} \\ \frac{\partial T}{\partial \dot{q}_3} \\ \vdots \\ \frac{\partial T}{\partial \dot{q}_N} \end{bmatrix} = \left\{ \frac{\partial T}{\partial \dot{q}} \right\} = \begin{bmatrix} \sum_{i=1}^{N} \Theta_{1i} \frac{\partial \bar{\boldsymbol{T}}}{\partial \dot{\gamma}_i} \\ \sum_{i=1}^{N} \Theta_{2i} \frac{\partial \bar{\boldsymbol{T}}}{\partial \dot{\gamma}_i} \\ \sum_{i=1}^{N} \Theta_{3i} \frac{\partial \bar{\boldsymbol{T}}}{\partial \dot{\gamma}_i} \\ \vdots \\ \sum_{i=1}^{N} \Theta_{Ni} \frac{\partial \bar{\boldsymbol{T}}}{\partial \dot{\gamma}_i} \end{bmatrix}$$

$$= \begin{bmatrix} \Theta_{11}\frac{\partial \bar{\boldsymbol{T}}}{\partial \dot{\gamma}_1} + \Theta_{12}\frac{\partial \bar{\boldsymbol{T}}}{\partial \dot{\gamma}_2} + \Theta_{13}\frac{\partial \bar{\boldsymbol{T}}}{\partial \dot{\gamma}_3} + \cdots + \Theta_{1N}\frac{\partial \bar{\boldsymbol{T}}}{\partial \dot{\gamma}_N} \\ \Theta_{21}\frac{\partial \bar{\boldsymbol{T}}}{\partial \dot{\gamma}_1} + \Theta_{22}\frac{\partial \bar{\boldsymbol{T}}}{\partial \dot{\gamma}_2} + \Theta_{23}\frac{\partial \bar{\boldsymbol{T}}}{\partial \dot{\gamma}_3} + \cdots + \Theta_{2N}\frac{\partial \bar{\boldsymbol{T}}}{\partial \dot{\gamma}_N} \\ \Theta_{31}\frac{\partial \bar{\boldsymbol{T}}}{\partial \dot{\gamma}_1} + \Theta_{32}\frac{\partial \bar{\boldsymbol{T}}}{\partial \dot{\gamma}_2} + \Theta_{33}\frac{\partial \bar{\boldsymbol{T}}}{\partial \dot{\gamma}_3} + \cdots + \Theta_{3N}\frac{\partial \bar{\boldsymbol{T}}}{\partial \dot{\gamma}_N} \\ \vdots \\ \Theta_{N1}\frac{\partial \bar{\boldsymbol{T}}}{\partial \dot{\gamma}_1} + \Theta_{N2}\frac{\partial \bar{\boldsymbol{T}}}{\partial \dot{\gamma}_2} + \Theta_{N3}\frac{\partial \bar{\boldsymbol{T}}}{\partial \dot{\gamma}_3} + \cdots + \Theta_{NN}\frac{\partial \bar{\boldsymbol{T}}}{\partial \dot{\gamma}_N} \end{bmatrix}$$

$$= \begin{bmatrix} \Theta_{11} & \Theta_{12} & \Theta_{13} & \cdots & \Theta_{1N} \\ \Theta_{21} & \Theta_{22} & \Theta_{23} & \cdots & \Theta_{2N} \\ \Theta_{31} & \Theta_{32} & \Theta_{33} & \cdots & \Theta_{3N} \\ \vdots & \vdots & \vdots & \cdots & \vdots \\ \Theta_{N1} & \Theta_{N2} & \Theta_{N3} & \cdots & \Theta_{NN} \end{bmatrix} \begin{bmatrix} \frac{\partial \bar{T}}{\partial \dot{\gamma}_1} \\ \frac{\partial \bar{T}}{\partial \dot{\gamma}_2} \\ \frac{\partial \bar{T}}{\partial \dot{\gamma}_3} \\ \vdots \\ \frac{\partial \bar{T}}{\partial \dot{\gamma}_N} \end{bmatrix} = [\vartheta] \left\{ \frac{\partial \bar{T}}{\partial \dot{\Gamma}} \right\} \tag{4.25}$$

From Eq. 4.25, it becomes apparent that the time derivative of $\{\partial T/\partial \dot{q}\}$ may be handily deduced to be:

$$\frac{d}{dt}\left\{\frac{\partial T}{\partial \dot{q}}\right\} = \frac{d}{dt}\left([\vartheta]\left\{\frac{\partial \bar{T}}{\partial \dot{\Gamma}}\right\}\right) = [\dot{\vartheta}]\left\{\frac{\partial \bar{T}}{\partial \dot{\Gamma}}\right\} + [\vartheta]\frac{d}{dt}\left\{\frac{\partial \bar{T}}{\partial \dot{\Gamma}}\right\} \tag{4.26}$$

As was noted previously, the coefficients Θ_{ij} are known functions only of the generalized coordinates $q_k;\ k = 1, 2, \ldots, N$, that is, $\Theta_{ij} = \Theta_{ij}(q_1, q_2, \ldots, q_N)$, and are not related to the generalized velocities $\dot{q}_k;\ k = 1, 2, \ldots, N$. This implies that the time derivative of any element of the matrix $[\vartheta]$, $\dot{\Theta}_{ij}$ becomes:

$$\frac{d\Theta_{ij}}{dt} = \frac{\partial \Theta_{ij}}{\partial q_1}\frac{dq_1}{dt} + \frac{\partial \Theta_{ij}}{\partial q_2}\frac{dq_2}{dt} + \frac{\partial \Theta_{ij}}{\partial q_3}\frac{dq_3}{dt} + \cdots + \frac{\partial \Theta_{ij}}{\partial q_N}\frac{dq_N}{dt} = \sum_{k=1}^{N}\frac{\partial \Theta_{ij}}{\partial q_k}\frac{dq_k}{dt} \tag{4.27}$$

Equation 4.27 may be written vectorially as:

$$\frac{d\Theta_{ij}}{dt} = \begin{bmatrix} \dot{q}_1 & \dot{q}_3 & \dot{q}_3 & \cdots & \dot{q}_N \end{bmatrix} \begin{bmatrix} \frac{\partial \Theta_{ij}}{\partial q_1} \\ \frac{\partial \Theta_{ij}}{\partial q_2} \\ \frac{\partial \Theta_{ij}}{\partial q_3} \\ \vdots \\ \frac{\partial \Theta_{ij}}{\partial q_N} \end{bmatrix}$$

$$= \{\dot{q}\}^T \left\{\frac{\partial \Theta_{ij}}{\partial q}\right\} = \{\dot{\Gamma}\}^T [\beta]^T \left\{\frac{\partial \Theta_{ij}}{\partial q}\right\} \tag{4.28}$$

where the result of Eq. 4.18, which stated that $\{\dot{q}\} = [\beta]\{\dot{\Gamma}\}$, was utilized in the above expression. Note that the term $d\Theta_{ij}/dt = \{\dot{\Gamma}\}^T [\beta]^T \left\{\frac{\partial \Theta_{ij}}{\partial q}\right\}$ is a scalar with the indices ij.

Hence every element $\dot{\Theta}_{ij}$ of the matrix $[\dot{\vartheta}]$ is of the form:

$$
[\dot{\vartheta}] = \begin{bmatrix}
\{\dot{\Gamma}\}^T [\beta]^T \left\{\frac{\partial \Theta_{11}}{\partial q}\right\} & \{\dot{\Gamma}\}^T [\beta]^T \left\{\frac{\partial \Theta_{12}}{\partial q}\right\} & \{\dot{\Gamma}\}^T [\beta]^T \left\{\frac{\partial \Theta_{13}}{\partial q}\right\} & \dots & \{\dot{\Gamma}\}^T [\beta]^T \left\{\frac{\partial \Theta_{1N}}{\partial q}\right\} \\
\{\dot{\Gamma}\}^T [\beta]^T \left\{\frac{\partial \Theta_{21}}{\partial q}\right\} & \{\dot{\Gamma}\}^T [\beta]^T \left\{\frac{\partial \Theta_{22}}{\partial q}\right\} & \{\dot{\Gamma}\}^T [\beta]^T \left\{\frac{\partial \Theta_{23}}{\partial q}\right\} & \dots & \{\dot{\Gamma}\}^T [\beta]^T \left\{\frac{\partial \Theta_{2N}}{\partial q}\right\} \\
\{\dot{\Gamma}\}^T [\beta]^T \left\{\frac{\partial \Theta_{31}}{\partial q}\right\} & \{\dot{\Gamma}\}^T [\beta]^T \left\{\frac{\partial \Theta_{32}}{\partial q}\right\} & \{\dot{\Gamma}\}^T [\beta]^T \left\{\frac{\partial \Theta_{33}}{\partial q}\right\} & \dots & \{\dot{\Gamma}\}^T [\beta]^T \left\{\frac{\partial \Theta_{3N}}{\partial q}\right\} \\
\vdots & \vdots & \vdots & \dots & \vdots \\
\{\dot{\Gamma}\}^T [\beta]^T \left\{\frac{\partial \Theta_{N1}}{\partial q}\right\} & \{\dot{\Gamma}\}^T [\beta]^T \left\{\frac{\partial \Theta_{N2}}{\partial q}\right\} & \{\dot{\Gamma}\}^T [\beta]^T \left\{\frac{\partial \Theta_{N3}}{\partial q}\right\} & \dots & \{\dot{\Gamma}\}^T [\beta]^T \left\{\frac{\partial \Theta_{NN}}{\partial q}\right\}
\end{bmatrix} \tag{4.29}
$$

Writing the above result in summary form, $[\dot{\vartheta}]$ turns out to be:

$$
[\dot{\vartheta}] = \left[\{\dot{\Gamma}\}^T [\beta]^T \left\{\frac{\partial \Theta_{ij}}{\partial q}\right\}_{\substack{i=1,2,\dots,N \\ j=1,2,\dots,N}} \right] \tag{4.30}
$$

Following a procedure similar to the one used to calculate $\{\partial T/\partial \dot{q}\}$, the derivative of T with respect to q for any index k turns out to be:

$$
\begin{aligned}
\frac{\partial T}{\partial q_k} &= \frac{\partial \bar{\boldsymbol{T}}}{\partial q_k} + \frac{\partial \bar{\boldsymbol{T}}}{\partial \dot{\gamma}_1}\frac{\partial \dot{\gamma}_1}{\partial q_k} + \frac{\partial \bar{\boldsymbol{T}}}{\partial \dot{\gamma}_2}\frac{\partial \dot{\gamma}_2}{\partial q_k} + \frac{\partial \bar{\boldsymbol{T}}}{\partial \dot{\gamma}_3}\frac{\partial \dot{\gamma}_3}{\partial q_k} + \dots + \frac{\partial \bar{\boldsymbol{T}}}{\partial \dot{\gamma}_N}\frac{\partial \dot{\gamma}_N}{\partial q_k} \\
&= \frac{\partial \bar{\boldsymbol{T}}}{\partial q_k} + \sum_{i=1}^{N} \frac{\partial \bar{\boldsymbol{T}}}{\partial \dot{\gamma}_i}\frac{\partial \dot{\gamma}_i}{\partial q_k}, \quad k = 1, 2, \dots, N
\end{aligned} \tag{4.31}
$$

Note that $\bar{\boldsymbol{T}}$ is a function of both the generalized coordinates q_k, $k = 1, 2, \dots, N$ and of the vector of quasi-velocities $\dot{\Gamma}$, that is:

$$
\bar{\boldsymbol{T}} = \bar{\boldsymbol{T}}(q, \dot{\Gamma}) = \bar{\boldsymbol{T}}(q_1, q_2, \dots, q_N, \dot{\gamma}_1, \dot{\gamma}_2, \dots, \dot{\gamma}_N)
$$

Additionally, the elements of $[\vartheta]$ are all functions of the generalized coordinates $q_1, q_2, \dots, q_N$, that is, any ij element of $[\vartheta]$ or $\Theta_{ij} = \Theta_{ij}(q_1, q_2, \dots, q_N)$, and so this implies, from Eq. 4.22, that:

$$
\begin{aligned}
\dot{\gamma}_i &= \Theta_{1i}(q_1, q2, q3, \ldots, q_N)\dot{q}_1 + \Theta_{2i}(q_1, q2, q3, \ldots, q_N)\dot{q}_2 \\
&\quad + \Theta_{3i}(q_1, q2, q3, \ldots, q_N)\dot{q}_3 \\
&\quad + \ldots + \Theta_{Ni}(q_1, q2, q3, \ldots, q_N)\dot{q}_N
\end{aligned} \tag{4.32}
$$

The partial derivative of the quasi-velocity $\dot{\gamma}_i$ with respect to the generalized coordinate q_k becomes:

$$
\frac{\partial \dot{\gamma}_i}{\partial q_k} = \frac{\partial \Theta_{1i}}{\partial q_k}\dot{q}_1 + \frac{\partial \Theta_{2i}}{\partial q_k}\dot{q}_2 + \frac{\partial \Theta_{3i}}{\partial q_k}\dot{q}_3 + \cdots + \frac{\partial \Theta_{Ni}}{\partial q_k}\dot{q}_N; \quad \frac{\partial \dot{\gamma}_i}{\partial q_k} = \sum_{j=1}^{N} \frac{\partial \Theta_{ji}}{\partial q_k}\dot{q}_j \tag{4.33}
$$

And so from Eq. 4.31, the expression $\partial T/\partial q_k,\ k = 1, 2, \ldots, N$ becomes:

$$
\begin{aligned}
\frac{\partial T}{\partial q_k} &= \frac{\partial \bar{\boldsymbol{T}}}{\partial q_k} + \sum_{i=1}^{N} \frac{\partial \bar{\boldsymbol{T}}}{\partial \dot{\gamma}_i}\frac{\partial \dot{\gamma}_i}{\partial q_k}, \quad k = 1, 2, \ldots, N \\
&= \frac{\partial \bar{\boldsymbol{T}}}{\partial q_k} + \sum_{i=1}^{N} \frac{\partial \bar{\boldsymbol{T}}}{\partial \dot{\gamma}_i} \sum_{j=1}^{N} \frac{\partial \Theta_{ji}}{\partial q_k}\dot{q}_j, \quad k = 1, 2, \ldots, N
\end{aligned} \tag{4.34}
$$

Expanding the double sum in Eq. 4.34 results in:

$$
\begin{aligned}
\sum_{i=1}^{N} \frac{\partial \bar{\boldsymbol{T}}}{\partial \dot{\gamma}_i} \sum_{j=1}^{N} \frac{\partial \Theta_{ji}}{\partial q_k}\dot{q}_j &= \frac{\partial \bar{\boldsymbol{T}}}{\partial \dot{\gamma}_1}\left(\frac{\partial \Theta_{11}}{\partial q_k}\dot{q}_1 + \frac{\partial \Theta_{21}}{\partial q_k}\dot{q}_2 + \frac{\partial \Theta_{31}}{\partial q_k}\dot{q}_3 + \cdots + \frac{\partial \Theta_{N1}}{\partial q_k}\dot{q}_N\right) \\
&\quad + \frac{\partial \bar{\boldsymbol{T}}}{\partial \dot{\gamma}_2}\left(\frac{\partial \Theta_{12}}{\partial q_k}\dot{q}_1 + \frac{\partial \Theta_{22}}{\partial q_k}\dot{q}_2 + \frac{\partial \Theta_{32}}{\partial q_k}\dot{q}_3 + \cdots + \frac{\partial \Theta_{N2}}{\partial q_k}\dot{q}_N\right) \\
&\quad + \frac{\partial \bar{\boldsymbol{T}}}{\partial \dot{\gamma}_3}\left(\frac{\partial \Theta_{13}}{\partial q_k}\dot{q}_1 + \frac{\partial \Theta_{23}}{\partial q_k}\dot{q}_2 + \frac{\partial \Theta_{33}}{\partial q_k}\dot{q}_3 + \cdots + \frac{\partial \Theta_{N3}}{\partial q_k}\dot{q}_N\right) \\
&\ \ \vdots \\
&\quad + \frac{\partial \bar{\boldsymbol{T}}}{\partial \dot{\gamma}_N}\left(\frac{\partial \Theta_{1N}}{\partial q_k}\dot{q}_1 + \frac{\partial \Theta_{2N}}{\partial q_k}\dot{q}_2 + \frac{\partial \Theta_{3N}}{\partial q_k}\dot{q}_3 + \cdots + \frac{\partial \Theta_{NN}}{\partial q_k}\dot{q}_N\right)
\end{aligned} \tag{4.35}
$$

Rewriting the preceding result in vector inner product form:

$$\begin{bmatrix} \left(\frac{\partial\Theta_{11}}{\partial q_k}\dot{q}_1 + \frac{\partial\Theta_{21}}{\partial q_k}\dot{q}_2 + \frac{\partial\Theta_{31}}{\partial q_k}\dot{q}_3 + \cdots + \frac{\partial\Theta_{N1}}{\partial q_k}\dot{q}_N\right) \\ \left(\frac{\partial\Theta_{12}}{\partial q_k}\dot{q}_1 + \frac{\partial\Theta_{22}}{\partial q_k}\dot{q}_2 + \frac{\partial\Theta_{32}}{\partial q_k}\dot{q}_3 + \cdots + \frac{\partial\Theta_{N2}}{\partial q_k}\dot{q}_N\right) \\ \left(\frac{\partial\Theta_{13}}{\partial q_k}\dot{q}_1 + \frac{\partial\Theta_{23}}{\partial q_k}\dot{q}_2 + \frac{\partial\Theta_{33}}{\partial q_k}\dot{q}_3 + \cdots + \frac{\partial\Theta_{N3}}{\partial q_k}\dot{q}_N\right) \\ \vdots \\ \left(\frac{\partial\Theta_{1N}}{\partial q_k}\dot{q}_1 + \frac{\partial\Theta_{2N}}{\partial q_k}\dot{q}_2 + \frac{\partial\Theta_{3N}}{\partial q_k}\dot{q}_3 + \cdots + \frac{\partial\Theta_{NN}}{\partial q_k}\dot{q}_N\right) \end{bmatrix}^T \begin{bmatrix} \frac{\partial\bar{\boldsymbol{T}}}{\partial\dot{\gamma}_1} \\ \frac{\partial\bar{\boldsymbol{T}}}{\partial\dot{\gamma}_2} \\ \frac{\partial\bar{\boldsymbol{T}}}{\partial\dot{\gamma}_3} \\ \vdots \\ \frac{\partial\bar{\boldsymbol{T}}}{\partial\dot{\gamma}_N} \end{bmatrix} \tag{4.36}$$

The leftmost vector in the preceding equation may also be rewritten as the inner product of two vectors as follows:

$$\left(\frac{\partial\Theta_{11}}{\partial q_k}\dot{q}_1 + \frac{\partial\Theta_{21}}{\partial q_k}\dot{q}_2 + \frac{\partial\Theta_{31}}{\partial q_k}\dot{q}_3 + \cdots + \frac{\partial\Theta_{N1}}{\partial q_k}\dot{q}_N\right) = \begin{bmatrix} \frac{\partial\Theta_{11}}{\partial q_k} & \frac{\partial\Theta_{21}}{\partial q_k} & \frac{\partial\Theta_{31}}{\partial q_k} & \ldots & \frac{\partial\Theta_{N1}}{\partial q_k} \end{bmatrix} \cdot \begin{bmatrix} \dot{q}_1 \\ \dot{q}_2 \\ \dot{q}_3 \\ \vdots \\ \dot{q}_N \end{bmatrix} \tag{4.37}$$

The same pattern repeats itself for the other rows, that is:

$$\begin{bmatrix} \left(\frac{\partial\Theta_{11}}{\partial q_k}\dot{q}_1 + \frac{\partial\Theta_{21}}{\partial q_k}\dot{q}_2 + \frac{\partial\Theta_{31}}{\partial q_k}\dot{q}_3 + \cdots + \frac{\partial\Theta_{N1}}{\partial q_k}\dot{q}_N\right) \\ \left(\frac{\partial\Theta_{12}}{\partial q_k}\dot{q}_1 + \frac{\partial\Theta_{22}}{\partial q_k}\dot{q}_2 + \frac{\partial\Theta_{32}}{\partial q_k}\dot{q}_3 + \cdots + \frac{\partial\Theta_{N2}}{\partial q_k}\dot{q}_N\right) \\ \left(\frac{\partial\Theta_{13}}{\partial q_k}\dot{q}_1 + \frac{\partial\Theta_{23}}{\partial q_k}\dot{q}_2 + \frac{\partial\Theta_{33}}{\partial q_k}\dot{q}_3 + \cdots + \frac{\partial\Theta_{N3}}{\partial q_k}\dot{q}_N\right) \\ \vdots \\ \left(\frac{\partial\Theta_{1N}}{\partial q_k}\dot{q}_1 + \frac{\partial\Theta_{2N}}{\partial q_k}\dot{q}_2 + \frac{\partial\Theta_{3N}}{\partial q_k}\dot{q}_3 + \cdots + \frac{\partial\Theta_{NN}}{\partial q_k}\dot{q}_N\right) \end{bmatrix}$$

$$= \begin{bmatrix} \frac{\partial \Theta_{11}}{\partial q_k} & \frac{\partial \Theta_{21}}{\partial q_k} & \frac{\partial \Theta_{31}}{\partial q_k} & \cdots & \frac{\partial \Theta_{N1}}{\partial q_k} \\ \frac{\partial \Theta_{12}}{\partial q_k} & \frac{\partial \Theta_{22}}{\partial q_k} & \frac{\partial \Theta_{32}}{\partial q_k} & \cdots & \frac{\partial \Theta_{N2}}{\partial q_k} \\ \frac{\partial \Theta_{13}}{\partial q_k} & \frac{\partial \Theta_{23}}{\partial q_k} & \frac{\partial \Theta_{33}}{\partial q_k} & \cdots & \frac{\partial \Theta_{N3}}{\partial q_k} \\ \vdots & \vdots & \vdots & \cdots & \vdots \\ \frac{\partial \Theta_{1N}}{\partial q_k} & \frac{\partial \Theta_{2N}}{\partial q_k} & \frac{\partial \Theta_{3N}}{\partial q_k} & \cdots & \frac{\partial \Theta_{NN}}{\partial q_k} \end{bmatrix} \begin{bmatrix} \dot{q}_1 \\ \dot{q}_2 \\ \dot{q}_3 \\ \vdots \\ \dot{q}_N \end{bmatrix} \tag{4.38}$$

Therefore the double sum of Eq. 4.34 may be written as a product of one row vector, one matrix, and one column vector, that is:

$$\sum_{i=1}^{N} \frac{\partial \bar{\boldsymbol{T}}}{\partial \dot{\gamma}_i} \sum_{j=1}^{N} \frac{\partial \Theta_{ji}}{\partial q_k} \dot{q}_j = \begin{bmatrix} \left(\frac{\partial \Theta_{11}}{\partial q_k} \dot{q}_1 + \frac{\partial \Theta_{21}}{\partial q_k} \dot{q}_2 + \frac{\partial \Theta_{31}}{\partial q_k} \dot{q}_3 + \cdots + \frac{\partial \Theta_{N1}}{\partial q_k} \dot{q}_N \right) \\ \left(\frac{\partial \Theta_{12}}{\partial q_k} \dot{q}_1 + \frac{\partial \Theta_{22}}{\partial q_k} \dot{q}_2 + \frac{\partial \Theta_{32}}{\partial q_k} \dot{q}_3 + \cdots + \frac{\partial \Theta_{N2}}{\partial q_k} \dot{q}_N \right) \\ \left(\frac{\partial \Theta_{13}}{\partial q_k} \dot{q}_1 + \frac{\partial \Theta_{23}}{\partial q_k} \dot{q}_2 + \frac{\partial \Theta_{33}}{\partial q_k} \dot{q}_3 + \cdots + \frac{\partial \Theta_{N3}}{\partial q_k} \dot{q}_N \right) \\ \vdots \\ \left(\frac{\partial \Theta_{1N}}{\partial q_k} \dot{q}_1 + \frac{\partial \Theta_{2N}}{\partial q_k} \dot{q}_2 + \frac{\partial \Theta_{3N}}{\partial q_k} \dot{q}_3 + \cdots + \frac{\partial \Theta_{NN}}{\partial q_k} \dot{q}_N \right) \end{bmatrix}^T \begin{bmatrix} \frac{\partial \bar{\boldsymbol{T}}}{\partial \dot{\gamma}_1} \\ \frac{\partial \bar{\boldsymbol{T}}}{\partial \dot{\gamma}_2} \\ \frac{\partial \bar{\boldsymbol{T}}}{\partial \dot{\gamma}_3} \\ \cdots \\ \frac{\partial \bar{\boldsymbol{T}}}{\partial \dot{\gamma}_N} \end{bmatrix}$$

$$= \begin{bmatrix} \frac{\partial \bar{\boldsymbol{T}}}{\partial \dot{\gamma}_1} & \frac{\partial \bar{\boldsymbol{T}}}{\partial \dot{\gamma}_2} & \frac{\partial \bar{\boldsymbol{T}}}{\partial \dot{\gamma}_3} & \cdots & \frac{\partial \bar{\boldsymbol{T}}}{\partial \dot{\gamma}_N} \end{bmatrix} \begin{bmatrix} \frac{\partial \Theta_{11}}{\partial q_k} & \frac{\partial \Theta_{21}}{\partial q_k} & \frac{\partial \Theta_{31}}{\partial q_k} & \cdots & \frac{\partial \Theta_{N1}}{\partial q_k} \\ \frac{\partial \Theta_{12}}{\partial q_k} & \frac{\partial \Theta_{22}}{\partial q_k} & \frac{\partial \Theta_{32}}{\partial q_k} & \cdots & \frac{\partial \Theta_{N2}}{\partial q_k} \\ \frac{\partial \Theta_{13}}{\partial q_k} & \frac{\partial \Theta_{23}}{\partial q_k} & \frac{\partial \Theta_{33}}{\partial q_k} & \cdots & \frac{\partial \Theta_{N3}}{\partial q_k} \\ \vdots & \vdots & \vdots & \cdots & \vdots \\ \frac{\partial \Theta_{1N}}{\partial q_k} & \frac{\partial \Theta_{2N}}{\partial q_k} & \frac{\partial \Theta_{3N}}{\partial q_k} & \cdots & \frac{\partial \Theta_{NN}}{\partial q_k} \end{bmatrix} \begin{bmatrix} \dot{q}_1 \\ \dot{q}_2 \\ \dot{q}_3 \\ \vdots \\ \dot{q}_N \end{bmatrix} \tag{4.39}$$

Hence summarizing, the term $\partial T/\partial q_k$ may be written succinctly as:

$$\frac{\partial T}{\partial q_k} = \frac{\partial \bar{\boldsymbol{T}}}{\partial q_k} + \left\{\frac{\partial \bar{\boldsymbol{T}}}{\partial \dot{\Gamma}}\right\}^T \left[\frac{\partial \vartheta}{\partial q_k}\right]^T \{\dot{q}\} = \frac{\partial \bar{\boldsymbol{T}}}{\partial q_k} + \{\dot{q}\}^T \left[\frac{\partial \vartheta}{\partial q_k}\right] \left\{\frac{\partial \bar{\boldsymbol{T}}}{\partial \dot{\Gamma}}\right\} \tag{4.40}$$

where $\{\dot{q}\}$ is an $N \times 1$ column vector, $\left[\frac{\partial \vartheta}{\partial q_k}\right]^T$ is an $N \times N$ matrix, and $\left\{\frac{\partial \bar{\boldsymbol{T}}}{\partial \dot{\Gamma}}\right\}^T$ is a $1 \times N$ row vector. Once again, it should be noted that the product $\left\{\frac{\partial \bar{\boldsymbol{T}}}{\partial \dot{\Gamma}}\right\}^T \left[\frac{\partial \vartheta}{\partial q_k}\right]^T \{\dot{q}\}$ or $\{\dot{q}\}^T \left[\frac{\partial \vartheta}{\partial q_k}\right] \left\{\frac{\partial \bar{\boldsymbol{T}}}{\partial \dot{\Gamma}}\right\}$ are scalars and so letting the index k vary over its range from $k = 1, 2, \ldots, N$, the column vector with entries $\partial T/\partial q_k$, $k = 1, 2, \ldots, N$ may be formed as follows:

$$\left\{\frac{\partial T}{\partial q}\right\} = \begin{bmatrix} \frac{\partial T}{\partial q_1} \\ \frac{\partial T}{\partial q_2} \\ \frac{\partial T}{\partial q_3} \\ \vdots \\ \frac{\partial T}{\partial q_N} \end{bmatrix} = \begin{bmatrix} \frac{\partial \bar{\boldsymbol{T}}}{\partial q_1} + \{\dot{q}\}^T \left[\frac{\partial \vartheta}{\partial q_1}\right] \left\{\frac{\partial \bar{\boldsymbol{T}}}{\partial \dot{\Gamma}}\right\} \\ \frac{\partial \bar{\boldsymbol{T}}}{\partial q_2} + \{\dot{q}\}^T \left[\frac{\partial \vartheta}{\partial q_2}\right] \left\{\frac{\partial \bar{\boldsymbol{T}}}{\partial \dot{\Gamma}}\right\} \\ \frac{\partial \bar{\boldsymbol{T}}}{\partial q_3} + \{\dot{q}\}^T \left[\frac{\partial \vartheta}{\partial q_3}\right] \left\{\frac{\partial \bar{\boldsymbol{T}}}{\partial \dot{\Gamma}}\right\} \\ \vdots \\ \frac{\partial \bar{\boldsymbol{T}}}{\partial q_N} + \{\dot{q}\}^T \left[\frac{\partial \vartheta}{\partial q_N}\right] \left\{\frac{\partial \bar{\boldsymbol{T}}}{\partial \dot{\Gamma}}\right\} \end{bmatrix} = \begin{bmatrix} \frac{\partial \bar{\boldsymbol{T}}}{\partial q_1} \\ \frac{\partial \bar{\boldsymbol{T}}}{\partial q_2} \\ \frac{\partial \bar{\boldsymbol{T}}}{\partial q_3} \\ \vdots \\ \frac{\partial \bar{\boldsymbol{T}}}{\partial q_N} \end{bmatrix} + \underbrace{\begin{bmatrix} \{\dot{q}\}^T \left[\frac{\partial \vartheta}{\partial q_1}\right] \\ \{\dot{q}\}^T \left[\frac{\partial \vartheta}{\partial q_2}\right] \\ \{\dot{q}\}^T \left[\frac{\partial \vartheta}{\partial q_3}\right] \\ \vdots \\ \{\dot{q}\}^T \left[\frac{\partial \vartheta}{\partial q_N}\right] \end{bmatrix}}_{[\eta]} \left\{\frac{\partial \bar{\boldsymbol{T}}}{\partial \dot{\Gamma}}\right\} \tag{4.41}$$

4.3 Lagrangian Dynamics with Quasi-Coordinates

The principle of virtual work led to the d'Alembert–Lagrange equations in matrix form. These equations were stated in Sect. 4.1, Eq. 4.12 and are repeated below for convenience:

$$\{\delta q\}^T \left(\frac{d}{dt}\left\{\frac{\partial T}{\partial \dot{q}}\right\} - \left\{\frac{\partial T}{\partial q}\right\} - \{Q\}\right) = 0$$

The following terms in the d'Alembert–Lagrange equations were derived in Sect. 4.2 (see Eq. 4.26):

$$\frac{d}{dt}\left\{\frac{\partial T}{\partial \dot{q}}\right\} = \frac{d}{dt}\left([\vartheta]\left\{\frac{\partial \bar{\boldsymbol{T}}}{\partial \dot{\Gamma}}\right\}\right) = [\dot{\vartheta}]\left\{\frac{\partial \bar{\boldsymbol{T}}}{\partial \dot{\Gamma}}\right\} + [\vartheta]\frac{d}{dt}\left\{\frac{\partial \bar{\boldsymbol{T}}}{\partial \dot{\Gamma}}\right\}$$

where, with the aid of Eq. 4.2, $\left[\dot{\vartheta}\right]\left\{\frac{\partial\bar{\boldsymbol{T}}}{\partial\dot{\Gamma}}\right\}$ can be shown to be:

$$
\begin{aligned}
&\left[\dot{\vartheta}\right]\left\{\frac{\partial\bar{\boldsymbol{T}}}{\partial\dot{\Gamma}}\right\} \\
&= \begin{bmatrix}
\{\dot{\Gamma}\}^T[\beta]^T\left\{\frac{\partial\Theta_{11}}{\partial q}\right\} & \{\dot{\Gamma}\}^T[\beta]^T\left\{\frac{\partial\Theta_{12}}{\partial q}\right\} & \{\dot{\Gamma}\}^T[\beta]^T\left\{\frac{\partial\Theta_{13}}{\partial q}\right\} & \cdots & \{\dot{\Gamma}\}^T[\beta]^T\left\{\frac{\partial\Theta_{1N}}{\partial q}\right\} \\
\{\dot{\Gamma}\}^T[\beta]^T\left\{\frac{\partial\Theta_{21}}{\partial q}\right\} & \{\dot{\Gamma}\}^T[\beta]^T\left\{\frac{\partial\Theta_{22}}{\partial q}\right\} & \{\dot{\Gamma}\}^T[\beta]^T\left\{\frac{\partial\Theta_{23}}{\partial q}\right\} & \cdots & \{\dot{\Gamma}\}^T[\beta]^T\left\{\frac{\partial\Theta_{2N}}{\partial q}\right\} \\
\{\dot{\Gamma}\}^T[\beta]^T\left\{\frac{\partial\Theta_{31}}{\partial q}\right\} & \{\dot{\Gamma}\}^T[\beta]^T\left\{\frac{\partial\Theta_{32}}{\partial q}\right\} & \{\dot{\Gamma}\}^T[\beta]^T\left\{\frac{\partial\Theta_{33}}{\partial q}\right\} & \cdots & \{\dot{\Gamma}\}^T[\beta]^T\left\{\frac{\partial\Theta_{3N}}{\partial q}\right\} \\
\vdots & \vdots & \vdots & \cdots & \vdots \\
\{\dot{\Gamma}\}^T[\beta]^T\left\{\frac{\partial\Theta_{N1}}{\partial q}\right\} & \{\dot{\Gamma}\}^T[\beta]^T\left\{\frac{\partial\Theta_{N2}}{\partial q}\right\} & \{\dot{\Gamma}\}^T[\beta]^T\left\{\frac{\partial\Theta_{N3}}{\partial q}\right\} & \cdots & \{\dot{\Gamma}\}^T[\beta]^T\left\{\frac{\partial\Theta_{NN}}{\partial q}\right\}
\end{bmatrix}
\begin{Bmatrix}
\frac{\partial\bar{\boldsymbol{T}}}{\partial\dot{\gamma}_1} \\
\frac{\partial\bar{\boldsymbol{T}}}{\partial\dot{\gamma}_2} \\
\frac{\partial\bar{\boldsymbol{T}}}{\partial\dot{\gamma}_3} \\
\vdots \\
\frac{\partial\bar{\boldsymbol{T}}}{\partial\dot{\gamma}_N}
\end{Bmatrix} \\
&= \left[\{\dot{\Gamma}\}^T[\beta]^T\left\{\frac{\partial\Theta_{ij}}{\partial q}\right\}_{\substack{i=1,2,\ldots,N \\ j=1,2,\ldots,N}}\right]\left\{\frac{\partial\bar{\boldsymbol{T}}}{\partial\dot{\Gamma}}\right\}
\end{aligned}
\tag{4.42}
$$

Furthermore the term $[\vartheta]\frac{d}{dt}\left\{\frac{\partial\bar{\boldsymbol{T}}}{\partial\dot{\Gamma}}\right\}$ was calculated to be:

$$
[\vartheta]\frac{d}{dt}\left\{\frac{\partial\bar{\boldsymbol{T}}}{\partial\dot{\Gamma}}\right\} =
\begin{bmatrix}
\Theta_{11} & \Theta_{12} & \Theta_{13} & \cdots & \Theta_{1N} \\
\Theta_{21} & \Theta_{22} & \Theta_{23} & \cdots & \Theta_{2N} \\
\Theta_{31} & \Theta_{32} & \Theta_{33} & \cdots & \Theta_{3N} \\
\vdots & \vdots & \vdots & \cdots & \vdots \\
\Theta_{N1} & \Theta_{N2} & \Theta_{N3} & \cdots & \Theta_{NN}
\end{bmatrix}
\begin{bmatrix}
\frac{d}{dt}\left(\frac{\partial\bar{\boldsymbol{T}}}{\partial\dot{\gamma}_1}\right) \\
\frac{d}{dt}\left(\frac{\partial\bar{\boldsymbol{T}}}{\partial\dot{\gamma}_2}\right) \\
\frac{d}{dt}\left(\frac{\partial\bar{\boldsymbol{T}}}{\partial\dot{\gamma}_3}\right) \\
\vdots \\
\frac{d}{dt}\left(\frac{\partial\bar{\boldsymbol{T}}}{\partial\dot{\gamma}_N}\right)
\end{bmatrix}
\tag{4.43}
$$

This implies that $\frac{d}{dt}\left\{\frac{\partial T}{\partial \dot{q}}\right\} = [\dot{\vartheta}]\left\{\frac{\partial \bar{T}}{\partial \dot{\Gamma}}\right\} + [\vartheta]\frac{d}{dt}\left\{\frac{\partial \bar{T}}{\partial \dot{\Gamma}}\right\}$, in shorthand form may be written as:

$$\frac{d}{dt}\left\{\frac{\partial T}{\partial \dot{q}}\right\} = \left[\{\dot{\Gamma}\}^T [\beta]^T \left\{\frac{\partial \Theta_{ij}}{\partial q}\right\}_{\substack{i=1,2,\ldots,N \\ j=1,2,\ldots,N}}\right]\left\{\frac{\partial \bar{\boldsymbol{T}}}{\partial \dot{\Gamma}}\right\} + [\vartheta]\frac{d}{dt}\left\{\frac{\partial \bar{\boldsymbol{T}}}{\partial \dot{\Gamma}}\right\} \tag{4.44}$$

In addition, from Eq. 4.41, $\left\{\frac{\partial T}{\partial q}\right\}$ was shown to be:

$$\left\{\frac{\partial T}{\partial q}\right\} = \begin{bmatrix} \frac{\partial T}{\partial q_1} \\ \frac{\partial T}{\partial q_2} \\ \frac{\partial T}{\partial q_3} \\ \vdots \\ \frac{\partial T}{\partial q_N} \end{bmatrix} = \begin{bmatrix} \frac{\partial \bar{\boldsymbol{T}}}{\partial q_1} + \{\dot{q}\}^T \left[\frac{\partial \vartheta}{\partial q_1}\right]\left\{\frac{\partial \bar{\boldsymbol{T}}}{\partial \dot{\Gamma}}\right\} \\ \frac{\partial \bar{\boldsymbol{T}}}{\partial q_2} + \{\dot{q}\}^T \left[\frac{\partial \vartheta}{\partial q_2}\right]\left\{\frac{\partial \bar{\boldsymbol{T}}}{\partial \dot{\Gamma}}\right\} \\ \frac{\partial \bar{\boldsymbol{T}}}{\partial q_3} + \{\dot{q}\}^T \left[\frac{\partial \vartheta}{\partial q_3}\right]\left\{\frac{\partial \bar{\boldsymbol{T}}}{\partial \dot{\Gamma}}\right\} \\ \vdots \\ \frac{\partial \bar{\boldsymbol{T}}}{\partial q_N} + \{\dot{q}\}^T \left[\frac{\partial \vartheta}{\partial q_N}\right]\left\{\frac{\partial \bar{\boldsymbol{T}}}{\partial \dot{\Gamma}}\right\} \end{bmatrix}$$

$$= \begin{bmatrix} \frac{\partial \bar{\boldsymbol{T}}}{\partial q_1} \\ \frac{\partial \bar{\boldsymbol{T}}}{\partial q_2} \\ \frac{\partial \bar{\boldsymbol{T}}}{\partial q_3} \\ \vdots \\ \frac{\partial \bar{\boldsymbol{T}}}{\partial q_N} \end{bmatrix} + \underbrace{\begin{bmatrix} \{\dot{q}\}^T \left[\frac{\partial \vartheta}{\partial q_1}\right] \\ \{\dot{q}\}^T \left[\frac{\partial \vartheta}{\partial q_2}\right] \\ \{\dot{q}\}^T \left[\frac{\partial \vartheta}{\partial q_3}\right] \\ \vdots \\ \{\dot{q}\}^T \left[\frac{\partial \vartheta}{\partial q_N}\right] \end{bmatrix}}_{\eta} \left\{\frac{\partial \bar{\boldsymbol{T}}}{\partial \dot{\Gamma}}\right\}$$

$$\Rightarrow \left\{\frac{\partial T}{\partial q}\right\} = \left\{\frac{\partial \bar{\boldsymbol{T}}}{\partial q}\right\} + [\eta]\left\{\frac{\partial \bar{\boldsymbol{T}}}{\partial \dot{\Gamma}}\right\} \tag{4.45}$$

Hence, the d'Alembert–Lagrange equation $\left(\frac{d}{dt}\left\{\frac{\partial T}{\partial \dot{q}}\right\} - \left\{\frac{\partial T}{\partial q}\right\} - \{Q\}\right) = 0$ may be written down in the following shorthand form as:

$$\left(\frac{d}{dt}\left\{\frac{\partial T}{\partial \dot{q}}\right\} - \left\{\frac{\partial T}{\partial q}\right\} - \{Q\}\right)$$

$$= [\vartheta]\frac{d}{dt}\left\{\frac{\partial \bar{\boldsymbol{T}}}{\partial \dot{\Gamma}}\right\} + \overbrace{\left[\left\{\dot{\Gamma}\right\}^T [\beta]^T \left\{\frac{\partial \Theta_{ij}}{\partial q}\right\}_{\substack{i=1,2,\ldots,N \\ j=1,2,\ldots,N}}\right]}^{[\dot{\vartheta}]} \left\{\frac{\partial \bar{\boldsymbol{T}}}{\partial \dot{\Gamma}}\right\}$$

$$- [\eta]\left\{\frac{\partial \bar{\boldsymbol{T}}}{\partial \dot{\Gamma}}\right\} - \left\{\frac{\partial \bar{\boldsymbol{T}}}{\partial q}\right\} - \{Q\} = 0 \tag{4.46}$$

The following expressions are defined to be:

$$\{N\} = [\beta]^T \{Q\}; \ [\Lambda] = \left[\left\{\dot{\Gamma}\right\}^T [\beta]^T \left\{\frac{\partial \Theta_{ij}}{\partial q}\right\}_{\substack{i=1,2,\ldots,N \\ j=1,2,\ldots,N}}\right] - [\eta] \tag{4.47}$$

Pre-multiplying the above d'Alembert–Lagrange equation (Eq. 4.46) by $[\beta]^T$ and using the definitions of $[\Lambda]$ and $\{N\}$, the result is:

$$[\beta]^T \left([\vartheta]\frac{d}{dt}\left\{\frac{\partial \bar{\boldsymbol{T}}}{\partial \dot{\Gamma}}\right\} + \left[\left\{\dot{\Gamma}\right\}^T [\beta]^T \left\{\frac{\partial \Theta_{ij}}{\partial q}\right\}_{\substack{i=1,2,\ldots,N \\ j=1,2,\ldots,N}}\right] \left\{\frac{\partial \bar{\boldsymbol{T}}}{\partial \dot{\Gamma}}\right\}\right.$$

$$\left. - [\eta]\left\{\frac{\partial \bar{\boldsymbol{T}}}{\partial \dot{\Gamma}}\right\} - \left\{\frac{\partial \bar{\boldsymbol{T}}}{\partial q}\right\}\right) = [\beta]^T \{Q\}$$

$$= [\beta]^T [\vartheta]\frac{d}{dt}\left\{\frac{\partial \bar{\boldsymbol{T}}}{\partial \dot{\Gamma}}\right\} + [\beta]^T [\Lambda]\left\{\frac{\partial \bar{\boldsymbol{T}}}{\partial \dot{\Gamma}}\right\} - [\beta]^T \left\{\frac{\partial \bar{\boldsymbol{T}}}{\partial q}\right\} = \{N\} \tag{4.48}$$

Recall that $[\beta]^T [\vartheta] = [I]$, and so the d'Alembert–Lagrange equation for quasi-coordinates may be written in the form:

$$\frac{d}{dt}\left\{\frac{\partial \bar{\boldsymbol{T}}}{\partial \dot{\Gamma}}\right\} + [\beta]^T [\Lambda]\left\{\frac{\partial \bar{\boldsymbol{T}}}{\partial \dot{\Gamma}}\right\} - [\beta]^T \left\{\frac{\partial \bar{\boldsymbol{T}}}{\partial q}\right\} = \{N\} \tag{4.49}$$

Example 1 [8, pp. 16–17]

A simple example involves the angular motion of a rigid body constrained to rotate about a point (see Meirovitch [23, pp. 159–160] or Cameron and Book [8, pp. 16–17]). Note that there are no constraints whatsoever on the rotational motion. The roll–pitch–yaw quasi-velocities w_x, w_y, w_z were introduced in order to simplify the equations and to illustrate the techniques involved in writing the d'Alembert–Lagrange equations of motion for non-holonomic joints. These quasi-velocities are non-holonomic in the sense that they are not the derivatives of any generalized coordinates (generalized angular velocities or Euler angular rates). Letting the body's angular rates represent the quasi-velocities, that is, $w_x = \dot{\gamma}_1, w_y = \dot{\gamma}_2, w_z = \dot{\gamma}_3$, while the Euler angular rates are defined to be the generalized velocities, i.e., $\dot{\phi} = \dot{q}_1, \dot{\theta} = \dot{q}_2, \dot{\psi} = \dot{q}_3$, the relationship between the Euler angular rates and the body's angular rates is simply:

$$\begin{bmatrix} w_x \\ w_y \\ w_z \end{bmatrix} = \begin{bmatrix} \dot{\phi} - \dot{\psi}\sin\theta \\ \dot{\theta}\cos\phi + \dot{\psi}\cos\theta\sin\phi \\ -\dot{\theta}\sin\phi + \dot{\psi}\cos\theta\cos\phi \end{bmatrix} = \underbrace{\begin{bmatrix} 1 & 0 & -\sin\theta \\ 0 & \cos\phi & \cos\theta\sin\phi \\ 0 & -\sin\phi & \cos\theta\cos\phi \end{bmatrix}}_{[\vartheta]^T} \begin{bmatrix} \dot{\phi} \\ \dot{\theta} \\ \dot{\psi} \end{bmatrix} \tag{4.50}$$

The kinetic energy $\bar{\boldsymbol{T}}$ is of the form:

$$\bar{\boldsymbol{T}} = \frac{1}{2}\left[\, w_x \; w_y \; w_z \,\right] \begin{bmatrix} I_x & 0 & 0 \\ 0 & I_y & 0 \\ 0 & 0 & I_z \end{bmatrix} \begin{bmatrix} w_x \\ w_y \\ w_z \end{bmatrix} = \frac{1}{2}\left(I_x w_x^2 + I_y w_y^2 + I_z w_z^2\right) \tag{4.51}$$

In outline form, the procedure for calculation of the d'Alembert–Lagrange equations for this problem is:

1. Write: $[\vartheta]$, $[\dot{\vartheta}]$, $\left[\frac{\partial\vartheta}{\partial\phi}\right]$, $\left[\frac{\partial\vartheta}{\partial\theta}\right]$, $\left[\frac{\partial\vartheta}{\partial\psi}\right]$

$$[\vartheta] = \begin{bmatrix} 1 & 0 & 0 \\ 0 & \cos\phi & -\sin\phi \\ -\sin\theta & \cos\theta\sin\phi & \cos\theta\cos\phi \end{bmatrix} \tag{4.52}$$

$$[\dot{\vartheta}] = \begin{bmatrix} 0 & 0 & 0 \\ 0 & -\dot{\phi}\sin\phi & -\dot{\phi}\cos\phi \\ -\dot{\theta}\cos\theta & -\dot{\theta}\sin\theta\sin\phi + \dot{\phi}\cos\phi\cos\theta & -\dot{\theta}\sin\theta\cos\phi - \dot{\phi}\sin\phi\cos\theta \end{bmatrix} \tag{4.53}$$

$$\left[\frac{\partial\vartheta}{\partial\phi}\right] = \begin{bmatrix} 0 & 0 & 0 \\ 0 & -\sin\phi & -\cos\phi \\ 0 & \cos\phi\cos\theta & -\sin\phi\cos\theta \end{bmatrix} \tag{4.54}$$

$$\left[\frac{\partial\vartheta}{\partial\theta}\right] = \begin{bmatrix} 0 & 0 & 0 \\ 0 & 0 & 0 \\ -\cos\theta & -\sin\phi\sin\theta & -\cos\phi\sin\theta \end{bmatrix} \tag{4.55}$$

$$\left[\frac{\partial\vartheta}{\partial\psi}\right] = \begin{bmatrix} 0\,0\,0 \\ 0\,0\,0 \\ 0\,0\,0 \end{bmatrix} \tag{4.56}$$

2. Write: $[\beta]$, $[\beta]^T$

$$[\beta] = \left([\vartheta]^T\right)^{-1} = \begin{bmatrix} 1 & \sin\phi\tan\theta & \cos\phi\tan\theta \\ 0 & \cos\phi & -\sin\phi \\ 0 & \sin\phi/\cos\theta & \cos\phi/\cos\theta \end{bmatrix}$$

$$\Rightarrow [\beta]^T = \begin{bmatrix} 1 & 0 & 0 \\ \sin\phi\tan\theta & \cos\phi & \sin\phi/\cos\theta \\ \cos\phi\tan\theta & -\sin\phi & \cos\phi/\cos\theta \end{bmatrix} \tag{4.57}$$

3. Write: $\{\dot{\Gamma}\}$

$$\{\dot{\Gamma}\} = \begin{bmatrix} w_x \\ w_y \\ w_z \end{bmatrix} = \begin{bmatrix} \dot{\phi} - \dot{\psi}\sin\theta \\ \dot{\theta}\cos\phi + \dot{\psi}\cos\theta\sin\phi \\ -\dot{\theta}\sin\phi + \dot{\psi}\cos\theta\cos\phi \end{bmatrix} \tag{4.58}$$

$$\begin{aligned} & w_z\cos\phi + w_y\sin\phi \\ &= \cos\phi\sin\phi + \dot{\psi}\cos\theta\sin^2\phi - \dot{\theta}\cos\phi\sin\phi + \dot{\psi}\cos\theta\cos^2\phi \\ &= \dot{\psi}\cos\theta \end{aligned} \tag{4.59}$$

4. Write: $[\eta]$

$$[\eta] = \begin{bmatrix} \{\dot{\Gamma}\}^T [\beta]^T [\partial\vartheta/\partial\phi] \\ \{\dot{\Gamma}\}^T [\beta]^T [\partial\vartheta/\partial\theta] \\ \{\dot{\Gamma}\}^T [\beta]^T [\partial\vartheta/\partial\psi] \end{bmatrix} \tag{4.60}$$

5. Write: $\{\dot{\Gamma}\}^T [\beta]^T$

$$\{\dot{\Gamma}\}^T [\beta]^T = \begin{bmatrix} w_x & w_y & w_z \end{bmatrix} \begin{bmatrix} 1 & 0 & 0 \\ \sin\phi\tan\theta & \cos\phi & \sin\phi/\cos\theta \\ \cos\phi\tan\theta & -\sin\phi & \cos\phi/\cos\theta \end{bmatrix}$$

$$= \begin{bmatrix} w_x + w_y \sin\phi\tan\theta + w_z \cos\phi\tan\theta & w_y\cos\phi - w_z\sin\phi & w_y\frac{\sin\phi}{\cos\theta} + w_z\frac{\cos\phi}{\cos\theta} \end{bmatrix}$$

$$= \begin{bmatrix} \dot{\phi} & \dot{\theta} & \dot{\psi} \end{bmatrix} \tag{4.61}$$

6. Write: $\{\dot{\Gamma}\}^T [\beta]^T [\partial\vartheta/\partial\phi]$; $\{\dot{\Gamma}\}^T [\beta]^T [\partial\vartheta/\partial\theta]$; $\{\dot{\Gamma}\}^T [\beta]^T [\partial\vartheta/\partial\psi]$; $[\eta]$

$$\begin{bmatrix} \dot{\phi} & \dot{\theta} & \dot{\psi} \end{bmatrix} \begin{bmatrix} \frac{\partial\vartheta}{\partial\phi} \end{bmatrix} = \begin{bmatrix} 0 & -\dot{\theta}\sin\phi + \dot{\psi}\cos\phi\cos\theta & -\dot{\theta}\cos\phi - \dot{\psi}\sin\phi\cos\theta \end{bmatrix}$$

$$= \begin{bmatrix} 0 & w_z & -w_y \end{bmatrix} \tag{4.62}$$

$$\begin{bmatrix} \dot{\phi} & \dot{\theta} & \dot{\psi} \end{bmatrix} \begin{bmatrix} \frac{\partial\vartheta}{\partial\theta} \end{bmatrix} = \begin{bmatrix} -\dot{\psi}\cos\theta & -\dot{\psi}\sin\theta\sin\phi & -\dot{\psi}\cos\phi\sin\theta \end{bmatrix}$$

$$\begin{bmatrix} \dot{\phi} & \dot{\theta} & \dot{\psi} \end{bmatrix} \begin{bmatrix} \frac{\partial\vartheta}{\partial\psi} \end{bmatrix} = \begin{bmatrix} 0 & 0 & 0 \end{bmatrix} \tag{4.63}$$

$$[\eta] = \begin{pmatrix} \begin{bmatrix} \dot{\phi} & \dot{\theta} & \dot{\psi} \end{bmatrix} \begin{bmatrix} \frac{\partial\vartheta}{\partial\phi} \end{bmatrix} \\ \begin{bmatrix} \dot{\phi} & \dot{\theta} & \dot{\psi} \end{bmatrix} \begin{bmatrix} \frac{\partial\vartheta}{\partial\theta} \end{bmatrix} \\ \begin{bmatrix} \dot{\phi} & \dot{\theta} & \dot{\psi} \end{bmatrix} \begin{bmatrix} \frac{\partial\vartheta}{\partial\theta} \end{bmatrix} \end{pmatrix} = \begin{bmatrix} 0 & -\dot{\theta}\sin\theta + \dot{\psi}\cos\phi\cos\theta & -\dot{\theta}\cos\phi - \dot{\psi}\sin\phi\cos\theta \\ -\dot{\psi}\cos\theta & -\dot{\psi}\sin\theta\sin\phi & -\dot{\psi}\cos\phi\sin\theta \\ 0 & 0 & 0 \end{bmatrix}$$

$$= \begin{bmatrix} 0 & w_z & -w_y \\ -\dot{\psi}\cos\theta & -\dot{\psi}\sin\theta\sin\phi & -\dot{\psi}\cos\phi\sin\theta \\ 0 & 0 & 0 \end{bmatrix}$$

$$
= \begin{bmatrix} 0 & w_z & -w_y \\ -(w_z \cos\phi + w_y \sin\phi) & -(w_z \cos\phi + w_y \sin\phi)\tan\theta\sin\phi & -(w_z \cos\phi + w_y \sin\phi)\tan\theta\cos\phi \\ 0 & 0 & 0 \end{bmatrix} \tag{4.64}
$$

7. Write: $[\Lambda] = \left[\dot{\vartheta}\right] - [\eta]$

$$
\begin{aligned}
[\Lambda] &= \left[\dot{\vartheta}\right] - [\eta] \\
&= \begin{bmatrix} 0 & 0 & 0 \\ 0 & -\dot{\phi}\sin\phi & -\dot{\phi}\cos\phi \\ -\dot{\theta}\cos\theta & -\dot{\theta}\sin\theta\sin\phi + \dot{\phi}\cos\phi\cos\theta & -\dot{\theta}\sin\theta\cos\phi - \dot{\phi}\sin\phi\cos\theta \end{bmatrix} \\
&\quad - \begin{bmatrix} 0 & w_z & -w_y \\ -\dot{\psi}\cos\theta & -\dot{\psi}\sin\theta\sin\phi & -\dot{\psi}\cos\theta\sin\phi \\ 0 & 0 & 0 \end{bmatrix} \\
&= \begin{bmatrix} 0 & -w_z & w_y \\ \dot{\psi}\cos\theta & \dot{\psi}\sin\theta\sin\phi - \dot{\phi}\sin\phi & \dot{\psi}\cos\theta\sin\phi - \dot{\phi}\cos\phi \\ -\dot{\theta}\cos\theta & -\dot{\theta}\sin\theta\sin\phi + \dot{\phi}\cos\phi\cos\theta & -\dot{\theta}\sin\theta\cos\phi - \dot{\phi}\sin\phi\cos\theta \end{bmatrix} \\
&= \begin{bmatrix} 0 & -w_z & w_y \\ w_z\cos\phi + w_y\sin\phi & \dot{\psi}\sin\theta\sin\phi - \dot{\phi}\sin\phi & \dot{\psi}\cos\theta\sin\phi - \dot{\phi}\cos\phi \\ -\dot{\theta}\cos\theta & -\dot{\theta}\sin\theta\sin\phi + \dot{\phi}\cos\phi\cos\theta & -\dot{\theta}\sin\theta\cos\phi - \dot{\phi}\sin\phi\cos\theta \end{bmatrix} \\
&= \begin{bmatrix} 0 & -w_z & w_y \\ (w_z\cos\phi + w_y\sin\phi) & \begin{matrix} -\dot{\phi}\sin\phi \\ +\sin\phi\tan\theta(w_z\cos\phi + w_y\sin\phi) \end{matrix} & \begin{matrix} -\dot{\phi}\cos\phi \\ +\cos\phi\tan\theta(w_z\cos\phi + w_y\sin\phi) \end{matrix} \\ -\dot{\theta}\cos\theta & -\dot{\theta}\sin\theta\sin\phi + \dot{\phi}\cos\phi\cos\theta & -\dot{\theta}\sin\theta\cos\phi - \dot{\phi}\sin\phi\cos\theta \end{bmatrix}
\end{aligned} \tag{4.65}
$$

8. Write: $[\beta]^T\,[\Lambda]$

$$[\beta]^T\,[\Lambda] = \begin{bmatrix} 1 & 0 & 0 \\ \sin\phi\tan\theta & \cos\phi & \sin\phi/\cos\theta \\ \cos\phi\tan\theta & -\sin\phi & \cos\phi/\cos\theta \end{bmatrix}$$

$$* \begin{bmatrix} 0 & -w_z & w_y \\ (w_z\cos\phi + w_y\sin\phi) & \begin{matrix} -\dot{\psi}\sin\psi \\ +\sin\phi\tan\theta(w_z\cos\phi + w_y\sin\phi) \end{matrix} & \begin{matrix} -\phi\cos\phi \\ +\cos\phi\tan\theta(w_z\cos\phi + w_y\sin\phi) \end{matrix} \\ -\dot{\theta}\cos\theta & -\dot{\theta}\sin\theta\sin\phi + \dot{\phi}\cos\phi\cos\theta & -\dot{\theta}\sin\theta\cos\phi - \dot{\phi}\sin\phi\cos\theta \end{bmatrix}$$

$$= \begin{bmatrix} 0 & -w_z & w_y \\ \begin{matrix} \cos\phi(w_z\cos\phi + w_y\sin\phi) \\ -\dot{\theta}\sin\phi \end{matrix} & \begin{matrix} \sin\phi\tan\theta(-w_z + w_z\cos^2\phi) \\ +\sin\phi\tan\theta(w_y\sin\phi\cos\phi - \dot{\theta}\sin\phi) \end{matrix} & \begin{matrix} w_y\tan\theta\sin\phi - \dot{\phi} \\ +w_z\cos^3\phi\tan\theta \\ -\dot{\theta}\tan\theta\cos\phi\sin\phi \\ +w_y\tan\theta\cos^2\phi\sin\phi \end{matrix} \\ \begin{matrix} -\sin\phi(w_z\cos\phi + w_y\sin\phi) \\ -\dot{\theta}\cos\phi \end{matrix} & \begin{matrix} \dot{\phi} - 2w_z\cos\phi\tan\theta \\ -w_y\sin\phi\tan\theta + w_z\cos^3\phi\tan\theta \\ -\dot{\theta}\cos\phi\sin\phi\tan\theta \\ +w_y\cos^2\phi\sin\phi\tan\theta \end{matrix} & \begin{matrix} w_y\cos^3\phi\tan\theta \\ -\dot{\theta}\cos^2\phi\tan\theta \\ -w_z\cos^2\phi\sin\phi\tan\theta \end{matrix} \end{bmatrix} \tag{4.66}$$

9. The following assumptions are used to simplify $[\beta]^T\,[\Lambda]$:

   (a) $\phi, \theta \approx 0$
   (b) $w_x \approx \dot{\phi};\; w_y \approx \dot{\theta};\; w_z \approx \dot{\psi}$

   The simplified version of $[\beta]^T\,[\Lambda]$ is the following:

$$\begin{bmatrix} 0 & -w_z & w_y \\ w_z & 0 & -w_x \\ -w_y & w_x & 0 \end{bmatrix} \tag{4.67}$$

10. Using the preceding result (the simplification in item 9), the matrix product $[\beta]^T [\Lambda] \left\{ \frac{\partial \bar{T}}{\partial \dot{\Gamma}} \right\}$ may be written as:

$$[\beta]^T [\Lambda] \begin{bmatrix} \frac{\partial \overline{T}}{\partial w_x} \\ \frac{\partial \overline{T}}{\partial w_y} \\ \frac{\partial \overline{T}}{\partial w_z} \end{bmatrix} = \begin{bmatrix} 0 & -w_z & w_y \\ w_z & 0 & -w_x \\ -w_y & w_x & 0 \end{bmatrix} \begin{bmatrix} \frac{\partial \overline{T}}{\partial w_x} \\ \frac{\partial \overline{T}}{\partial w_y} \\ \frac{\partial \overline{T}}{\partial w_z} \end{bmatrix} = \begin{bmatrix} -w_z \frac{\partial \overline{T}}{\partial w_y} + w_y \frac{\partial \overline{T}}{\partial w_z} \\ w_z \frac{\partial \overline{T}}{\partial w_x} - w_x \frac{\partial \overline{T}}{\partial w_z} \\ -w_y \frac{\partial \overline{T}}{\partial w_x} + w_x \frac{\partial \overline{T}}{\partial w_y} \end{bmatrix} \tag{4.68}$$

11. The matrix product $-[\beta]^T \left[\frac{\partial \overline{T}}{\partial \phi} \; \frac{\partial \overline{T}}{\partial \theta} \; \frac{\partial \overline{T}}{\partial \psi} \right]^T$ may be shown to be:

$$-\begin{bmatrix} 1 & 0 & 0 \\ \sin\phi \tan\theta & \cos\phi & \sin\phi / \cos\theta \\ \cos\phi \tan\theta & -\sin\phi & \cos\phi / \cos\theta \end{bmatrix} \begin{bmatrix} \frac{\partial \overline{T}}{\partial \phi} \\ \frac{\partial \overline{T}}{\partial \theta} \\ \frac{\partial \overline{T}}{\partial \psi} \end{bmatrix}$$

$$= \begin{bmatrix} -\frac{\partial \overline{T}}{\partial \phi} \\ -\sin\phi \tan\theta \frac{\partial \overline{T}}{\partial \phi} - \cos\phi \frac{\partial \overline{T}}{\partial \theta} - \sin\phi / \cos\theta \frac{\partial \overline{T}}{\partial \psi} \\ -\cos\phi \tan\theta \frac{\partial \overline{T}}{\partial \phi} + \sin\phi \frac{\partial \overline{T}}{\partial \theta} - \cos\phi / \cos\theta \frac{\partial \overline{T}}{\partial \psi} \end{bmatrix} \tag{4.69}$$

12. Hence, combining $\left[\frac{d}{dt}\left(\frac{\partial \overline{T}}{\partial w_x}\right) \frac{d}{dt}\left(\frac{\partial \overline{T}}{\partial w_y}\right) \frac{d}{dt}\left(\frac{\partial \overline{T}}{\partial w_z}\right) \right]^T$ with the results in items 10 (Eqs. 4.68) and 11 (Eq. 4.69), respectively, we have the following set of d'Alembert–Lagrange differential equations which describe the rotational joint's motion as follows:

$$\begin{aligned} &\frac{d}{dt}\left(\frac{\partial \overline{T}}{\partial w_x}\right) - w_z \frac{\partial \overline{T}}{\partial w_y} + w_y \frac{\partial \overline{T}}{\partial w_z} - \frac{\partial \overline{T}}{\partial \phi} = \tau_1 \\ &\frac{d}{dt}\left(\frac{\partial \overline{T}}{\partial w_y}\right) + w_z \frac{\partial \overline{T}}{\partial w_x} - w_x \frac{\partial \overline{T}}{\partial w_z} - \sin\phi \tan\theta \frac{\partial \overline{T}}{\partial \phi} - \cos\phi \frac{\partial \overline{T}}{\partial \theta} - \frac{\sin\phi}{\cos\theta} \frac{\partial \overline{T}}{\partial \psi} = \tau_2 \\ &\frac{d}{dt}\left(\frac{\partial \overline{T}}{\partial w_z}\right) - w_y \frac{\partial \overline{T}}{\partial w_x} + w_x \frac{\partial \overline{T}}{\partial w_y} - \cos\phi \tan\theta \frac{\partial \overline{T}}{\partial \phi} + \sin\phi \frac{\partial \overline{T}}{\partial \theta} - \frac{\cos\phi}{\cos\theta} \frac{\partial \overline{T}}{\partial \psi} = \tau_3 \end{aligned} \tag{4.70}$$

where τ_1, τ_2, and τ_3 are the input torques in the body's x, y, and z directions, respectively, required to maintain the rotation of the joint.

Example 2 [8, pp. 14–15]

Consider the common tricycle which typifies a wide variety of three-wheeled mobile robots (see Fig. 4.1). This vehicle is subject to two non-holonomic constraints. The velocity of the front wheel is in the direction that it is pointing and this constitutes the first of the two non-holonomic constraints, i.e., the front wheel doesn't slide. The second constraint is related to the rear wheels which also don't slide from side to side—there is no side slip of the rear wheels. The system has four degrees of freedom, three of which describe the tricycle's position: x, y, θ. The additional degree of freedom is the steering angle ϕ. The four degrees of freedom are reduced by the two non-holonomic constraints to two. The two independent degrees of freedom are the velocity of the front wheel v and the front wheel's turning rate $\dot{\phi}$. The turning rate of the rear wheels $\dot{\theta}$ is dependent upon the forward velocity v and the front wheel's steering angle ϕ, that is, $\dot{\theta} = (v \sin \phi)/l$. The upshot of the fact that the system has only two degrees of freedom is that solely the d'Alembert–Lagrange equations of motion with the derivatives dv/dt and $d^2\phi/dt^2$ are of interest, i.e.,

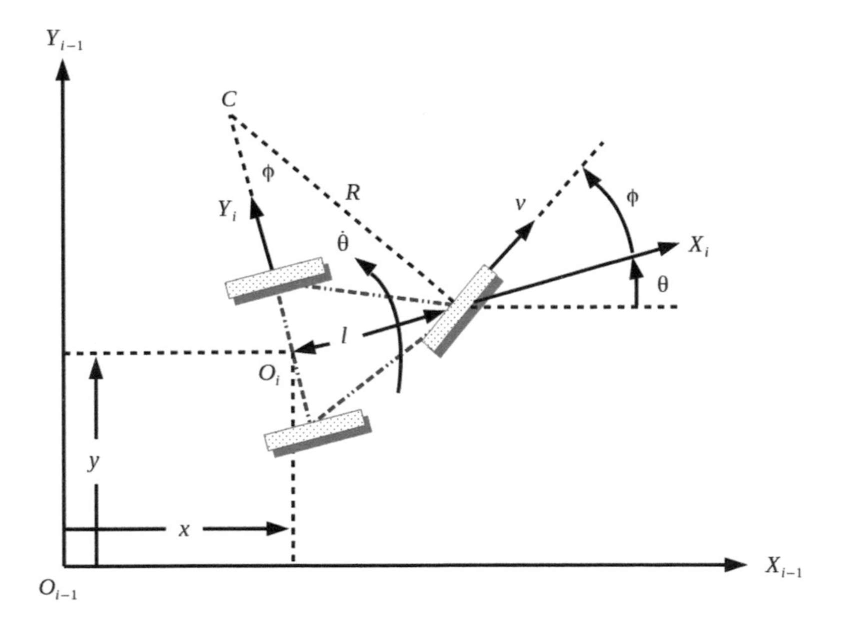

Fig. 4.1 Tricycle geometry and notation

factors in the d'Alembert–Lagrange equations with the term $d\theta/dt$ may be ignored. The equations which describe the tricycle's motion are the following:

$$\begin{aligned}\dot{x} &= v\ \cos\phi\ \cos\theta \\ \dot{y} &= v\ \cos\phi\ \sin\theta \\ \dot{\theta} &= \frac{v}{l}\sin\phi\end{aligned} \tag{4.71}$$

The velocity v may be solved for as follows:

$$\begin{aligned}\dot{x}\cos\theta &= v\ \cos\phi\ \cos^2\theta \\ \dot{y}\sin\theta &= v\ \cos\phi\ \sin^2\theta \\ \Rightarrow v &= \dot{x}\frac{\cos\theta}{\cos\phi} + \dot{y}\frac{\sin\theta}{\cos\phi}\end{aligned} \tag{4.72}$$

The two constraints are: front and rear wheels don't slip which implies that velocities in the Y_i and R directions (perpendicular to v), respectively, are zero. This may be written as (see Fig. 4.2):

$$v_{Yi} = \dot{y}\cos\theta - \dot{x}\sin\theta = 0;\ \ v_R \approx l\dot{\theta}\cos\theta\cos\phi - \dot{x}\sin\phi = 0 \tag{4.73}$$

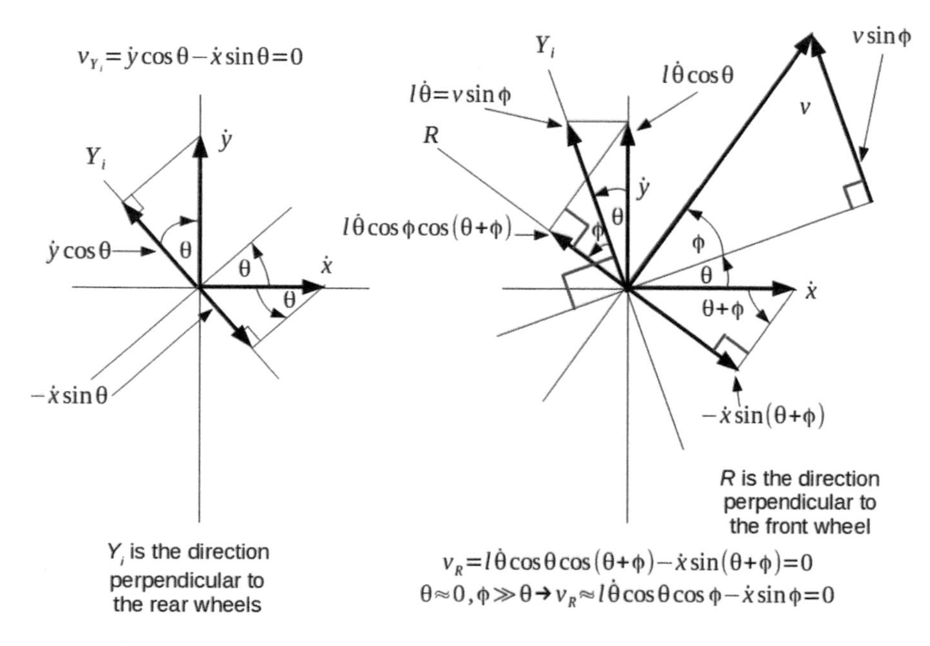

Fig. 4.2 Geometry for velocities perpendicular to both front and rear wheels

The equation relating the vector of quasi-velocities $\{\dot{\Gamma}\}$ to the vector of generalized velocities $\{\dot{q}\}$ is derived from the non-holonomic constraints $v_{Yi} = 0$, $v_R = 0$ and the forward velocity of the front wheel $v = \dot{x}(\cos\theta/\cos\phi) + \dot{y}(\sin\theta/\cos\phi)$, and may be written as:

$$\{\dot{\Gamma}\} = \begin{bmatrix} v \\ v_{Yi} \\ v_R \\ \dot{\phi} \end{bmatrix}; \ \{\dot{q}\} = \begin{bmatrix} \dot{x} \\ \dot{y} \\ \dot{\theta} \\ \dot{\phi} \end{bmatrix}$$

$$\{\dot{\Gamma}\} = \begin{bmatrix} v \\ v_{Yi} \\ v_R \\ \dot{\phi} \end{bmatrix} = \underbrace{\begin{bmatrix} \frac{\cos\theta}{\cos\phi} & \frac{\sin\theta}{\cos\phi} & 0 & 0 \\ -\sin\theta & \cos\theta & 0 & 0 \\ -\sin\phi & 0 & l\cos\theta\cos\phi & 0 \\ 0 & 0 & 0 & 1 \end{bmatrix}}_{[\vartheta]^T} \underbrace{\begin{bmatrix} \dot{x} \\ \dot{y} \\ \dot{\theta} \\ \dot{\phi} \end{bmatrix}}_{\{\dot{q}\}} \tag{4.74}$$

The inverse relationship $\{\dot{q}\} = [\beta]\{\dot{\Gamma}\}$ becomes:

$$\begin{bmatrix} \dot{x} \\ \dot{y} \\ \dot{\theta} \\ \dot{\phi} \end{bmatrix} = \underbrace{\begin{bmatrix} \cos\phi\cos\theta & -\sin\theta & 0 & 0 \\ \cos\phi\sin\theta & \cos\theta & 0 & 0 \\ \frac{\sin\phi}{l} & -\frac{\tan\phi\tan\theta}{l} & \frac{1}{l\cos\phi\cos\theta} & 0 \\ 0 & 0 & 0 & 1 \end{bmatrix}}_{[\beta]} \begin{bmatrix} v \\ v_{Yi} \\ v_R \\ \dot{\phi} \end{bmatrix} \tag{4.75}$$

The Lagrangian $\overline{\mathcal{L}}$ (assuming that there is no potential energy, only kinetic energy) may be written as:

$$\overline{\mathcal{L}} = \frac{1}{2}m\left(v^2 + v_{Yi}^2 + v_R^2\right) + \frac{1}{2}I_\phi\dot{\phi}^2 + \frac{1}{2}I_\theta\dot{\theta}^2 \tag{4.76}$$

where m is the mass of the tricycle, I_ϕ is the moment of inertia of the front wheel assembly at the joint connecting the front and rear wheels, relative to the steering angle rate $\dot{\phi}$, and I_θ is the moment of inertia of the rear wheel assembly at the joint connecting the front and rear wheels, relative to the tracking angle rate of the rear wheels $\dot{\theta}$. Although the Lagrangian $\overline{\mathcal{L}}$ contains the $\dot{\theta}$ term, in the sequel, it will be seen that $\partial\overline{\mathcal{L}}/\partial\dot{\theta}$ doesn't enter into any of the calculations. Recalling the discussion prior to the introduction of the example on pp. 181, the d'Alembert–Lagrange equations for quasi-velocities are:

$$\frac{d}{dt}\left\{\frac{\partial \overline{\mathcal{L}}}{\partial \dot{\Gamma}}\right\} + [\beta]^T [\Lambda] \left\{\frac{\partial \overline{\mathcal{L}}}{\partial \dot{\Gamma}}\right\} - [\beta]^T \left\{\frac{\partial \overline{\mathcal{L}}}{\partial q}\right\} = \{N\} \tag{4.77}$$

where

$$[\Lambda] = \left[\dot{\vartheta}\right] - [\eta]$$

$$\left\{\frac{\partial \bar{T}}{\partial \dot{\Gamma}}\right\} = \begin{bmatrix} \frac{\partial \bar{T}}{\partial v} \\ \frac{\partial \bar{T}}{\partial v_{Yi}} \\ \frac{\partial \bar{T}}{\partial v_R} \\ \frac{\partial \bar{T}}{\partial \dot{\phi}} \end{bmatrix}; \quad \left\{\frac{\partial \bar{T}}{\partial q}\right\} = \begin{bmatrix} \frac{\partial \bar{T}}{\partial x} \\ \frac{\partial \bar{T}}{\partial y} \\ \frac{\partial \bar{T}}{\partial \theta} \\ \frac{\partial \bar{T}}{\partial \phi} \end{bmatrix} \tag{4.78}$$

Differentiation with respect to time of $[\vartheta]$ may be written as:

$$\frac{d}{dt}[\vartheta] = \frac{d}{dt}\begin{bmatrix} \frac{\cos\theta}{\cos\phi} & -\sin\theta & -\sin\phi & 0 \\ \frac{\sin\theta}{\cos\phi} & \cos\theta & 0 & 0 \\ 0 & 0 & l\cos\theta\cos\phi & 0 \\ 0 & 0 & 0 & 1 \end{bmatrix}$$

$$= \begin{bmatrix} \left(\frac{\dot{\phi}\cos\theta\sin\phi}{\cos^2\phi} - \frac{\dot{\theta}\sin\theta}{\cos\phi}\right) & -\dot{\theta}\cos\theta & -\dot{\phi}\cos\phi & 0 \\ \left(\frac{\dot{\phi}\sin\theta\sin\phi}{\cos^2\phi} + \frac{\dot{\theta}\cos\theta}{\cos\phi}\right) & -\dot{\theta}\sin\theta & 0 & 0 \\ 0 & 0 & -\left(l\dot{\phi}\cos\theta\sin\phi + l\dot{\theta}\cos\phi\sin\theta\right) & 0 \\ 0 & 0 & 0 & 0 \end{bmatrix} \tag{4.79}$$

The matrix $[\eta]$ is composed as follows:

$$[\eta] = \begin{bmatrix} \left\{\dot{\Gamma}\right\}^T [\beta]^T \left[\frac{\partial \vartheta}{\partial x}\right] \\ \left\{\dot{\Gamma}\right\}^T [\beta]^T \left[\frac{\partial \vartheta}{\partial y}\right] \\ \left\{\dot{\Gamma}\right\}^T [\beta]^T \left[\frac{\partial \vartheta}{\partial \theta}\right] \\ \left\{\dot{\Gamma}\right\}^T [\beta]^T \left[\frac{\partial \vartheta}{\partial \phi}\right] \end{bmatrix} \tag{4.80}$$

The term $\left\{\dot{\Gamma}\right\}^T [\beta]^T$ is of the form:

$$\left\{\dot{\Gamma}\right\}^T [\beta]^T = \begin{bmatrix} v & v_{Yi} & v_R & \dot{\phi} \end{bmatrix} \begin{bmatrix} \cos\phi\cos\theta & \cos\phi\sin\theta & \frac{\sin\phi}{l} & 0 \\ -\sin\theta & \cos\theta & -\frac{\tan\phi\tan\theta}{l} & 0 \\ 0 & 0 & \frac{1}{l\cos\phi\cos\theta} & 0 \\ 0 & 0 & 0 & 1 \end{bmatrix}$$

$$= \begin{bmatrix} (v\cos\phi\cos\theta - v_{Yi}\sin\theta) \\ (v\cos\phi\sin\theta + v_{Yi}\cos\theta) \\ \left(\frac{v\sin\theta}{l} - \frac{v_{Yi}\tan\theta\tan\phi}{l} + \frac{v_R}{l\cos\theta\cos\phi}\right) \\ \dot{\phi} \end{bmatrix}^T \tag{4.81}$$

The remaining elements which make up the matrix $[\eta]$ are:

$$\left[\frac{\partial \vartheta}{\partial x}\right] = \begin{bmatrix} 0&0&0&0 \\ 0&0&0&0 \\ 0&0&0&0 \\ 0&0&0&0 \end{bmatrix}; \quad \left[\frac{\partial \vartheta}{\partial y}\right] = \begin{bmatrix} 0&0&0&0 \\ 0&0&0&0 \\ 0&0&0&0 \\ 0&0&0&0 \end{bmatrix} \tag{4.82}$$

$$\left[\frac{\partial\vartheta}{\partial\theta}\right]=\begin{bmatrix}-\frac{\sin\theta}{\cos\phi} & -\cos\theta & 0 & 0\\ \frac{\cos\theta}{\cos\phi} & -\sin\theta & 0 & 0\\ 0 & 0 & -l\cos\phi\sin\theta & 0\\ 0 & 0 & 0 & 0\end{bmatrix} \tag{4.83}$$

$$\left[\frac{\partial\vartheta}{\partial\phi}\right]=\begin{bmatrix}\frac{\cos\theta\sin\phi}{\cos^2\phi} & 0 & -\cos\phi & 0\\ \frac{\sin\phi\sin\theta}{\cos^2\phi} & 0 & 0 & 0\\ 0 & 0 & -l\cos\theta\sin\phi & 0\\ 0 & 0 & 0 & 0\end{bmatrix} \tag{4.84}$$

Carrying out the prescribed operations results in:

$$\left\{\dot{\Gamma}\right\}^T[\beta]^T\left[\frac{\partial\vartheta}{\partial x}\right]=\left\{\dot{\Gamma}\right\}^T[\beta]^T\begin{bmatrix}0&0&0&0\\0&0&0&0\\0&0&0&0\\0&0&0&0\end{bmatrix}=\begin{bmatrix}0&0&0&0\end{bmatrix} \tag{4.85}$$

$$\left\{\dot{\Gamma}\right\}^T[\beta]^T\left[\frac{\partial\vartheta}{\partial y}\right]=\left\{\dot{\Gamma}\right\}^T[\beta]^T\begin{bmatrix}0&0&0&0\\0&0&0&0\\0&0&0&0\\0&0&0&0\end{bmatrix}=\begin{bmatrix}0&0&0&0\end{bmatrix} \tag{4.86}$$

$$\left\{\dot{\Gamma}\right\}^T[\beta]^T\left[\frac{\partial\vartheta}{\partial\theta}\right]$$

$$=\begin{bmatrix}(v\cos\phi\cos\theta-v_{Yi}\sin\theta)\\ (v\cos\phi\sin\theta+v_{Yi}\cos\theta)\\ \left(\frac{v\sin\theta}{l}-\frac{v_{Yi}\tan\theta\tan\phi}{l}+\frac{v_R}{l\cos\theta\cos\phi}\right)\\ \dot{\phi}\end{bmatrix}^T\begin{bmatrix}-\frac{\sin\theta}{\cos\phi} & -\cos\theta & 0 & 0\\ \frac{\cos\theta}{\cos\phi} & -\sin\theta & 0 & 0\\ 0 & 0 & -l\cos\phi\sin\theta & 0\\ 0 & 0 & 0 & 0\end{bmatrix}$$

$$= \begin{bmatrix} \frac{\cos\theta(v_{Yi}\ \cos\theta + v\ \cos\phi\ \sin\theta)}{\cos\phi} + \frac{\sin\theta(v_{Yi}\ \sin\theta - v\ \cos\phi\ \cos\theta)}{\cos\phi} \\ \cos\theta\ (v_{Yi}\ \sin\theta - v\ \cos\phi\ \cos\theta) - \sin\theta\ (v_{Yi}\ \cos\theta + v\ \cos\phi\ \sin\theta) \\ -l\ \cos\phi\ \sin\theta\ \left(\frac{v\ \sin\phi}{l} + \frac{v_R}{l\ \cos\phi\ \cos\theta} - \frac{v_{Yi}\ \sin\phi\ \sin\theta}{l\ \cos\phi\ \cos\theta}\right) \\ 0 \end{bmatrix}^T$$

$$= \left[\frac{v_{Yi}}{\cos\phi}\ -v\cos\phi\ \tan\theta\ (v_{Yi}\ \sin\phi\ \sin\theta - v_R) - v\ \cos\phi\ \sin\theta\ \sin\phi\ 0\right] \tag{4.87}$$

$$\{\dot{\Gamma}\}^T\,[\beta]^T \left[\frac{\partial\vartheta}{\partial\phi}\right]$$

$$= \begin{bmatrix} (v\cos\phi\cos\theta - v_{Yi}\sin\theta) \\ (v\cos\phi\sin\theta + v_{Yi}\cos\theta) \\ \left(\frac{v\sin\theta}{l} - \frac{v_{Yi}\tan\theta\tan\phi}{l} + \frac{v_R}{l\cos\theta\cos\phi}\right) \\ \dot{\phi} \end{bmatrix}^T \begin{bmatrix} \frac{\cos\theta\sin\phi}{\cos^2\phi} & 0 & -\cos\phi & 0 \\ \frac{\sin\phi\sin\theta}{\cos^2\phi} & 0 & 0 & 0 \\ 0 & 0 & -l\cos\theta\sin\phi & 0 \\ 0 & 0 & 0 & 0 \end{bmatrix}$$

$$= \begin{bmatrix} \frac{\sin\phi\ \sin\theta\ (v_{Yi}\ \cos\theta + v\ \cos\phi\ \sin\theta)}{\cos^2\phi} - \frac{\cos\theta\ \sin\phi\ (v_{Yi}\ \sin\theta - v\ \cos\phi\ \cos\theta)}{\cos^2\phi} \\ 0 \\ \cos\phi\ (v_{Yi}\ \sin\theta - v\ \cos\phi\ \cos\theta) - l\ \cos\theta\ \sin\phi\ \left(\frac{v\ \sin\phi}{l} + \frac{v_R}{l\ \cos\phi\ \cos\theta} - \frac{v_{Yi}\ \sin\phi\ \sin\theta}{l\ \cos\phi\ \cos\theta}\right) \\ 0 \end{bmatrix}^T$$

$$= \left[v\ \tan\phi\ 0\ v_{Yi}\frac{\sin\theta}{\cos\phi} - v_R\tan\phi - v\cos\theta\ 0\right] \tag{4.88}$$

Combining the above results, the matrix $[\eta]$ becomes:

$$[\eta] = \begin{bmatrix} \{\dot{\Gamma}\}^T [\beta]^T \left[\frac{\partial \vartheta}{\partial x}\right] \\ \{\dot{\Gamma}\}^T [\beta]^T \left[\frac{\partial \vartheta}{\partial y}\right] \\ \{\dot{\Gamma}\}^T [\beta]^T \left[\frac{\partial \vartheta}{\partial \theta}\right] \\ \{\dot{\Gamma}\}^T [\beta]^T \left[\frac{\partial \vartheta}{\partial \phi}\right] \end{bmatrix}$$

$$= \begin{bmatrix} 0 & 0 & 0 & 0 \\ 0 & 0 & 0 & 0 \\ \frac{v_{Yi}}{\cos\phi} & -v\cos\phi\,\tan\theta & (v_{Yi}\,\sin\phi\,\sin\theta - v_R) - v\,\cos\phi\,\sin\theta\,\sin\phi & 0 \\ v\,\tan\phi & 0 & v_{Yi}\frac{\sin\theta}{\cos\phi} - v_R\tan\phi - v\cos\theta & 0 \end{bmatrix} \tag{4.89}$$

The matrix $[\Lambda] = \left[\dot{\vartheta}\right] - [\eta]$ may be written as:

$$[\Lambda] = \left[\dot{\vartheta}\right] - [\eta]$$

$$= \begin{bmatrix} \left(\frac{\dot{\phi}\cos\theta\sin\phi}{\cos^2\phi} - \frac{\dot{\theta}\sin\theta}{\cos\phi}\right) & -\dot{\theta}\cos\theta & -\dot{\phi}\cos\phi & 0 \\ \left(\frac{\dot{\phi}\sin\theta\sin\phi}{\cos^2\phi} + \frac{\dot{\theta}\cos\theta}{\cos\phi}\right) & -\dot{\theta}\sin\theta & 0 & 0 \\ 0 & 0 & -\left(l\dot{\phi}\cos\theta\sin\phi + l\dot{\theta}\cos\phi\sin\theta\right) & 0 \\ 0 & 0 & 0 & 0 \end{bmatrix}$$

$$- \begin{bmatrix} 0 & 0 & 0 & 0 \\ 0 & 0 & 0 & 0 \\ \frac{v_{Yi}}{\cos\phi} & -v\cos\phi\,\tan\theta & (v_{Yi}\,\sin\phi\,\sin\theta - v_R) - v\,\cos\phi\,\sin\theta\,\sin\phi & 0 \\ \frac{v\,\sin\phi}{\cos\phi} & 0 & v_{Yi}\frac{\sin\theta}{\cos\phi} - v_R\tan\phi - v\cos\theta & 0 \end{bmatrix}$$

$$= \begin{bmatrix} \left(\frac{\dot{\phi}\cos\theta\sin\phi}{\cos^2\phi} - \frac{\dot{\theta}\sin\theta}{\cos\phi}\right) & -\dot{\theta}\cos\theta & -\dot{\phi}\cos\phi & 0 \\ \left(\frac{\dot{\phi}\sin\theta\sin\phi}{\cos^2\phi} + \frac{\dot{\theta}\cos\theta}{\cos\phi}\right) & -\dot{\theta}\sin\theta & 0 & 0 \\ -\frac{v_{Yi}}{\cos\phi} & v\cos\phi & \begin{array}{c} -\tan\theta\ (v_{Yi}\ \sin\phi\ \sin\theta - v_R) + v\ \cos\phi\ \sin\theta\ \sin\phi \\ -\left(l\dot{\phi}\cos\theta\sin\phi + l\dot{\theta}\cos\phi\sin\theta\right) \end{array} & 0 \\ -v\tan\phi & 0 & -v_{Yi}\frac{\sin\theta}{\cos\phi} + v_R\tan\phi + v\cos\theta & 0 \end{bmatrix} \tag{4.90}$$

Using the fact that $v_R = v_{Yi} = 0$, the matrix $[\Lambda]$ may be simplified to be:

$$[\Lambda] = \begin{bmatrix} \left(\frac{\dot{\phi}\cos\theta\sin\phi}{\cos^2\phi} - \frac{\dot{\theta}\sin\theta}{\cos\phi}\right) & -\dot{\theta}\cos\theta & -\dot{\phi}\cos\phi & 0 \\ \left(\frac{\dot{\phi}\sin\theta\sin\phi}{\cos^2\phi} + \frac{\dot{\theta}\cos\theta}{\cos\phi}\right) & -\dot{\theta}\sin\theta & 0 & 0 \\ 0 & v\cos\phi & \begin{array}{c} v\ \cos\phi\ \sin\theta\ \sin\phi \\ -\left(l\dot{\phi}\cos\theta\sin\phi + l\dot{\theta}\cos\phi\sin\theta\right) \end{array} & 0 \\ -\frac{v\ \sin\phi}{\cos\phi} & 0 & v\cos\theta & 0 \end{bmatrix} \tag{4.91}$$

Because the only independent generalized coordinates are v and ϕ, the dynamic equations of interest will be those related to v and ϕ. The coefficients of interest can then be obtained by pre-multiplying the matrix $[\Lambda]$, by the appropriate rows of the $[\beta]^T$ matrix. Since our interest lies only in v and ϕ, the matrix $[\Lambda]$ should be pre-multiplied by the first and fourth rows of $[\beta]^T$, respectively, to obtain the desired coefficients. We assume the following:

1. $\cos\phi\sin\phi \approx 0$
2. $\cos\phi\sin\theta \approx 0$
3. $v_{Yi} = 0$, $v_R = 0$
4. $\dot{\theta}$ may be ignored in the dynamic equations
5. The only dynamics of interest are related to v and ϕ

The product of $\left[\beta^T\right]_{1st\ row} * [\Lambda]$ becomes:

$$\left[\cos\phi\cos\theta\ \cos\phi\sin\theta\ \frac{\sin\phi}{l}\ 0\right]$$

$$* \begin{bmatrix} \left(\frac{\dot{\phi}\cos\theta\sin\phi}{\cos^2\phi} - \frac{\dot{\theta}\sin\theta}{\cos\phi}\right) & -\dot{\theta}\cos\theta & -\dot{\phi}\cos\phi & 0 \\ \left(\frac{\dot{\phi}\sin\theta\sin\phi}{\cos^2\phi} + \frac{\dot{\theta}\cos\theta}{\cos\phi}\right) & -\dot{\theta}\sin\theta & 0 & 0 \\ 0 & v\cos\phi & \begin{matrix} v\cos\phi\ \sin\theta\ \sin\phi \\ -\left(l\dot{\phi}\cos\theta\sin\phi + l\dot{\theta}\cos\phi\sin\theta\right) \end{matrix} & 0 \\ -\frac{v\ \sin\phi}{\cos\phi} & 0 & v\cos\theta & 0 \end{bmatrix}$$

$$= \begin{bmatrix} \cos\phi\cos\theta\left(\frac{\dot{\phi}\cos\theta\sin\phi}{\cos^2\phi} - \frac{\dot{\theta}\sin\theta}{\cos\phi}\right) + \cos\phi\sin\theta\left(\frac{\dot{\phi}\sin\theta\sin\phi}{\cos^2\phi} + \frac{\dot{\theta}\cos\theta}{\cos\phi}\right) \\ -\dot{\theta}\cos\phi\cos^2\theta - \dot{\theta}\cos\phi\sin^2\theta + \frac{\sin\phi}{l}v\cos\phi \\ -\dot{\phi}\cos^2\phi\cos\theta + \frac{\sin\phi}{l}\begin{bmatrix} v\ \cos\phi\ \sin\theta\ \sin\phi \\ -\left(l\dot{\phi}\cos\theta\sin\phi + l\dot{\theta}\cos\phi\sin\theta\right) \end{bmatrix} \\ 0 \end{bmatrix}^T$$

$$= \left[\ \dot{\phi}\tan\phi\ \ -\dot{\theta}\cos\phi + \tfrac{v}{l}\sin\phi\cos\phi\ \ \tfrac{v\cos\phi\sin^2\phi\sin\theta}{l} - \dot{\phi}\cos\theta - \dot{\theta}\cos\phi\sin\phi\sin\theta\ \ 0\right]$$

$$\approx \left[\dot{\phi}\tan\phi\ \ 0\ \ -\dot{\phi}\cos\theta\ \ 0\right] \tag{4.92}$$

Hence, the result of $\left[\beta^T\right]_{1st\,row} * [\Lambda]\left\{\frac{\partial\overline{\mathcal{L}}}{\partial\dot{\Gamma}}\right\}$ becomes:

$$\left[\beta^T\right]_{1st\,row} * [\Lambda]\left\{\frac{\partial\overline{\mathcal{L}}}{\partial\dot{\Gamma}}\right\} = \left[\dot{\phi}\tan\phi\ \ 0\ \ -\dot{\phi}\cos\theta\ \ 0\right]\begin{bmatrix} \frac{\partial\overline{\mathcal{L}}}{\partial v} \\ \frac{\partial\overline{\mathcal{L}}}{\partial v_{Yi}} \\ \frac{\partial\overline{\mathcal{L}}}{\partial v_R} \\ \frac{\partial\overline{\mathcal{L}}}{\partial\dot{\phi}} \end{bmatrix}$$

$$= \dot{\phi}\tan\phi\frac{\partial\overline{\mathcal{L}}}{\partial v} - \dot{\phi}\cos\theta\frac{\partial\overline{\mathcal{L}}}{\partial v_R} \tag{4.93}$$

Note that all terms related to $\dot{\theta}$ are not taken into account, that is, $\dot{\theta}\cos\phi\sin\phi\sin\theta \approx 0$, since the only independent quasi-velocity and generalized coordinates are v and $\dot{\phi}$, respectively. Using exactly the same procedure as above, the term $[\beta^T]_{4th\,row} * [\Lambda]$ may be shown to be:

$$\begin{bmatrix} 0 & 0 & 0 & 1 \end{bmatrix}$$

$$* \begin{bmatrix} \left(\frac{\dot{\phi}\cos\theta\sin\phi}{\cos^2\phi} - \frac{\dot{\theta}\sin\theta}{\cos\phi}\right) & -\dot{\theta}\cos\theta & -\dot{\phi}\cos\phi & 0 \\ \left(\frac{\dot{\phi}\sin\theta\sin\phi}{\cos^2\phi} + \frac{\dot{\theta}\cos\theta}{\cos\phi}\right) & -\dot{\theta}\sin\theta & 0 & 0 \\ 0 & v\cos\phi & \begin{array}{c} v\cos\phi\,\sin\theta\,\sin\phi \\ -\left(l\dot{\phi}\cos\theta\sin\phi + l\dot{\theta}\cos\phi\sin\theta\right) \end{array} & 0 \\ -\frac{v\sin\phi}{\cos\phi} & 0 & v\cos\theta & 0 \end{bmatrix}$$

$$= \begin{bmatrix} -v\tan\phi & 0 & v\cos\theta & 0 \end{bmatrix} \tag{4.94}$$

Hence, the result of $[\beta^T]_{4th\,row} * [\Lambda]\left\{\frac{\partial\overline{\mathcal{L}}}{\partial\Gamma}\right\}$ becomes:

$$\left[\beta^T\right]_{4th\,row} * [\Lambda]\left\{\frac{\partial\overline{\mathcal{L}}}{\partial\dot{\Gamma}}\right\} = \begin{bmatrix} -v\tan\phi & 0 & v\cos\theta & 0 \end{bmatrix} \begin{bmatrix} \frac{\partial\overline{\mathcal{L}}}{\partial v} \\ \frac{\partial\overline{\mathcal{L}}}{\partial v_{Yi}} \\ \frac{\partial\overline{\mathcal{L}}}{\partial v_R} \\ \frac{\partial\overline{\mathcal{L}}}{\partial\dot{\phi}} \end{bmatrix}$$

$$= -v\tan\phi\frac{\partial\overline{\mathcal{L}}}{\partial v} + v\cos\theta\frac{\partial\overline{\mathcal{L}}}{\partial v_R} \tag{4.95}$$

The additional terms in the d'Alembert–Lagrange equations

$$\frac{d}{dt}\left\{\frac{\partial\overline{\mathcal{L}}}{\partial\dot{\Gamma}}\right\} + [\beta]^T[\Lambda]\left\{\frac{\partial\overline{\mathcal{L}}}{\partial\dot{\Gamma}}\right\} - [\beta]^T\left\{\frac{\partial\overline{\mathcal{L}}}{\partial q}\right\} = \{N\} \tag{4.96}$$

are: $\frac{d}{dt}\left\{\frac{\partial\overline{\mathcal{L}}}{\partial\Gamma}\right\}$ and $-[\beta]^T\left\{\frac{\partial\overline{\mathcal{L}}}{\partial q}\right\}$, respectively. For the $-[\beta]^T\left\{\frac{\partial\overline{\mathcal{L}}}{\partial q}\right\}$ terms, only the first and fourth rows of $[\beta]^T$ are required and the results are:

$$-\left[\beta^T\right]_{1st\,row}\begin{bmatrix}\frac{\partial\overline{\mathcal{L}}}{\partial x}\\ \frac{\partial\overline{\mathcal{L}}}{\partial y}\\ \frac{\partial\overline{\mathcal{L}}}{\partial\theta}\\ \frac{\partial\overline{\mathcal{L}}}{\partial\phi}\end{bmatrix}=-\left[\cos\phi\cos\theta\ \cos\phi\sin\theta\ \frac{\sin\phi}{l}\ 0\right]\begin{bmatrix}\frac{\partial\bar{L}}{\partial x}\\ \frac{\partial\overline{\mathcal{L}}}{\partial y}\\ \frac{\partial\overline{\mathcal{L}}}{\partial\theta}\\ \frac{\partial\overline{\mathcal{L}}}{\partial\phi}\end{bmatrix}$$

$$=-\cos\phi\cos\theta\frac{\partial\overline{\mathcal{L}}}{\partial x}-\cos\phi\sin\theta\frac{\partial\bar{L}}{\partial y}-\frac{\sin\phi}{l}\frac{\partial\overline{\mathcal{L}}}{\partial\dot{\theta}}$$

$$-\left[\beta^T\right]_{4th\,row}\begin{bmatrix}\frac{\partial\overline{\mathcal{L}}}{\partial x}\\ \frac{\partial\overline{\mathcal{L}}}{\partial y}\\ \frac{\partial\overline{\mathcal{L}}}{\partial\theta}\\ \frac{\partial\overline{\mathcal{L}}}{\partial\phi}\end{bmatrix}=-\left[0\,0\,0\,1\right]\begin{bmatrix}\frac{\partial\overline{\mathcal{L}}}{\partial x}\\ \frac{\partial\overline{\mathcal{L}}}{\partial y}\\ \frac{\partial\overline{\mathcal{L}}}{\partial\theta}\\ \frac{\partial\overline{\mathcal{L}}}{\partial\phi}\end{bmatrix}=-\frac{\partial\overline{\mathcal{L}}}{\partial\phi}\tag{4.97}$$

Finally, the relevant terms of $\frac{d}{dt}\left\{\frac{\partial\overline{\mathcal{L}}}{\partial\Gamma}\right\}$, that is, the first and fourth terms of the vector are:

$$\frac{d}{dt}\left\{\frac{\partial\overline{\mathcal{L}}}{\partial v}\right\};\ \frac{d}{dt}\left\{\frac{\partial\overline{\mathcal{L}}}{\partial\dot{\phi}}\right\}$$

Combining all of the foregoing results, the relevant d'Alembert–Lagrange equations become:

$$\begin{aligned}&\frac{d}{dt}\left\{\frac{\partial\bar{L}}{\partial v}\right\}-\cos\phi\cos\theta\frac{\partial\bar{L}}{\partial x}-\cos\phi\sin\theta\frac{\partial\bar{L}}{\partial y}-\frac{\sin\phi}{l}\frac{\partial\bar{L}}{\partial\dot{\theta}}\\ &+\dot{\phi}\tan\phi\frac{\partial\bar{L}}{\partial v}-\dot{\phi}\cos\theta\frac{\partial\bar{L}}{\partial v_R}=F_v\\ &\frac{d}{dt}\left\{\frac{\partial\bar{L}}{\partial\dot{\phi}}\right\}-\frac{\partial\bar{L}}{\partial\phi}-v\tan\phi\frac{\partial\bar{L}}{\partial v}+v\cos\theta\frac{\partial\bar{L}}{\partial v_R}=\tau\end{aligned}\tag{4.98}$$

where F_v is the external generalized force applied to the tricycle and τ is the external generalized torque.

4.4 Lagrangian Dynamics with Quasi-Coordinates: Prof. Ranjan Vepa's Approach

The following section is based on the approach by Prof. Ranjan Vepa (see Vepa [50, pp. 69–72]).

For most systems where quasi-velocities are involved, the kinetic energy is a function of both the generalized coordinates ψ, θ, ϕ and the quasi-velocities $[p_b, q_b, r_b]^T$. Hence there are two approaches to writing the Lagrangian dynamics. In the first approach, the quasi-velocities are written in terms of the generalized velocities and generalized coordinates, that is:

$$\begin{bmatrix} p_b \\ q_b \\ r_b \end{bmatrix} = \begin{bmatrix} \dot{\phi} - \dot{\psi} \sin\theta \\ \dot{\theta} \cos\phi + \dot{\psi} \cos\theta \sin\phi \\ -\dot{\theta} \sin\phi + \dot{\psi} \cos\theta \cos\phi \end{bmatrix} \tag{4.99}$$

These terms are then substituted into the energy equations and subsequently the system's Lagrangian dynamics are derived. In the second approach, the mixed quasi-velocities and generalized coordinates are retained but the Lagrangian dynamics must be modified accordingly. The method to be developed in the sequel, in order to obtain the Lagrangian dynamics for the mixed quasi-velocities and generalized coordinates, was adapted from the work of Ranjan Vepa [50, pp. 69–72]. The time derivative of a rotation matrix will be required in the sequel and so it is now introduced.

Time Derivative of a Rotation Matrix (See Britting [6, pp. 16–17], and Noureldin et al. [26, pp. 37–38, 43–45])

Consider a time varying matrix $T_{a/b}(t)$ which represents the Euler transformation between the coordinate frames F_a and F_b. That is, a vector in frame F_b is transformed (rotated) to a vector in frame F_a by $T_{a/b}(t)$. Frame F_b rotates with angular velocity Ω relative to F_a, which we may regard as fixed. At time t, the F_a and F_b frames are related through the Euler rotation matrix, $T_{a/b}(t)$. During the next instant of time, Δt, frame F_b rotates to a new orientation such that the Euler rotation matrix at $t + \Delta t$ is given by $T_{a/b}(t + \Delta t)$. By definition, the time rate of change of $\dot{T}_{a/b}(t)$ is given by:

$$\dot{T}_{a/b}(t) = \lim_{\Delta t \to 0} \frac{\Delta T_{a/b}(t)}{\Delta t} = \lim_{\Delta t \to 0} \frac{T_{a/b}(t + \Delta t) - T_{a/b}(t)}{\Delta t} \tag{4.100}$$

Take, for example, the rotation matrix from the body frame to the inertial axis system, that is:

$$T_{I/B}(t) = \begin{bmatrix} \cos\theta\cos\psi & \sin\phi\sin\theta\cos\psi - \cos\phi\sin\psi & \cos\phi\sin\theta\cos\psi + \sin\phi\sin\psi \\ \cos\theta\sin\psi & \sin\phi\sin\theta\sin\psi + \cos\phi\cos\psi & \cos\phi\sin\theta\sin\psi - \sin\phi\cos\psi \\ -\sin\theta & \sin\phi\cos\theta & \cos\phi\cos\theta \end{bmatrix} \tag{4.101}$$

Allowing for very small angular rotations from time t to time $t + \Delta t$, such that $\phi' = \phi + \Delta\phi, \psi' = \psi + \Delta\psi, \theta' = \theta + \Delta\theta$, the rotation matrix $T_{I/B}(t + \Delta t)$ becomes:

$$T_{I/B}(t + \Delta t)$$

$$= \begin{bmatrix} \cos\theta'\cos\psi' & \sin\phi'\sin\theta'\cos\psi' - \cos\phi'\sin\psi' & \cos\phi'\sin\theta'\cos\psi' + \sin\phi'\sin\psi' \\ \cos\theta'\sin\psi' & \sin\phi'\sin\theta'\sin\psi' + \cos\phi'\cos\psi' & \cos\phi'\sin\theta'\sin\psi' - \sin\phi'\cos\psi' \\ -\sin\theta' & \sin\phi'\cos\theta' & \cos\phi'\cos\theta' \end{bmatrix} \tag{4.102}$$

Furthermore assuming that the Δ angles are very small, so that $\Delta\phi \to 0, \Delta\psi \to 0, \Delta\theta \to 0$, and $\cos\Delta\phi = \cos\Delta\theta = \cos\Delta\psi = 1, \sin\Delta\phi = \Delta\phi, \sin\Delta\psi = \Delta\psi, \sin\Delta\theta = \Delta\theta;\ \Delta\phi\Delta\psi = \Delta\phi\Delta\theta = \Delta\psi\Delta\theta \approx 0$, we have:

1. $T_{I/B}(t + \Delta t)(1,1) = \boldsymbol{\cos\psi\cos\theta} - \Delta\psi\cos\theta\sin\psi - \Delta\theta\cos\psi\sin\theta$
2. $T_{I/B}(t + \Delta t)(1,2) = \boldsymbol{\cos\psi\sin\phi\sin\theta - \cos\phi\sin\psi} + \Delta\phi\sin\phi\sin\psi - \Delta\psi\cos\phi\cos\psi + \Delta\phi\cos\phi\cos\psi\sin\theta + \Delta\theta\cos\psi\cos\theta\sin\phi - \Delta\psi\sin\phi\sin\psi\sin\theta$
3. $T_{I/B}(t + \Delta t)(1,3) = \boldsymbol{\sin\phi\sin\psi + \cos\phi\cos\psi\sin\theta} + \Delta\phi\cos\phi\sin\psi + \Delta\psi\cos\psi\sin\phi + \Delta\theta\cos\phi\cos\psi\cos\theta - \Delta\phi\cos\psi\sin\phi\sin\theta - \Delta\psi\cos\phi\sin\psi\sin\theta$
4. $T_{I/B}(t + \Delta t)(2,1) = \boldsymbol{\cos\theta\sin\psi} + \Delta\psi\cos\psi\cos\theta - \Delta\theta\sin\psi\sin\theta$
5. $T_{I/B}(t + \Delta t)(2,2) = \boldsymbol{\cos\phi\cos\psi + \sin\phi\sin\psi\sin\theta} - \Delta\phi\cos\psi\sin\phi - \Delta\psi\cos\phi\sin\psi + \Delta\phi\cos\phi\sin\psi\sin\theta + \Delta\psi\cos\psi\sin\phi\sin\theta + \Delta\theta\cos\theta\sin\phi\sin\psi$
6. $T_{I/B}(t + \Delta t)(2,3) = \boldsymbol{\cos\phi\sin\psi\sin\theta - \cos\psi\sin\phi} + \Delta\psi\sin\phi\sin\psi - \Delta\phi\cos\phi\cos\psi + \Delta\psi\cos\phi\cos\psi\sin\theta + \Delta\theta\cos\phi\cos\theta\sin\psi - \Delta\phi\sin\phi\sin\psi\sin\theta$
7. $T_{I/B}(t + \Delta t)(3,1) = \boldsymbol{-\sin\theta} - \Delta\theta\cos\theta$
8. $T_{I/B}(t + \Delta t)(3,2) = \boldsymbol{\cos\theta\sin\phi} + \Delta\phi\cos\phi\cos\theta - \Delta\theta\sin\phi\sin\theta$
9. $T_{I/B}(t + \Delta t)(3,3) = \boldsymbol{\cos\phi\cos\theta} - \Delta\phi\cos\theta\sin\phi - \Delta\theta\cos\phi\sin\theta$

Note that the bold items in the above list constitute the elements of the $T_{I/B}(t)$ matrix. Hence subtracting the matrix $T_{I/B}(t)$ from $T_{I/B}(t + \Delta t)$, dividing throughout by Δt and letting $\lim_{\Delta t \to 0}$, we are left with:

$$\lim_{\Delta t \to 0} \frac{T_{I/B}(t+\Delta t) - T_{I/B}(t)}{\Delta t} = \lim_{\Delta t \to 0} \frac{\Delta T_{I/B}(t)}{\Delta t} = \frac{dT_{I/B}(t)}{dt}$$

Postulating that $T_{I/B}(t+\Delta t) = T_{I/B}(t)[I + Y(t)]$ allows us to solve for $T_{I/B}(t + \Delta t) - T_{I/B}(t)$ as follows:

$$T_{I/B}(t+\Delta t) = T_{I/B}(t)[I + Y(t)] \Rightarrow Y(t) = T_{I/B}^{T}(t)\left(T_{I/B}(t+\Delta t) - T_{I/B}(t)\right)$$
$$\Rightarrow T_{I/B}(t+\Delta t) - T_{I/B}(t) = T_{I/B}(t)Y(t)$$

For example, assuming that $Y(t) = \begin{bmatrix} y_{11} & y_{12} & y_{13} \\ y_{21} & y_{22} & y_{23} \\ y_{31} & y_{32} & y_{33} \end{bmatrix}$, then:

$$\begin{aligned}
T_{I/B}(t+\Delta t)(1,1) - T_{I/B}(t)(1,1) &= -\Delta\psi \cos\theta \sin\psi - \Delta\theta \cos\psi \sin\theta \\
&= T_{I/B}(t)(1,1)y_{11} + T_{I/B}(t)(1,2)y_{21} + T_{I/B}(t)(1,3)y_{31} \\
&= y_{11}\cos\psi\cos\theta + y_{21}(\cos\psi\sin\phi\sin\theta - \cos\phi\sin\psi) \\
&\quad + y_{31}(\sin\phi\sin\psi + \cos\phi\cos\psi\sin\theta)
\end{aligned}$$

$$= \cos\psi\sin\theta\underbrace{[y_{21}\sin\phi + y_{31}\cos\phi]}_{-\Delta\theta} + \cos\theta\sin\psi\underbrace{\left[y_{31}\frac{\sin\phi}{\cos\theta} - y_{21}\frac{\cos\phi}{\cos\theta}\right]}_{-\Delta\psi}$$
$$+ y_{11}\cos\psi\cos\theta$$
$$\Rightarrow y_{11} = 0;\quad y_{21}\sin\phi + y_{31}\cos\phi = -\Delta\theta;\quad y_{31}\sin\phi - y_{21}\cos\phi = -\Delta\psi\cos\theta$$

The coefficients y_{21}, y_{31} may be determined as follows:

$$\begin{aligned}
&y_{31}\cos^2\phi + y_{21}\sin\phi\cos\phi = -\Delta\theta\cos\phi \\
&y_{31}\sin^2\phi - y_{21}\cos\phi\sin\phi = -\Delta\psi\cos\theta\sin\phi \\
&\Rightarrow y_{31} = -\Delta\psi\cos\theta\sin\phi - \Delta\theta\cos\phi \\
&y_{21}\sin\phi + y_{31}\cos\phi = -\Delta\theta \Rightarrow y_{21}\sin\phi = -y_{31}\cos\phi - \Delta\theta \\
&= -\Delta\theta + \Delta\theta\cos^2\phi + \Delta\psi\cos\theta\cos\phi\sin\phi \\
&\Rightarrow y_{21}\sin\phi = \Delta\theta(\cos^2\theta - 1) + \Delta\psi\cos\theta\cos\phi\sin\phi \\
&\Rightarrow y_{21} = -\Delta\theta\sin\phi + \Delta\psi\cos\theta\cos\phi
\end{aligned}$$

A similar procedure may be carried out for the evaluation of y_{12}, y_{22}, y_{32} and y_{13}, y_{23}, y_{33}, thus resulting in the $Y(t)$ matrix, which is of the form:

$$Y(t) = \begin{bmatrix} 0 & \Delta\theta \sin\phi - \Delta\psi \cos\phi \cos\theta & \Delta\theta \cos\phi + \Delta\psi \cos\theta \sin\phi \\ \Delta\psi \cos\phi \cos\theta - \Delta\theta \sin\phi & 0 & \Delta\psi \sin\theta - \Delta\phi \\ -\Delta\theta \cos\phi - \Delta\psi \cos\theta \sin\phi & \Delta\phi - \dot{\psi} \sin\theta & 0 \end{bmatrix} \tag{4.103}$$

Finally dividing $T_{I/B}(t + \Delta t) - T_{I/B}(t)$ throughout by Δt and letting $\lim_{\Delta t \to 0}$ results in:

$$\lim_{\Delta t \to 0} \frac{T_{I/B}(t + \Delta t) - T_{I/B}(t)}{\Delta t} = \lim_{\Delta t \to 0} \frac{\Delta T_{I/B}(t)}{\Delta t} = \frac{dT_{I/B}(t)}{dt} = T_{I/B}(t) \lim_{\Delta t \to 0} Y(t)$$

$$= T_{I/B}(t) \begin{bmatrix} 0 & \dot{\theta} \sin\phi - \dot{\psi} \cos\phi \cos\theta & \dot{\theta} \cos\phi + \dot{\psi} \cos\theta \sin\phi \\ \dot{\psi} \cos\phi \cos\theta - \dot{\theta} \sin\phi & 0 & \dot{\psi} \sin\theta - \dot{\phi} \\ -\dot{\theta} \cos\phi - \dot{\psi} \cos\theta \sin\phi & \dot{\phi} - \dot{\psi} \sin\theta & 0 \end{bmatrix}$$

$$= T_{I/B}(t) \begin{bmatrix} 0 & -r_b & q_b \\ r_b & 0 & -p_b \\ -q_b & p_b & 0 \end{bmatrix} = T_{I/B}(t)\hat{\Omega}^T \tag{4.104}$$

where $\dot{\psi} = \lim_{\Delta t \to 0} \frac{\Delta\psi}{\Delta t}$; $\dot{\phi} = \lim_{\Delta t \to 0} \frac{\Delta\phi}{\Delta t}$; $\dot{\theta} = \lim_{\Delta t \to 0} \frac{\Delta\theta}{\Delta t}$.

The time derivative of $T_{I/B}(t)$ may be obtained as well from purely geometric considerations. Since the rotation at time $t + \Delta t$ consists of the rotation up to time t, followed by the small rotation $I + \Delta\theta_b$ from time t to time $t + \Delta t$, the total rotation from time $t = 0$ to time $t + \Delta t$ may be written as: $T_{I/B}(t + \Delta t) = T_{I/B}(t)(I + \Delta\theta_b)$. As may be seen from Fig. 4.3, $\Delta\theta_b$ is given by:

$$\Delta\theta_b = \begin{bmatrix} 0 & -\Delta\theta_{Yaw} & \Delta\theta_{Pitch} \\ \Delta\theta_{Yaw} & 0 & -\Delta\theta_{Roll} \\ -\Delta\theta_{Pitch} & \Delta\theta_{Roll} & 0 \end{bmatrix} = \begin{bmatrix} 0 & -\Delta\psi & \Delta\theta \\ \Delta\psi & 0 & -\Delta\phi \\ -\Delta\theta & \Delta\phi & 0 \end{bmatrix} \tag{4.105}$$

Note that $\Delta\theta_{Roll}, \Delta\theta_{Pitch}, \Delta\theta_{Yaw}$ are the small rotation angles through which frame F_b has rotated during the time Δt. Dividing $\Delta\theta_b$ by Δt and and letting $\Delta t \to 0$ results in:

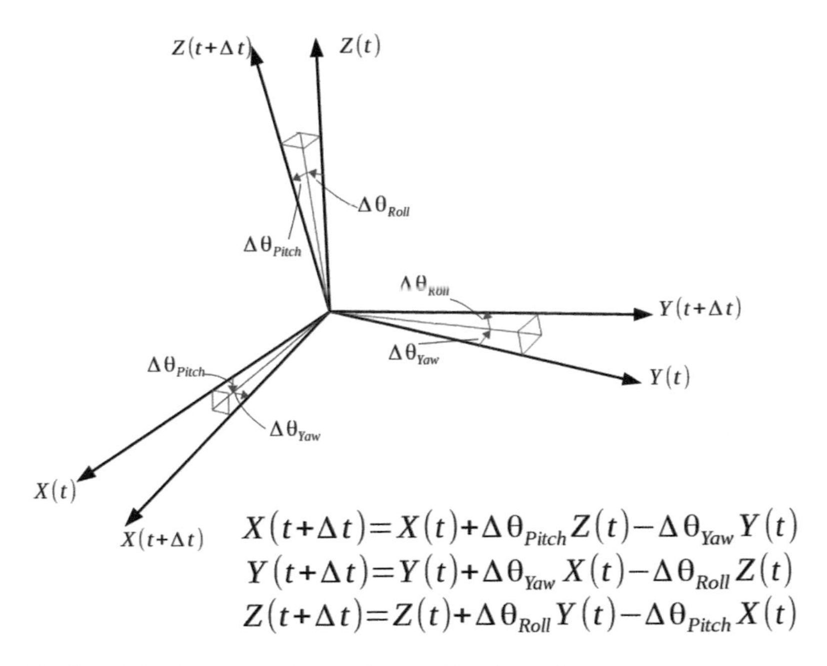

Fig. 4.3 Time derivative of a rotation matrix—small angle rotations

$$\dot{T}_{a/b}(t) = \lim_{\Delta t \to 0} T_{a/b}(t)\frac{\Delta\theta_b}{\Delta t} = T_{a/b}(t) \lim_{\Delta t \to 0} \begin{bmatrix} 0 & -\frac{\Delta\theta_{Yaw}}{\Delta t} & \frac{\Delta\theta_{Pitch}}{\Delta t} \\ \frac{\Delta\theta_{Yaw}}{\Delta t} & 0 & -\frac{\Delta\theta_{Roll}}{\Delta t} \\ -\frac{\Delta\theta_{Pitch}}{\Delta t} & \frac{\Delta\theta_{Roll}}{\Delta t} & 0 \end{bmatrix}$$

$$= T_{a/b}(t) \begin{bmatrix} 0 & -r_b & q_b \\ r_b & 0 & -p_b \\ -q_b & p_b & 0 \end{bmatrix} = T_{a/b}(t)\hat{\Omega}^T = -T_{a/b}(t)\hat{\Omega}; \;\; \hat{\Omega}^T = -\hat{\Omega} \qquad (4.106)$$

where p_b, q_b, r_b are the roll, pitch, and yaw angular velocities of frame F_b with respect to frame F_a, that is, $\Omega = \begin{bmatrix} p_b \\ q_b \\ r_b \end{bmatrix}$. In the limit as $\Delta t \to 0$, $\frac{\Delta\theta_b}{\Delta t}$ is the skew-symmetric form of the vector angular velocity of the F_b frame relative to the F_a frame during time Δt. This skew-symmetric matrix is the matrix equivalent of a vector cross product between vector Ω and velocity vector $V = \begin{bmatrix} v_x \\ v_y \\ v_z \end{bmatrix}$, that is,

$\lim_{\Delta t \to 0} \frac{\Delta \theta_b}{\Delta t} \begin{bmatrix} v_x \\ v_y \\ v_z \end{bmatrix} = -\Omega \times \begin{bmatrix} v_x \\ v_y \\ v_z \end{bmatrix}$.

Summarizing then, the derivative with respect to time of the Euler rotation matrix $T_{a/b}$ is:

$$\boxed{\dot{T}_{a/b} = T_{a/b} \hat{\Omega}^T = -T_{a/b} \hat{\Omega}} \tag{4.107}$$

Transformation for Translational Motion

In the sequel, whenever the intent is clear from the context of the discussion, the symbols $\{\cdot\}$ and $[\cdot]$ shall be dropped.

The basic idea is to transform the Lagrangian $\mathcal{L}$ into $\overline{\mathcal{L}}$ in such a way that the latter contains the velocities in body-fixed axes u, v, w, translations x_b, y_b, z_b, and body angular rates p_b, q_b, r_b. The Lagrangian $\mathcal{L}$ is a function of the inertial location and velocities of the system under consideration (the generalized inertial coordinates and generalized inertial velocities), that is, $\mathcal{L} = \mathcal{L}(x_I, y_I, z_I, u_I, v_I, w_I) = \mathcal{L}\left(X_I, \dot{X}_I\right)$, while the modified Lagrangian contains the system's location and velocities in body axis coordinates. In other words, $\overline{\mathcal{L}} = \overline{\mathcal{L}}(x_b, y_b, z_b, u, v, w) = \overline{\mathcal{L}}\left(X_B, \dot{X}_B\right)$. The body-centered and inertial velocities and positions are intimately related via Euler transformations $T_{B/I}$ and $T_{I/B}$ which transform from inertial to body-centered coordinates and vice versa. Stated more precisely, the transformation from body-centered velocities to velocities along inertial axes is: $\begin{bmatrix} u_I \\ v_I \\ w_I \end{bmatrix} = T_{I/B} \begin{bmatrix} u \\ v \\ w \end{bmatrix}$. The partial derivatives of $\mathcal{L}$ with respect to u_I, v_I, and w_I, for example, may be written in the following fashion:

$$\begin{aligned}
\frac{\partial \mathcal{L}}{\partial u_I} &= \frac{\partial \overline{\mathcal{L}}}{\partial u} \frac{\partial u}{\partial u_I} + \frac{\partial \overline{\mathcal{L}}}{\partial v} \frac{\partial v}{\partial u_I} + \frac{\partial \overline{\mathcal{L}}}{\partial w} \frac{\partial w}{\partial u_I} \\
\frac{\partial \mathcal{L}}{\partial v_I} &= \frac{\partial \overline{\mathcal{L}}}{\partial u} \frac{\partial u}{\partial v_I} + \frac{\partial \overline{\mathcal{L}}}{\partial v} \frac{\partial v}{\partial v_I} + \frac{\partial \overline{\mathcal{L}}}{\partial w} \frac{\partial w}{\partial v_I} \\
\frac{\partial \mathcal{L}}{\partial v_I} &= \frac{\partial \overline{\mathcal{L}}}{\partial u} \frac{\partial u}{\partial v_I} + \frac{\partial \overline{\mathcal{L}}}{\partial v} \frac{\partial v}{\partial v_I} + \frac{\partial \overline{\mathcal{L}}}{\partial w} \frac{\partial w}{\partial v_I}
\end{aligned} \tag{4.108}$$

In matrix form, the partial derivative of $\mathcal{L}$ with respect to the velocity vector in inertial coordinates V_I may be written as:

$$\begin{bmatrix} \frac{\partial \mathcal{L}}{\partial u_I} \\ \frac{\partial \mathcal{L}}{\partial v_I} \\ \frac{\partial \mathcal{L}}{\partial w_I} \end{bmatrix} = \begin{bmatrix} \frac{\partial u}{\partial u_I} & \frac{\partial v}{\partial u_I} & \frac{\partial w}{\partial u_I} \\ \frac{\partial u}{\partial v_I} & \frac{\partial v}{\partial v_I} & \frac{\partial w}{\partial v_I} \\ \frac{\partial u}{\partial w_I} & \frac{\partial v}{\partial w_I} & \frac{\partial w}{\partial w_I} \end{bmatrix} \begin{bmatrix} \frac{\partial \overline{\mathcal{L}}}{\partial u} \\ \frac{\partial \overline{\mathcal{L}}}{\partial v} \\ \frac{\partial \overline{\mathcal{L}}}{\partial w} \end{bmatrix} \tag{4.109}$$

Lagrange's equation in generalized inertial coordinates for the translational degrees of freedom may be written in the form:

$$\frac{d}{dt}\left(\frac{\partial \mathcal{L}}{\partial \dot{X}_I}\right) - \frac{\partial \mathcal{L}}{\partial X_I} = Q_I;\ \dot{X}_I \equiv V_I \Rightarrow \frac{d}{dt}\left(\frac{\partial \mathcal{L}}{\partial V_I}\right) - \frac{\partial \mathcal{L}}{\partial X_I} = Q_I \tag{4.110}$$

In Sect. 4.1, the transformation between the velocity vector in body coordinates and the corresponding velocity vector in inertial coordinates was shown to be:

$$T_{I/B} = \begin{bmatrix} \frac{\partial u}{\partial u_I} & \frac{\partial v}{\partial u_I} & \frac{\partial w}{\partial u_I} \\ \frac{\partial u}{\partial v_I} & \frac{\partial v}{\partial v_I} & \frac{\partial w}{\partial v_I} \\ \frac{\partial u}{\partial w_I} & \frac{\partial v}{\partial w_I} & \frac{\partial w}{\partial w_I} \end{bmatrix} \tag{4.111}$$

where $T_{I/B}$ is the direction cosine matrix (or Euler transformation matrix) which rotates the velocity vector in body coordinates, $[u, v, w]^T$, into the velocity vector in inertial coordinates $[u_I, v_I, w_I]^T$. The foregoing relationship may be written succinctly in the form: $\frac{\partial \mathcal{L}}{\partial V_I} = T_{I/B}\left\{\frac{\partial \overline{\mathcal{L}}}{\partial V_B}\right\}$, where $V_I = \begin{bmatrix} u_I \\ v_I \\ w_I \end{bmatrix}$ and $V_B = \begin{bmatrix} u \\ v \\ w \end{bmatrix}$.

Similarly, for position the following relationship holds:

$$\begin{bmatrix} \frac{\partial \mathcal{L}}{\partial x_I} \\ \frac{\partial \mathcal{L}}{\partial y_I} \\ \frac{\partial \mathcal{L}}{\partial z_I} \end{bmatrix} = \begin{bmatrix} \frac{\partial x_b}{\partial x_I} & \frac{\partial y_b}{\partial x_I} & \frac{\partial z_b}{\partial x_I} \\ \frac{\partial x_b}{\partial y_I} & \frac{\partial y_b}{\partial y_I} & \frac{\partial z_b}{\partial y_I} \\ \frac{\partial x_b}{\partial z_I} & \frac{\partial y_b}{\partial z_I} & \frac{\partial z_b}{\partial z_I} \end{bmatrix} \begin{bmatrix} \frac{\partial \overline{\mathcal{L}}}{\partial x_b} \\ \frac{\partial \overline{\mathcal{L}}}{\partial y_b} \\ \frac{\partial \overline{\mathcal{L}}}{\partial z_b} \end{bmatrix} \tag{4.112}$$

which may be written in compact form as: $\frac{\partial \mathcal{L}}{\partial X_I} = T_{I/B}\frac{\partial \overline{\mathcal{L}}}{\partial X_B}$. The preceding results imply that:

$$\frac{\partial \mathcal{L}}{\partial V_I} = T_{I/B}\frac{\partial \overline{\mathcal{L}}}{\partial V_B} \Rightarrow T_{B/I}\frac{\partial \mathcal{L}}{\partial V_I} = \frac{\partial \overline{\mathcal{L}}}{\partial V_B}$$
$$\frac{\partial \mathcal{L}}{\partial X_I} = T_{I/B}\frac{\partial \overline{\mathcal{L}}}{\partial X_B} \Rightarrow T_{B/I}\frac{\partial \mathcal{L}}{\partial X_I} = \frac{\partial \overline{\mathcal{L}}}{\partial X_B} \tag{4.113}$$

where $X_I = [x_I, y_I, z_I]^T$ is a position vector in inertial coordinates and $X_B = [x_b, y_b, z_b]^T$ is the corresponding position vector in body (or rotating) coordinates. Similarly V_I, V_B are the velocity vectors in inertial and body coordinate frames, respectively. The derivative with respect to time of an Euler rotation matrix $T_{I/B}$ takes the following form:

$$\frac{dT_{I/B}}{dt} = T_{I/B}(t)\underbrace{\begin{bmatrix} 0 & -r_b & q_b \\ r_b & 0 & -p_b \\ -q_b & p_b & 0 \end{bmatrix}}_{\hat{\Omega}^T \;=\; \omega\times} = T_{I/B}(t)\hat{\Omega}^T = -T_{I/B}(t)\hat{\Omega};\ \hat{\Omega}^T = -\hat{\Omega} \tag{4.114}$$

Hence the time derivative of $T_{B/I}\frac{\partial \mathcal{L}}{\partial V_I}$ may be written as:

$$\frac{d}{dt}\left[T_{B/I}\frac{\partial \mathcal{L}}{\partial V_I}\right] = \dot{T}_{B/I}\frac{\partial \mathcal{L}}{\partial V_I} + T_{B/I}\frac{d}{dt}\frac{\partial \mathcal{L}}{\partial V_I} = \frac{d}{dt}\left(\frac{\partial \overline{\mathcal{L}}}{\partial V_B}\right) \tag{4.115}$$

The rotation matrix $T_{I/B}$ has the property that the product $T_{I/B}T_{B/I} = I$, that is, the inverse of $T_{I/B}$ implies rotation in exactly the opposite direction and in the opposite sequence. This implies that:

$$\dot{T}_{B/I}T_{I/B} + T_{B/I}\dot{T}_{I/B} = 0 \Rightarrow \dot{T}_{B/I}T_{I/B} = -T_{B/I}\dot{T}_{I/B}$$
$$\Rightarrow \dot{T}_{B/I} = -T_{B/I}\dot{T}_{I/B}T_{B/I} = T_{B/I}T_{I/B}\hat{\Omega}T_{B/I} = \hat{\Omega}T_{B/I} \tag{4.116}$$

Using the fact that $\dot{T}_{B/I} = \hat{\Omega} T_{B/I}$ and substituting this result into the equation for the time derivative of $\frac{\partial \overline{\mathcal{L}}}{\partial V_B}$ leads to:

$$\begin{gathered}
\frac{d}{dt}\left[\frac{\partial \overline{\mathcal{L}}}{\partial V_B}\right] = \hat{\Omega} T_{B/I}\left(\frac{\partial \mathcal{L}}{\partial V_I}\right) + T_{B/I}\frac{d}{dt}\left(\frac{\partial \mathcal{L}}{\partial V_I}\right) \\
\Rightarrow T_{B/I}\frac{d}{dt}\left(\frac{\partial \mathcal{L}}{\partial V_I}\right) - \frac{d}{dt}\left[\frac{\partial \overline{\mathcal{L}}}{\partial V_B}\right] - \hat{\Omega} T_{B/I}\left(\frac{\partial \mathcal{L}}{\partial V_I}\right) \\
= \frac{d}{dt}\left[\frac{\partial \overline{\mathcal{L}}}{\partial V_B}\right] - \hat{\Omega} T_{B/I} T_{I/B}\left(\frac{\partial \overline{\mathcal{L}}}{\partial V_B}\right) \\
\boxed{\Rightarrow T_{B/I}\frac{d}{dt}\left(\frac{\partial \mathcal{L}}{\partial V_I}\right) = \frac{d}{dt}\left[\frac{\partial \overline{\mathcal{L}}}{\partial V_B}\right] - \hat{\Omega}\left(\frac{\partial \overline{\mathcal{L}}}{\partial V_B}\right)}
\end{gathered} \tag{4.117}$$

Furthermore, we know that:

$$\boxed{\frac{\partial \mathcal{L}}{\partial X_I} = T_{I/B}\frac{\partial \overline{\mathcal{L}}}{\partial X_B} \Rightarrow \frac{\partial \overline{\mathcal{L}}}{\partial X_B} = T_{B/I}\frac{\partial \mathcal{L}}{\partial X_I} \Rightarrow -\frac{\partial \overline{\mathcal{L}}}{\partial X_B} = -T_{B/I}\frac{\partial \mathcal{L}}{\partial X_I}} \tag{4.118}$$

Combining the terms in $T_{B/I}\left[\frac{d}{dt}\left(\frac{\partial \mathcal{L}}{\partial \dot{X}_I}\right) - \frac{\partial \mathcal{L}}{\partial X_I}\right] - T_{B/I} Q_I = 0$ results in the transformed translational Lagrangian dynamic equation which is:

$$\begin{gathered}
\boxed{T_{B/I}\left[\frac{d}{dt}\left(\frac{\partial \mathcal{L}}{\partial \dot{X}_I}\right) - \frac{\partial \mathcal{L}}{\partial X_I} - Q_I\right] = \frac{d}{dt}\left(\frac{\partial \overline{\mathcal{L}}}{\partial V_B}\right) - \hat{\Omega}\frac{\partial \overline{\mathcal{L}}}{\partial V_B} - \frac{\partial \overline{\mathcal{L}}}{\partial X_B} - T_{B/I} Q_I = 0} \\
\boxed{\Rightarrow T_{B/I}\left[\frac{d}{dt}\left(\frac{\partial \mathcal{L}}{\partial \dot{X}_I}\right) - \frac{\partial \mathcal{L}}{\partial X_I} - T_{B/I} Q_I\right] = \frac{d}{dt}\left(\frac{\partial \overline{\mathcal{L}}}{\partial V_B}\right) + \hat{\Omega}^T\frac{\partial \overline{\mathcal{L}}}{\partial V_B} - \frac{\partial \overline{\mathcal{L}}}{\partial X_B} - \underbrace{T_{B/I} Q_I}_{F_B} = 0}
\end{gathered} \tag{4.119}$$

since $\hat{\Omega}^T = -\hat{\Omega}$. In addition, $\dot{X}_I \equiv V_I$. Note that in the development of the transformation of the translational equations of motion, it was tacitly assumed that the origin of the body-centered coordinate system is at the body's center of mass. The same assumption will be used in developing the rotational equations of motion in the sequel.

Example 3

The transformed equations for translational motion shall be applied to a body with six degrees of freedom (where the potential energy due to gravity has not been taken into account). The Lagrangian $\overline{\mathcal{L}}$ for translational motion is:

$$\overline{\mathcal{L}} = \frac{1}{2}m(u^2 + v^2 + w^2) \tag{4.120}$$

where u, v, and w are the velocities of the body along the body-centered coordinate axes, with the center of mass at the origin of this axis system. Carrying out the operation indicated by $\frac{d}{dt}\left(\frac{\partial \overline{\mathcal{L}}}{\partial V_B}\right) + \hat{\Omega}^T \frac{\partial \overline{\mathcal{L}}}{\partial V_B} - \frac{\partial \overline{\mathcal{L}}}{\partial X_B} - T_{B/I} Q_I = 0$ results in:

$$m\begin{bmatrix} \dot{u} \\ \dot{v} \\ \dot{w} \end{bmatrix} + m\begin{bmatrix} 0 & -r_b & q_b \\ r_b & 0 & -p_b \\ -q_b & p_b & 0 \end{bmatrix}\begin{bmatrix} u \\ v \\ w \end{bmatrix} = \begin{bmatrix} F_{Bx} \\ F_{By} \\ F_{Bz} \end{bmatrix}$$

$$\begin{aligned} \Rightarrow m(\dot{u} - r_b v + q_b w) &= F_{Bx} \\ m(\dot{v} + r_b u - p_b w) &= F_{By} \\ m(\dot{w} + p_b v - q_b u) &= F_{Bz} \end{aligned} \tag{4.121}$$

Transformation for Rotational Motion
For the rotational dynamics case, we use the following definitions:

1. $\Phi \triangleq [\phi, \theta, \psi]^T$
2. $\dot{\Phi} \triangleq [\dot{\phi}, \dot{\theta}, \dot{\psi}]^T$
3. $Q_B \triangleq \left[p_b\left(\Phi, \dot{\Phi}\right), q_b\left(\Phi, \dot{\Phi}\right), r_b\left(\Phi, \dot{\Phi}\right)\right]^T = \omega$
4. $p_b = p_b\left(\phi, \theta, \psi, \dot{\phi}, \dot{\theta}, \dot{\psi}\right) = p_b\left(\Phi, \dot{\Phi}\right)$
5. $q_b = q_b\left(\phi, \theta, \psi, \dot{\phi}, \dot{\theta}, \dot{\psi}\right) = q_b\left(\Phi, \dot{\Phi}\right)$
6. $r_b = r_b\left(\phi, \theta, \psi, \dot{\phi}, \dot{\theta}, \dot{\psi}\right) = r_b\left(\Phi, \dot{\Phi}\right)$
7. $\mathcal{L} = \mathcal{L}\left(\phi, \theta, \psi, \dot{\phi}, \dot{\theta}, \dot{\psi}\right) = \mathcal{L}\left(\Phi, \dot{\Phi}\right)$
8. $\overline{\mathcal{L}} = \overline{\mathcal{L}}\left(p_b\left(\phi, \theta, \psi, \dot{\phi}, \dot{\theta}, \dot{\psi}\right), q_b\left(\phi, \theta, \psi, \dot{\phi}, \dot{\theta}, \dot{\psi}\right), r_b\left(\phi, \theta, \psi, \dot{\phi}, \dot{\theta}, \dot{\psi}\right)\right) = \overline{\mathcal{L}}(\Phi, Q_B)$

The relationship between the body's angular rates and the Euler angular rates may be written in the following manner:

$$\begin{aligned} p_b &= \dot{\phi} - \dot{\psi}\sin\theta;\ \ q_b = \dot{\theta}\cos\phi + \dot{\psi}\sin\phi\cos\theta; \\ r_b &= -\dot{\theta}\sin\phi + \dot{\psi}\cos\phi\cos\theta \\ \dot{\phi} &= p_b + (q_b\sin\phi + r_b\cos\phi)\tan\theta;\ \ \dot{\theta} = q_b\cos\phi - r_b\sin\phi; \\ \dot{\psi} &= \frac{(q_b\sin\phi + r_b\cos\phi)}{\cos\theta} \end{aligned} \tag{4.122}$$

With the definitions of $\omega = Q_B \triangleq [p_b, q_b, r_b]^T$ and $\Phi \triangleq [\phi, \theta, \psi]^T$, the transpose of the Jacobian matrix J, that is, $J^T = \partial Q_B/\partial \Phi$, turns out to be:

$$\frac{\partial Q_B}{\partial \Phi} = \begin{bmatrix} \frac{\partial p_b}{\partial \phi} & \frac{\partial q_b}{\partial \phi} & \frac{\partial r_b}{\partial \phi} \\ \frac{\partial p_b}{\partial \theta} & \frac{\partial q_b}{\partial \theta} & \frac{\partial r_b}{\partial \theta} \\ \frac{\partial p_b}{\partial \psi} & \frac{\partial q_b}{\partial \psi} & \frac{\partial r_b}{\partial \psi} \end{bmatrix} = \begin{bmatrix} \frac{\partial}{\partial \phi}\left[\, p_b \; q_b \; r_b \,\right] \\ \frac{\partial}{\partial \theta}\left[\, p_b \; q_b \; r_b \,\right] \\ \frac{\partial}{\partial \psi}\left[\, p_b \; q_b \; r_b \,\right] \end{bmatrix}$$

$$= \begin{bmatrix} 0 & -\dot{\theta}\sin\phi + \dot{\psi}\cos\phi\cos\theta & -\dot{\theta}\cos\phi - \dot{\psi}\sin\phi\cos\theta \\ -\dot{\psi}\cos\theta & -\dot{\psi}\sin\phi\sin\theta & -\dot{\psi}\cos\phi\sin\theta \\ 0 & 0 & 0 \end{bmatrix} \quad (4.123)$$

$$\frac{\partial Q_B}{\partial \dot{\Phi}} = \begin{bmatrix} \frac{\partial p_b}{\partial \dot{\phi}} & \frac{\partial q_b}{\partial \dot{\phi}} & \frac{\partial r_b}{\partial \dot{\phi}} \\ \frac{\partial p_b}{\partial \dot{\theta}} & \frac{\partial q_b}{\partial \dot{\theta}} & \frac{\partial r_b}{\partial \dot{\theta}} \\ \frac{\partial p_b}{\partial \dot{\psi}} & \frac{\partial q_b}{\partial \dot{\psi}} & \frac{\partial r_b}{\partial \dot{\psi}} \end{bmatrix} = \begin{bmatrix} 1 & 0 & 0 \\ 0 & \cos\phi & -\sin\phi \\ -\sin\theta & \sin\phi\cos\theta & \cos\phi\cos\theta \end{bmatrix} = M^T$$

(4.124)

The inverse matrix of M, i.e., $L = M^{-1}$, is defined to be:

$$M = \begin{bmatrix} 1 & 0 & -\sin\theta \\ 0 & \cos\phi & \sin\phi\cos\theta \\ 0 & -\sin\phi & \cos\phi\cos\theta \end{bmatrix}; \quad L = M^{-1} = \begin{bmatrix} 1 & \sin\phi\tan\theta & \cos\phi\tan\theta \\ 0 & \cos\phi & -\sin\phi \\ 0 & \frac{\sin\phi}{\cos\theta} & \frac{\cos\phi}{\cos\theta} \end{bmatrix}$$

(4.125)

Since the Lagrangian $\mathcal{L}$ and transformed Lagrangian $\overline{\mathcal{L}}$ are related, the following operation is justified:

$$\frac{\partial \mathcal{L}}{\partial \dot{\phi}} = \frac{\partial \overline{\mathcal{L}}}{\partial p_b}\frac{\partial p_b}{\partial \dot{\phi}} + \frac{\partial \overline{\mathcal{L}}}{\partial q_b}\frac{\partial q_b}{\partial \dot{\phi}} + \frac{\partial \overline{\mathcal{L}}}{\partial r_b}\frac{\partial r_b}{\partial \dot{\phi}}$$

$$\frac{\partial \mathcal{L}}{\partial \dot{\theta}} = \frac{\partial \overline{\mathcal{L}}}{\partial p_b}\frac{\partial p_b}{\partial \dot{\theta}} + \frac{\partial \overline{\mathcal{L}}}{\partial q_b}\frac{\partial q_b}{\partial \dot{\theta}} + \frac{\partial \overline{\mathcal{L}}}{\partial r_b}\frac{\partial r_b}{\partial \dot{\theta}}$$

$$\frac{\partial \mathcal{L}}{\partial \dot{\psi}} = \frac{\partial \overline{\mathcal{L}}}{\partial p_b}\frac{\partial p_b}{\partial \dot{\psi}} + \frac{\partial \overline{\mathcal{L}}}{\partial q_b}\frac{\partial q_b}{\partial \dot{\psi}} + \frac{\partial \overline{\mathcal{L}}}{\partial r_b}\frac{\partial r_b}{\partial \dot{\psi}}$$

(4.126)

Writing the above in matrix form, the partial derivative of $\mathcal{L}$ with respect to the Euler angular rates, $\dot{\Phi}$, may be written as:

$$\frac{\partial \mathcal{L}}{\partial \dot{\Phi}} = \frac{\partial Q_B}{\partial \dot{\Phi}} \frac{\partial \overline{\mathcal{L}}}{\partial Q_B} \Rightarrow \begin{bmatrix} \frac{\partial \mathcal{L}}{\partial \dot{\phi}} \\ \frac{\partial \mathcal{L}}{\partial \dot{\Phi}} \\ \frac{\partial \mathcal{L}}{\partial \dot{\psi}} \end{bmatrix} = \underbrace{\begin{bmatrix} \frac{\partial p_b}{\partial \dot{\phi}} & \frac{\partial q_b}{\partial \dot{\phi}} & \frac{\partial r_b}{\partial \dot{\phi}} \\ \frac{\partial p_b}{\partial \dot{\theta}} & \frac{\partial q_b}{\partial \dot{\theta}} & \frac{\partial r_b}{\partial \dot{\theta}} \\ \frac{\partial p_b}{\partial \dot{\psi}} & \frac{\partial q_b}{\partial \dot{\psi}} & \frac{\partial r_b}{\partial \dot{\psi}} \end{bmatrix}}_{\frac{\partial Q_B}{\partial \dot{\Phi}}} \underbrace{\begin{bmatrix} \frac{\partial \overline{\mathcal{L}}}{\partial p_b} \\ \frac{\partial \overline{\mathcal{L}}}{\partial q_b} \\ \frac{\partial \overline{\mathcal{L}}}{\partial r_b} \end{bmatrix}}_{\frac{\partial \overline{\mathcal{L}}}{\partial Q_B}} = M^T \begin{bmatrix} \frac{\partial \overline{\mathcal{L}}}{\partial p_b} \\ \frac{\partial \overline{\mathcal{L}}}{\partial q_b} \\ \frac{\partial \overline{\mathcal{L}}}{\partial r_b} \end{bmatrix} \tag{4.127}$$

Symbolically, the partial derivative of the Lagrangian $\mathcal{L}$ with respect to the vector of Euler angles Φ may be written in terms of the partial derivatives of the modified Lagrangian $\overline{\mathcal{L}}$, with respect to Φ and the vector of body angular rates Q_B, that is:

$$\frac{\partial \mathcal{L}}{\partial \Phi} = \frac{\partial \overline{\mathcal{L}}}{\partial \Phi} + \frac{\partial Q_B}{\partial \Phi} \frac{\partial \overline{\mathcal{L}}}{\partial Q_B} \tag{4.128}$$

To recap, we shall use the following definitions in the sequel:

1. $\partial \overline{\mathcal{L}}/\partial Q_B \triangleq \begin{bmatrix} \frac{\partial \overline{\mathcal{L}}}{\partial p_b} & \frac{\partial \overline{\mathcal{L}}}{\partial q_b} & \frac{\partial \overline{\mathcal{L}}}{\partial r_b} \end{bmatrix}^T$

2. $\partial Q_B/\partial \Phi \triangleq \begin{bmatrix} \frac{\partial p_b}{\partial \phi} & \frac{\partial q_b}{\partial \phi} & \frac{\partial r_b}{\partial \phi} \\ \frac{\partial p_b}{\partial \theta} & \frac{\partial q_b}{\partial \theta} & \frac{\partial r_b}{\partial \theta} \\ \frac{\partial p_b}{\partial \psi} & \frac{\partial q_b}{\partial \psi} & \frac{\partial r_b}{\partial \psi} \end{bmatrix} = \begin{bmatrix} \frac{\partial}{\partial \phi} \left[p_b \; q_b \; r_b \right] \\ \frac{\partial}{\partial \theta} \left[p_b \; q_b \; r_b \right] \\ \frac{\partial}{\partial \psi} \left[p_b \; q_b \; r_b \right] \end{bmatrix}$

$$= \begin{bmatrix} 0 & -\dot{\theta} \sin\phi + \dot{\psi} \cos\phi \cos\theta & -\dot{\theta} \cos\phi - \dot{\psi} \sin\phi \cos\theta \\ -\dot{\psi} \cos\theta & -\dot{\psi} \sin\phi \sin\theta & -\dot{\psi} \cos\phi \sin\theta \\ 0 & 0 & 0 \end{bmatrix}$$

3. $\partial Q_B/\partial \dot{\Phi} \triangleq \begin{bmatrix} \frac{\partial p_b}{\partial \dot{\phi}} & \frac{\partial q_b}{\partial \dot{\phi}} & \frac{\partial r_b}{\partial \dot{\phi}} \\ \frac{\partial p_b}{\partial \dot{\theta}} & \frac{\partial q_b}{\partial \dot{\theta}} & \frac{\partial r_b}{\partial \dot{\theta}} \\ \frac{\partial p_b}{\partial \dot{\psi}} & \frac{\partial q_b}{\partial \dot{\psi}} & \frac{\partial r_b}{\partial \dot{\psi}} \end{bmatrix} = \begin{bmatrix} 1 & 0 & 0 \\ 0 & \cos\phi & -\sin\phi \\ -\sin\theta & \sin\phi\cos\theta & \cos\phi\cos\theta \end{bmatrix} = M^T$

4. $M = \begin{bmatrix} 1 & 0 & -\sin\theta \\ 0 & \cos\phi & \sin\phi\cos\theta \\ 0 & -\sin\phi & \cos\phi\cos\theta \end{bmatrix}; \quad L = M^{-1} = \begin{bmatrix} 1 & \sin\phi\tan\theta & \cos\phi\tan\theta \\ 0 & \cos\phi & -\sin\phi \\ 0 & \frac{\sin\phi}{\cos\theta} & \frac{\cos\phi}{\cos\theta} \end{bmatrix}$

From the relationship between the body's angular rates and the Euler angular rates we can deduce the following:

$$\begin{bmatrix} p_b \\ q_b \\ r_b \end{bmatrix} = \begin{bmatrix} 1 & 0 & -\sin\theta \\ 0 & \cos\phi & \sin\phi\cos\theta \\ 0 & -\sin\phi & \cos\phi\cos\theta \end{bmatrix} \begin{bmatrix} \dot{\phi} \\ \dot{\theta} \\ \dot{\psi} \end{bmatrix} \Rightarrow Q_B = Md\Phi/dt \Rightarrow d\Phi/dt = LQ_B \tag{4.129}$$

The d'Alembert–Lagrange equation for the rotational modes is of the following form:

$$K\left[\frac{d}{dt}\left(\frac{\partial \mathcal{L}}{\partial \dot{\Phi}}\right) - \frac{\partial \mathcal{L}}{\partial \Phi} - Q_\tau\right] = 0 \tag{4.130}$$

where K is a constant. The time derivative of $\partial\mathcal{L}/\partial\dot{\Phi}$ is calculated as follows (see Eq. 4.127):

$$\begin{aligned} &\frac{\partial \mathcal{L}}{\partial \dot{\Phi}} = M^T \frac{\partial \overline{\mathcal{L}}}{\partial Q_B} \Rightarrow \frac{d}{dt}\left(\frac{\partial \mathcal{L}}{\partial \dot{\Phi}}\right) = \dot{M}^T \frac{\partial \overline{\mathcal{L}}}{\partial Q_B} + M^T \frac{d}{dt}\left(\frac{\partial \overline{\mathcal{L}}}{\partial Q_B}\right) \\ &\Rightarrow \left(M^T\right)^{-1} \frac{d}{dt}\left(\frac{\partial \mathcal{L}}{\partial \dot{\Phi}}\right) = \left(M^T\right)^{-1} \dot{M}^T \left(\frac{\partial \overline{\mathcal{L}}}{\partial Q_B}\right) + \frac{d}{dt}\left(\frac{\partial \overline{\mathcal{L}}}{\partial Q_B}\right) \\ &\Rightarrow L^T \frac{d}{dt}\left(\frac{\partial \mathcal{L}}{\partial \dot{\Phi}}\right) = L^T \dot{M}^T \left(\frac{\partial \overline{\mathcal{L}}}{\partial Q_B}\right) + \frac{d}{dt}\left(\frac{\partial \overline{\mathcal{L}}}{\partial Q_B}\right) \end{aligned} \tag{4.131}$$

The term $\partial\mathcal{L}/\partial\Phi$ was shown to be:

$$\frac{\partial \mathcal{L}}{\partial \Phi} = \frac{\partial \overline{\mathcal{L}}}{\partial \Phi} + \frac{\partial Q_B}{\partial \Phi}\frac{\partial \overline{\mathcal{L}}}{\partial Q_B} \tag{4.132}$$

Therefore the d'Alembert–Lagrange equation for the rotational modes, that is, $K\left[\frac{d}{dt}\left(\frac{\partial \mathcal{L}}{\partial \dot{\Phi}}\right) - \frac{\partial \mathcal{L}}{\partial \Phi} - Q_\tau\right] = 0$, may be written as follows:

$$L^T \frac{d}{dt}\left(\frac{\partial \mathcal{L}}{\partial \dot{\Phi}}\right) = L^T\left[\frac{\partial \mathcal{L}}{\partial \Phi} + Q_\tau\right] = L^T\left[\frac{\partial \overline{\mathcal{L}}}{\partial \Phi} + \frac{\partial Q_B}{\partial \Phi}\frac{\partial \overline{\mathcal{L}}}{\partial Q_B} + Q_\tau\right] \tag{4.133}$$

Using this latter result, the transformed d'Alembert–Lagrange equation for the rotational modes may be written as:

$$\begin{aligned}
& L^T \frac{d}{dt}\left(\frac{\partial \mathcal{L}}{\partial \dot{\Phi}}\right) = L^T \dot{M}^T \left(\frac{\partial \overline{\mathcal{L}}}{\partial Q_B}\right) + \frac{d}{dt}\left(\frac{\partial \overline{\mathcal{L}}}{\partial Q_B}\right) \\
& \Rightarrow L^T \left[\frac{\partial \overline{\mathcal{L}}}{\partial \Phi} + \frac{\partial Q_B}{\partial \Phi}\frac{\partial \overline{\mathcal{L}}}{\partial Q_B} + Q_\tau\right] = L^T \dot{M}^T \left(\frac{\partial \overline{\mathcal{L}}}{\partial Q_B}\right) + \frac{d}{dt}\left(\frac{\partial \overline{\mathcal{L}}}{\partial Q_B}\right) \\
& \Rightarrow \frac{d}{dt}\left(\frac{\partial \overline{\mathcal{L}}}{\partial Q_B}\right) - L^T \frac{\partial \overline{\mathcal{L}}}{\partial \Phi} + L^T \dot{M}^T \left(\frac{\partial \overline{\mathcal{L}}}{\partial Q_B}\right) - L^T \frac{\partial Q_B}{\partial \Phi}\frac{\partial \overline{\mathcal{L}}}{\partial Q_B} = L^T Q_\tau \\
& \Rightarrow \frac{d}{dt}\left(\frac{\partial \overline{\mathcal{L}}}{\partial Q_B}\right) - L^T \frac{\partial \overline{\mathcal{L}}}{\partial \Phi} + L^T \left(\dot{M}^T - \frac{\partial Q_B}{\partial \Phi}\right)\left(\frac{\partial \overline{\mathcal{L}}}{\partial Q_B}\right) = L^T Q_\tau
\end{aligned} \tag{4.134}$$

Previously the value of $\partial Q_B/\partial \Phi$ was shown to be:

$$\frac{\partial Q_B}{\partial \Phi} = \begin{bmatrix} 0 & -\dot{\theta}\sin\phi + \dot{\psi}\cos\phi\cos\theta & -\dot{\theta}\cos\phi - \dot{\psi}\sin\phi\cos\theta \\ -\dot{\psi}\cos\theta & -\dot{\psi}\sin\phi\sin\theta & -\dot{\psi}\cos\phi\sin\theta \\ 0 & 0 & 0 \end{bmatrix} \tag{4.135}$$

The calculated time derivative of M^T is:

$$\begin{aligned}
M^T &= \begin{bmatrix} 1 & 0 & 0 \\ 0 & \cos\phi & -\sin\phi \\ -\sin\theta & \sin\phi\cos\theta & \cos\phi\cos\theta \end{bmatrix} \\
\Rightarrow \dot{M}^T &= \begin{bmatrix} 0 & 0 & 0 \\ 0 & -\dot{\phi}\sin\phi & -\dot{\phi}\cos\phi \\ -\dot{\theta}\cos\theta & \dot{\phi}\cos\phi\cos\theta - \dot{\theta}\sin\phi\sin\theta & -\dot{\phi}\sin\phi\cos\theta - \dot{\theta}\cos\phi\sin\theta \end{bmatrix}
\end{aligned} \tag{4.136}$$

Hence using the relationship $\begin{bmatrix} p_b \\ q_b \\ r_b \end{bmatrix} = \begin{bmatrix} \dot{\phi} - \dot{\psi}\sin\theta \\ \dot{\theta}\cos\phi + \dot{\psi}\cos\theta\sin\phi \\ -\dot{\theta}\sin\phi + \dot{\psi}\cos\theta\cos\phi \end{bmatrix}$, the expression $\left(\dot{M}^T - \frac{\partial Q_B}{\partial \Phi}\right)$ turns out to be:

$$\left(\dot{M}^T - \frac{\partial Q_B}{\partial \Phi}\right)$$

$$= \begin{bmatrix} 0 & 0 & 0 \\ 0 & -\dot{\phi}\sin\phi & -\dot{\phi}\cos\phi \\ -\dot{\theta}\cos\theta & \dot{\phi}\cos\phi\cos\theta - \dot{\theta}\sin\phi\sin\theta & -\dot{\phi}\sin\phi\cos\theta - \dot{\theta}\cos\phi\sin\theta \end{bmatrix}$$

$$- \begin{bmatrix} 0 & -\dot{\theta}\sin\phi + \dot{\psi}\cos\phi\cos\theta & -\dot{\theta}\cos\phi - \dot{\psi}\sin\phi\cos\theta \\ -\dot{\psi}\cos\theta & -\dot{\psi}\sin\phi\sin\theta & -\dot{\psi}\cos\phi\sin\theta \\ 0 & 0 & 0 \end{bmatrix}$$

$$= \begin{bmatrix} 0 & \dot{\theta}\sin\phi - \dot{\psi}\cos\phi\cos\theta & \dot{\theta}\cos\phi + \dot{\psi}\sin\phi\cos\theta \\ \dot{\psi}\cos\theta & \dot{\psi}\sin\phi\sin\theta - \dot{\phi}\sin\phi & \dot{\psi}\cos\phi\sin\theta - \dot{\phi}\cos\phi \\ -\dot{\theta}\cos\theta & \dot{\phi}\cos\phi\cos\theta - \dot{\theta}\sin\phi\sin\theta & -\dot{\phi}\sin\phi\cos\theta - \dot{\theta}\cos\phi\sin\theta \end{bmatrix}$$

$$= \begin{bmatrix} 0 & -r_b & q_b \\ r_b\cos\phi + q_b\sin\phi & -p_b\sin\phi & -p_b\cos\phi \\ r_b\sin\phi\cos\theta - q_b\cos\phi\cos\theta & p_b\cos\phi\cos\theta + r_b\sin\theta & -p_b\sin\phi\cos\theta - q_b\sin\theta \end{bmatrix} \tag{4.137}$$

The latter expression may be rewritten in the form of a matrix product:

$$\begin{bmatrix} 0 & -r_b & q_b \\ r_b \cos\phi + q_b \sin\phi & -p_b \sin\phi & -p_b \cos\phi \\ r_b \sin\phi\cos\theta - q_b \cos\phi\cos\theta & p_b \cos\phi\cos\theta + r_b \sin\theta & -p_b \sin\phi\cos\theta - q_b \sin\theta \end{bmatrix}$$

$$= \underbrace{\begin{bmatrix} 1 & 0 & 0 \\ 0 & \cos\phi & -\sin\phi \\ -\sin\theta & \sin\phi\cos\theta & \cos\phi\cos\theta \end{bmatrix}}_{M^T} \underbrace{\begin{bmatrix} 0 & -r_b & q_b \\ r_b & 0 & -p_b \\ -q_b & p_b & 0 \end{bmatrix}}_{\hat{\Omega}^T} \tag{4.138}$$

The d'Alembert–Lagrange equation using quasi-velocities for rotational motion then becomes:

$$\frac{d}{dt}\left(\frac{\partial\overline{\mathcal{L}}}{\partial Q_B}\right) - L^T\frac{\partial\overline{\mathcal{L}}}{\partial\Phi} + L^T\left(\dot{M}^T - \frac{\partial Q_B}{\partial\Phi}\right)\left(\frac{\partial\overline{\mathcal{L}}}{\partial Q_B}\right) = L^T Q_\tau$$

$$\Rightarrow \frac{d}{dt}\left(\frac{\partial\overline{\mathcal{L}}}{\partial Q_B}\right) - L^T\frac{\partial\overline{\mathcal{L}}}{\partial\Phi} + L^T\left(M^T\hat{\Omega}^T\right)\left(\frac{\partial\overline{\mathcal{L}}}{\partial Q_B}\right) = L^T Q_\tau$$

$$\boxed{\begin{gathered} \Rightarrow \frac{d}{dt}\left\{\frac{\partial\overline{\mathcal{L}}}{\partial Q_B}\right\} - L^T\frac{\partial\overline{\mathcal{L}}}{\partial\Phi} + \hat{\Omega}^T\left(\frac{\partial\overline{\mathcal{L}}}{\partial Q_B}\right) = L^T Q_\tau;\ L^T M^T = I \\ \hat{\Omega}^T = \begin{bmatrix} 0 & -r_b & q_b \\ r_b & 0 & -p_b \\ -q_b & p_b & 0 \end{bmatrix} \end{gathered}} \tag{4.139}$$

Example 4: Equations of Rotational Motion

The transformed equations for rotational motion will be applied to a body with six degrees of freedom. The Lagrangian $\overline{\mathcal{L}}$ for rotational motion is:

$$\overline{\mathcal{L}} = \frac{1}{2}(I_x p_b^2 + I_y q_b^2 + I_z r_b^2) \tag{4.140}$$

where I_x, I_y, and I_z are the bodies' moments of inertia at the center of mass, p_b, q_b, and r_b are the rotational velocities of the body along the body-centered coordinate axes, with the center of mass at the origin of this axis system. Carrying out the operation indicated by $\frac{d}{dt}\left(\frac{\partial \overline{\mathcal{L}}}{\partial Q_B}\right) - L^T \frac{\partial \overline{\mathcal{L}}}{\partial \Phi} + \hat{\Omega}^T \left(\frac{\partial \overline{\mathcal{L}}}{\partial Q_B}\right) = L^T Q_\tau$ results in:

$$\begin{bmatrix} I_x \dot{p}_b \\ I_y \dot{q}_b \\ I_z \dot{r}_b \end{bmatrix} + \begin{bmatrix} 0 & -r_b & q_b \\ r_b & 0 & -p_b \\ -q_b & p_b & 0 \end{bmatrix} \begin{bmatrix} I_x p_b \\ I_y q_b \\ I_z r_b \end{bmatrix} = \begin{bmatrix} \tau_{Bx} \\ \tau_{By} \\ \tau_{Bz} \end{bmatrix}$$

$$\begin{aligned} \Rightarrow I_x \dot{p}_b + q_b r_b (I_z - I_y) &= \tau_{Bx} \\ I_y \dot{p}_b + p_b r_b (I_x - I_z) &= \tau_{By} \\ I_z \dot{p}_b + p_b q_b (I_y - I_x) &= \tau_{Bz} \end{aligned} \tag{4.141}$$

4.5 Lagrangian Dynamics in Quasi-Coordinates—Vepa's Approach—Origin Not at Mass Center

The following section is based on the approach by Prof. Ranjan Vepa (see Vepa [50, pp. 73–75]).
The velocity and position vectors at point P, and at the center of mass C are defined as follows:

$$V_p = \begin{bmatrix} u \\ v \\ w \end{bmatrix}; \; V_c = \begin{bmatrix} u_c \\ v_c \\ w_c \end{bmatrix}; \; R_p - R_c = \begin{bmatrix} x_p - x_c \\ y_p - y_c \\ z_p - z_c \end{bmatrix} \tag{4.142}$$

V_p is the velocity vector at point P, while V_c is the velocity vector at the center of mass point C, respectively. $R_p - R_c$ is the distance from the center of mass to any point P. The velocity components with respect to the point P are related to the velocity of the center of mass and the coordinates of point P by the relation:

$$V_p = V_c + \omega \times (R_p - R_c)$$

$$\begin{bmatrix} u \\ v \\ w \end{bmatrix} = \begin{bmatrix} u_c \\ v_c \\ w_c \end{bmatrix} + \begin{bmatrix} 0 & -r_b & q_b \\ r_b & 0 & -p_b \\ -q_b & p_b & 0 \end{bmatrix} \begin{bmatrix} x_p - x_c \\ y_p - y_c \\ z_p - z_c \end{bmatrix}$$

$$\begin{bmatrix} u \\ v \\ w \end{bmatrix} = \begin{bmatrix} u_c + q_b(z_p - z_c) - r_b(y_p - y_c) \\ v_c + r_b(x_p - x_c) - p_b(z_p - z_c) \\ w_c + p_b(y_p - y_c) - q_b(x_p - x_c) \end{bmatrix} \tag{4.143}$$

The results of the foregoing equation imply that:

$$\frac{\partial u}{\partial u_c} = 1; \quad \frac{\partial v}{\partial v_c} = 1; \quad \frac{\partial w}{\partial w_c} = 1$$

$$\begin{aligned} \frac{\partial u}{\partial v_c} &= \frac{\partial u}{\partial w_c} = 0 \\ \frac{\partial v}{\partial u_c} &= \frac{\partial v}{\partial w_c} = 0 \\ \frac{\partial w}{\partial u_c} &= \frac{\partial w}{\partial v_c} = 0 \end{aligned} \tag{4.144}$$

The velocity at the center of mass may also be related to the velocity of the rotating and moving coordinates centered at O and the distance from the point O to the center of mass at point C, as follows:

$$V_c = V_o + \omega \times R_c = V_o + \overline{V}_c$$

$$\begin{bmatrix} u_c \\ v_c \\ w_c \end{bmatrix} = \begin{bmatrix} \dot{x}_c \\ \dot{y}_c \\ \dot{z}_c \end{bmatrix} + \begin{bmatrix} 0 & -r_b & q_b \\ r_b & 0 & -p_b \\ -q_b & p_b & 0 \end{bmatrix} \begin{bmatrix} x_c \\ y_c \\ z_c \end{bmatrix}$$

$$\begin{bmatrix} u_c \\ v_c \\ w_c \end{bmatrix} = \begin{bmatrix} \dot{x}_c + q_b z_c - r_b y_c \\ \dot{y}_c + r_b x_c - p_b z_c \\ \dot{z}_c + p_b y_c - q_b x_c \end{bmatrix} \tag{4.145}$$

Given the Lagrangian $\overline{\mathcal{L}}$, expressed in terms of the body-centered velocity components $V_p = \begin{bmatrix} u & v & w \end{bmatrix}^T$, the partial derivatives of $\overline{\mathcal{L}}$ with respect to u, v, and w, taking into account the fact that u, v, and w are intimately related to u_c, v_c, and w_c, respectively, may be written in the following fashion:

$$\begin{aligned} \frac{\partial \overline{\mathcal{L}}}{\partial u} &= \frac{\partial \overline{\mathcal{L}}}{\partial u_c}\frac{\partial u_c}{\partial u} + \frac{\partial \overline{\mathcal{L}}}{\partial v_c}\frac{\partial v_c}{\partial u} + \frac{\partial \overline{\mathcal{L}}}{\partial w_c}\frac{\partial w_c}{\partial u} \\ \frac{\partial \overline{\mathcal{L}}}{\partial v} &= \frac{\partial \overline{\mathcal{L}}}{\partial u_c}\frac{\partial u_c}{\partial v} + \frac{\partial \overline{\mathcal{L}}}{\partial v_c}\frac{\partial v_c}{\partial v} + \frac{\partial \overline{\mathcal{L}}}{\partial w_c}\frac{\partial w_c}{\partial v} \\ \frac{\partial \overline{\mathcal{L}}}{\partial w} &= \frac{\partial \overline{\mathcal{L}}}{\partial u_c}\frac{\partial u_c}{\partial w} + \frac{\partial \overline{\mathcal{L}}}{\partial v_c}\frac{\partial v_c}{\partial w} + \frac{\partial \overline{\mathcal{L}}}{\partial w_c}\frac{\partial w_c}{\partial w} \end{aligned} \tag{4.146}$$

While the vector V_p is written in body-centered coordinates, it should be remembered that it is the velocity of point P with respect to the inertial coordinate frame X, Y, Z. In matrix form, the partial derivative of $\overline{\mathcal{L}}$ with respect to the velocity vector V_p may be written as:

$$\begin{bmatrix} \frac{\partial \overline{\mathcal{L}}}{\partial u} \\ \frac{\partial \overline{\mathcal{L}}}{\partial v} \\ \frac{\partial \overline{\mathcal{L}}}{\partial w} \end{bmatrix} = \begin{bmatrix} \frac{\partial u_c}{\partial u} & \frac{\partial v_c}{\partial u} & \frac{\partial w_c}{\partial u} \\ \frac{\partial u_c}{\partial v} & \frac{\partial v_c}{\partial v} & \frac{\partial w_c}{\partial v} \\ \frac{\partial u_c}{\partial w} & \frac{\partial v_c}{\partial w} & \frac{\partial w_c}{\partial w} \end{bmatrix} \begin{bmatrix} \frac{\partial \overline{\mathcal{L}}}{\partial u_c} \\ \frac{\partial \overline{\mathcal{L}}}{\partial v_c} \\ \frac{\partial \overline{\mathcal{L}}}{\partial w_c} \end{bmatrix} \tag{4.147}$$

$$\begin{bmatrix} \frac{\partial u_c}{\partial u} & \frac{\partial v_c}{\partial u} & \frac{\partial w_c}{\partial u} \\ \frac{\partial u_c}{\partial v} & \frac{\partial v_c}{\partial v} & \frac{\partial w_c}{\partial v} \\ \frac{\partial u_c}{\partial w} & \frac{\partial v_c}{\partial w} & \frac{\partial w_c}{\partial w} \end{bmatrix} = \begin{bmatrix} 1 & 0 & 0 \\ 0 & 1 & 0 \\ 0 & 0 & 1 \end{bmatrix} \Rightarrow \begin{bmatrix} \frac{\partial \overline{\mathcal{L}}}{\partial u} \\ \frac{\partial \overline{\mathcal{L}}}{\partial v} \\ \frac{\partial \overline{\mathcal{L}}}{\partial w} \end{bmatrix} = \begin{bmatrix} \frac{\partial \overline{\mathcal{L}}}{\partial u_c} \\ \frac{\partial \overline{\mathcal{L}}}{\partial v_c} \\ \frac{\partial \overline{\mathcal{L}}}{\partial w_c} \end{bmatrix} \tag{4.148}$$

Similarly the Lagrangian $\overline{\mathcal{L}}$, in terms of the body-centered position components $R_p = \begin{bmatrix} x_p & y_p & z_p \end{bmatrix}^T$, and the partial derivatives of $\overline{\mathcal{L}}$ with respect to the coordinates of the center of mass, x_c, y_c, and z_c, for example, may be written in the following fashion:

$$\begin{aligned}\frac{\partial \overline{\mathcal{L}}}{\partial x_p} &= \frac{\partial \overline{\mathcal{L}}}{\partial x_c}\frac{\partial x_c}{\partial x_p} + \frac{\partial \overline{\mathcal{L}}}{\partial y_c}\frac{\partial y_c}{\partial x_p} + \frac{\partial \overline{\mathcal{L}}}{\partial z_c}\frac{\partial z_c}{\partial x_p} \\ \frac{\partial \overline{\mathcal{L}}}{\partial y_p} &= \frac{\partial \overline{\mathcal{L}}}{\partial x_c}\frac{\partial x_c}{\partial y_p} + \frac{\partial \overline{\mathcal{L}}}{\partial y_c}\frac{\partial y_c}{\partial y_p} + \frac{\partial \overline{\mathcal{L}}}{\partial z_c}\frac{\partial z_c}{\partial y_p} \\ \frac{\partial \overline{\mathcal{L}}}{\partial z_p} &= \frac{\partial \overline{\mathcal{L}}}{\partial x_c}\frac{\partial x_c}{\partial z_p} + \frac{\partial \overline{\mathcal{L}}}{\partial y_c}\frac{\partial y_c}{\partial z_p} + \frac{\partial \overline{\mathcal{L}}}{\partial z_c}\frac{\partial z_c}{\partial z_p}\end{aligned} \tag{4.149}$$

The matrix of partial derivatives of $\overline{\mathcal{L}}$ with respect to the position vector R_p is:

$$\begin{bmatrix} \frac{\partial \overline{\mathcal{L}}}{\partial x_p} \\ \frac{\partial \overline{\mathcal{L}}}{\partial y_p} \\ \frac{\partial \overline{\mathcal{L}}}{\partial z_p} \end{bmatrix} = \begin{bmatrix} \frac{\partial x_c}{\partial x_p} & \frac{\partial y_c}{\partial x_p} & \frac{\partial z_c}{\partial x_p} \\ \frac{\partial x_c}{\partial y_p} & \frac{\partial y_c}{\partial y_p} & \frac{\partial z_c}{\partial y_p} \\ \frac{\partial x_c}{\partial z_p} & \frac{\partial y_c}{\partial z_p} & \frac{\partial z_c}{\partial z_p} \end{bmatrix} \begin{bmatrix} \frac{\partial \overline{\mathcal{L}}}{\partial x_c} \\ \frac{\partial \overline{\mathcal{L}}}{\partial y_c} \\ \frac{\partial \overline{\mathcal{L}}}{\partial z_c} \end{bmatrix} \tag{4.150}$$

Since R_p and R_c are related via a constant vector K, that is, $R_p = K + R_c \Rightarrow x_p = x_c + k_x;\; y_p = y_c + k_y;\; z_p = z_c + k_z$, and both are within the rotating coordinate axis system, the partial derivatives between the components of R_p and R_c turn out to be:

$$\begin{aligned} &\frac{\partial x_c}{\partial x_p} = 1; \; \frac{\partial y_c}{\partial x_p} = 0; \; \frac{\partial z_c}{\partial x_p} = 0 \\ &\frac{\partial x_c}{\partial y_p} = 0; \; \frac{\partial y_c}{\partial y_p} = 1; \; \frac{\partial z_c}{\partial y_p} = 0 \\ &\frac{\partial x_c}{\partial z_p} = 0; \; \frac{\partial y_c}{\partial z_p} = 0; \; \frac{\partial z_c}{\partial z_p} = 1 \end{aligned} \tag{4.151}$$

Rewriting Eq. 4.151 as a matrix:

$$\begin{bmatrix} \frac{\partial x_c}{\partial x_p} & \frac{\partial y_c}{\partial x_p} & \frac{\partial z_c}{\partial x_p} \\ \frac{\partial x_c}{\partial y_p} & \frac{\partial y_c}{\partial y_p} & \frac{\partial z_c}{\partial y_p} \\ \frac{\partial x_c}{\partial z_p} & \frac{\partial y_c}{\partial z_p} & \frac{\partial z_c}{\partial z_p} \end{bmatrix} = \begin{bmatrix} 1\,0\,0 \\ 0\,1\,0 \\ 0\,0\,1 \end{bmatrix} \Rightarrow \begin{bmatrix} \frac{\partial \overline{\mathcal{L}}}{\partial x_p} \\ \frac{\partial \overline{\mathcal{L}}}{\partial y_p} \\ \frac{\partial \overline{\mathcal{L}}}{\partial z_p} \end{bmatrix} = \begin{bmatrix} \frac{\partial \overline{\mathcal{L}}}{\partial x_c} \\ \frac{\partial \overline{\mathcal{L}}}{\partial y_c} \\ \frac{\partial \overline{\mathcal{L}}}{\partial z_c} \end{bmatrix} \tag{4.152}$$

For the rotational modes, the partial derivative of the Lagrangian $\mathcal{L}$ with respect to the component parts of the vector $\omega = \left[\, p_b \; q_b \; r_b \,\right]^T$ may be written symbolically as (Eq. 4.128 is similar in form):

$$\left\{\frac{\partial \mathcal{L}}{\partial \omega}\right\} = \left\{\frac{\partial \overline{\mathcal{L}}}{\partial \omega}\right\} + \left[\frac{\partial V_c}{\partial \omega}\right]\left\{\frac{\partial \overline{\mathcal{L}}}{\partial V_c}\right\} \tag{4.153}$$

where $V_c = \left[\, u_c \; v_c \; w_c \,\right]$. Rewriting using the foregoing definitions of the vectors ω and V_c, the partial derivatives of the Lagrangian $\mathcal{L}$ with respect to the component parts of the vector $\omega = \left[\, p_b \; q_b \; r_b \,\right]$ and with respect to $\overline{\mathcal{L}}$ become:

$$\begin{aligned}
\frac{\partial \mathcal{L}}{\partial p_b} &= \frac{\partial \overline{\mathcal{L}}}{\partial p_b} + \frac{\partial u_c}{\partial p_b}\frac{\partial \overline{\mathcal{L}}}{\partial u_c} + \frac{\partial v_c}{\partial p_b}\frac{\partial \overline{\mathcal{L}}}{\partial v_c} + \frac{\partial w_c}{\partial p_b}\frac{\partial \overline{\mathcal{L}}}{\partial w_c} \\
\frac{\partial \mathcal{L}}{\partial q_b} &= \frac{\partial \overline{\mathcal{L}}}{\partial q_b} + \frac{\partial u_c}{\partial q_b}\frac{\partial \overline{\mathcal{L}}}{\partial u_c} + \frac{\partial v_c}{\partial q_b}\frac{\partial \overline{\mathcal{L}}}{\partial v_c} + \frac{\partial w_c}{\partial q_b}\frac{\partial \overline{\mathcal{L}}}{\partial w_c} \\
\frac{\partial \mathcal{L}}{\partial r_b} &= \frac{\partial \overline{\mathcal{L}}}{\partial r_b} + \frac{\partial u_c}{\partial r_b}\frac{\partial \overline{\mathcal{L}}}{\partial u_c} + \frac{\partial v_c}{\partial r_b}\frac{\partial \overline{\mathcal{L}}}{\partial v_c} + \frac{\partial w_c}{\partial r_b}\frac{\partial \overline{\mathcal{L}}}{\partial w_c}
\end{aligned} \tag{4.154}$$

As a matrix, this latter result is:

$$\begin{bmatrix} \frac{\partial \mathcal{L}}{\partial p_b} \\ \frac{\partial \mathcal{L}}{\partial q_b} \\ \frac{\partial \mathcal{L}}{\partial r_b} \end{bmatrix} = \begin{bmatrix} \frac{\partial \overline{\mathcal{L}}}{\partial p_b} \\ \frac{\partial \overline{\mathcal{L}}}{\partial q_b} \\ \frac{\partial \overline{\mathcal{L}}}{\partial r_b} \end{bmatrix} + \begin{bmatrix} \frac{\partial u_c}{\partial p_b} & \frac{\partial v_c}{\partial p_b} & \frac{\partial w_c}{\partial p_b} \\ \frac{\partial u_c}{\partial q_b} & \frac{\partial v_c}{\partial q_b} & \frac{\partial w_c}{\partial q_b} \\ \frac{\partial u_c}{\partial r_b} & \frac{\partial v_c}{\partial r_b} & \frac{\partial w_c}{\partial r_b} \end{bmatrix} \begin{bmatrix} \frac{\partial \overline{\mathcal{L}}}{\partial u_c} \\ \frac{\partial \overline{\mathcal{L}}}{\partial v_c} \\ \frac{\partial \overline{\mathcal{L}}}{\partial w_c} \end{bmatrix} \tag{4.155}$$

It was previously shown that the velocity at the center of mass with respect to an inertial coordinate system is:

$$V_c = V_o + \omega \times R_c = V_o - R_c \times \omega \tag{4.156}$$

$$\begin{bmatrix} u_c \\ v_c \\ w_c \end{bmatrix} = \begin{bmatrix} \dot{x}_c + q_b z_c - r_b y_c \\ \dot{y}_c + r_b x_c - p_b z_c \\ \dot{z}_c + p_b y_c - q_b x_c \end{bmatrix} \tag{4.157}$$

Therefore the foregoing matrix of partial derivatives of the velocities of V_c with respect to ω becomes:

$$\begin{bmatrix} \frac{\partial u_c}{\partial p_b} & \frac{\partial v_c}{\partial p_b} & \frac{\partial w_c}{\partial p_b} \\ \frac{\partial u_c}{\partial q_b} & \frac{\partial v_c}{\partial q_b} & \frac{\partial w_c}{\partial q_b} \\ \frac{\partial u_c}{\partial r_b} & \frac{\partial v_c}{\partial r_b} & \frac{\partial w_c}{\partial r_b} \end{bmatrix} = \begin{bmatrix} 0 & -z_c & y_c \\ z_c & 0 & -x_c \\ -y_c & x_c & 0 \end{bmatrix} \triangleq R_c \times$$

$$\Rightarrow \begin{bmatrix} \frac{\partial \mathcal{L}}{\partial p_b} \\ \frac{\partial \mathcal{L}}{\partial q_b} \\ \frac{\partial \mathcal{L}}{\partial r_b} \end{bmatrix} = \begin{bmatrix} \frac{\partial \overline{\mathcal{L}}}{\partial p_b} \\ \frac{\partial \overline{\mathcal{L}}}{\partial q_b} \\ \frac{\partial \overline{\mathcal{L}}}{\partial r_b} \end{bmatrix} + R_c \times \begin{bmatrix} \frac{\partial \overline{\mathcal{L}}}{\partial u_c} \\ \frac{\partial \overline{\mathcal{L}}}{\partial v_c} \\ \frac{\partial \overline{\mathcal{L}}}{\partial w_c} \end{bmatrix} \Rightarrow \left\{ \frac{\partial \mathcal{L}}{\partial \omega} \right\} = \left\{ \frac{\partial \overline{\mathcal{L}}}{\partial \omega} \right\} + R_c \times \left\{ \frac{\partial \overline{\mathcal{L}}}{\partial V_c} \right\} \tag{4.158}$$

Hence the time derivative of $\left\{ \frac{\partial \mathcal{L}}{\partial \omega} \right\}$ in rotating x, y, z coordinates becomes:

$$\begin{aligned} \frac{d}{dt}\left\{ \frac{\partial \mathcal{L}}{\partial \omega} \right\} &= \frac{d}{dt}\left\{ \frac{\partial \overline{\mathcal{L}}}{\partial \omega} \right\} + \frac{d}{dt}\left(R_c \times \left\{ \frac{\partial \overline{\mathcal{L}}}{\partial V_c} \right\} \right) = \frac{d}{dt}\left\{ \frac{\partial \overline{\mathcal{L}}}{\partial \omega} \right\} \\ &+ \frac{dR_c}{dt} \times \left\{ \frac{\partial \overline{\mathcal{L}}}{\partial V_c} \right\} + R_c \times \frac{d}{dt}\left\{ \frac{\partial \overline{\mathcal{L}}}{\partial V_c} \right\} \end{aligned}$$

$$\begin{aligned} \Rightarrow \frac{d}{dt}\begin{bmatrix} \frac{\partial \mathcal{L}}{\partial p_b} \\ \frac{\partial \mathcal{L}}{\partial q_b} \\ \frac{\partial \mathcal{L}}{\partial r_b} \end{bmatrix} &= \frac{d}{dt}\begin{bmatrix} \frac{\partial \overline{\mathcal{L}}}{\partial p_b} \\ \frac{\partial \overline{\mathcal{L}}}{\partial q_b} \\ \frac{\partial \overline{\mathcal{L}}}{\partial r_b} \end{bmatrix} + \left(\frac{d}{dt}\begin{bmatrix} 0 & -z_c & y_c \\ z_c & 0 & -x_c \\ -y_c & x_c & 0 \end{bmatrix} \right) \begin{bmatrix} \frac{\partial \overline{\mathcal{L}}}{\partial u_c} \\ \frac{\partial \overline{\mathcal{L}}}{\partial v_c} \\ \frac{\partial \overline{\mathcal{L}}}{\partial w_c} \end{bmatrix} \\ &+ \begin{bmatrix} 0 & -z_c & y_c \\ z_c & 0 & -x_c \\ -y_c & x_c & 0 \end{bmatrix} \frac{d}{dt} \begin{bmatrix} \frac{\partial \overline{\mathcal{L}}}{\partial u_c} \\ \frac{\partial \overline{\mathcal{L}}}{\partial v_c} \\ \frac{\partial \overline{\mathcal{L}}}{\partial w_c} \end{bmatrix} \end{aligned} \tag{4.159}$$

The vector product $\omega \times \left\{\frac{\partial \mathcal{L}}{\partial \omega}\right\}$ is (see Eqs. 4.153 and 4.158):

$$\omega \times \left\{\frac{\partial \mathcal{L}}{\partial \omega}\right\} = \omega \times \left\{\frac{\partial \overline{\mathcal{L}}}{\partial \omega}\right\} + \omega \times \left\{R_c \times \left(\frac{\partial \overline{\mathcal{L}}}{\partial V_c}\right)\right\} \tag{4.160}$$

Therefore the sum of $\frac{d}{dt}\left\{\frac{\partial \mathcal{L}}{\partial \omega}\right\} + \omega \times \left\{\frac{\partial \mathcal{L}}{\partial \omega}\right\}$, which is the time derivative of $\frac{d}{dt}\left\{\frac{\partial \mathcal{L}}{\partial \omega}\right\}$ in the inertial X, Y, Z frame, becomes:

$$\frac{d}{dt}\left\{\frac{\partial \mathcal{L}}{\partial \omega}\right\} + \omega \times \left\{\frac{\partial \mathcal{L}}{\partial \omega}\right\} = \frac{d}{dt}\left\{\frac{\partial \overline{\mathcal{L}}}{\partial \omega}\right\} + \frac{d}{dt}\left(R_c \times \left\{\frac{\partial \overline{\mathcal{L}}}{\partial V_c}\right\}\right) + \omega \times \left\{R_c \times \left(\frac{\partial \overline{\mathcal{L}}}{\partial V_c}\right)\right\} + \omega \times \left\{\frac{\partial \overline{\mathcal{L}}}{\partial \omega}\right\} \tag{4.161}$$

The derivative of a vector Q in inertial coordinates is $\dot{Q}_{inertial} = \dot{Q}_{rotating} + \omega \times Q$, and so:

$$\underbrace{\left\{V_c \times \left\{\frac{\partial \overline{\mathcal{L}}}{\partial V_c}\right\}\right\}}_{\dot{Q}_{inertial}} = \underbrace{\frac{d}{dt}\left\{R_c \times \left\{\frac{\partial \overline{\mathcal{L}}}{\partial V_c}\right\}\right\}}_{\dot{Q}_{rotating}} + \underbrace{\omega \times \left\{R_c \times \left\{\frac{\partial \overline{\mathcal{L}}}{\partial V_c}\right\}\right\}}_{\omega \times Q} \tag{4.162}$$

Hence the time derivative of $\frac{d}{dt}\left\{\frac{\partial \mathcal{L}}{\partial \omega}\right\}$ in inertial coordinates is:

$$\frac{d}{dt}\left\{\frac{\partial \mathcal{L}}{\partial \omega}\right\} + \omega \times \left\{\frac{\partial \mathcal{L}}{\partial \omega}\right\} = \frac{d}{dt}\left\{\frac{\partial \overline{\mathcal{L}}}{\partial \omega}\right\} + \omega \times \left\{\frac{\partial \overline{\mathcal{L}}}{\partial \omega}\right\} + \left\{V_c \times \left\{\frac{\partial \overline{\mathcal{L}}}{\partial V_c}\right\}\right\}$$

$$= \frac{d}{dt}\begin{bmatrix} \frac{\partial \overline{\mathcal{L}}}{\partial p_b} \\ \frac{\partial \overline{\mathcal{L}}}{\partial q_b} \\ \frac{\partial \overline{\mathcal{L}}}{\partial r_b} \end{bmatrix} + \begin{bmatrix} 0 & -r_b & q_b \\ r_b & 0 & -p_b \\ -q_b & p_b & 0 \end{bmatrix}\begin{bmatrix} \frac{\partial \overline{\mathcal{L}}}{\partial p_b} \\ \frac{\partial \overline{\mathcal{L}}}{\partial q_b} \\ \frac{\partial \overline{\mathcal{L}}}{\partial r_b} \end{bmatrix} + \begin{bmatrix} 0 & -w_c & v_c \\ w_c & 0 & -u_c \\ -v_c & u_c & 0 \end{bmatrix}\begin{bmatrix} \frac{\partial \overline{\mathcal{L}}}{\partial u_c} \\ \frac{\partial \overline{\mathcal{L}}}{\partial v_c} \\ \frac{\partial \overline{\mathcal{L}}}{\partial w_c} \end{bmatrix} \tag{4.163}$$

Transformation for Translational Motion

The transformed translational d'Alembert–Lagrange dynamic equations, when the origin is at the center of mass, were previously shown to be (see Eq. 4.119):

$$\frac{d}{dt}\begin{bmatrix} \frac{\partial \overline{\mathcal{L}}}{\partial u} \\ \frac{\partial \overline{\mathcal{L}}}{\partial v} \\ \frac{\partial \overline{\mathcal{L}}}{\partial w} \end{bmatrix} + \begin{bmatrix} 0 & -r_b & q_b \\ r_b & 0 & -p_b \\ -q_b & p_b & 0 \end{bmatrix} \begin{bmatrix} \frac{\partial \overline{\mathcal{L}}}{\partial u} \\ \frac{\partial \overline{\mathcal{L}}}{\partial v} \\ \frac{\partial \overline{\mathcal{L}}}{\partial w} \end{bmatrix} - \begin{bmatrix} \frac{\partial \overline{\mathcal{L}}}{\partial x_p} \\ \frac{\partial \overline{\mathcal{L}}}{\partial y_p} \\ \frac{\partial \overline{\mathcal{L}}}{\partial z_p} \end{bmatrix} = F_B \tag{4.164}$$

We have shown above that:

$$\begin{bmatrix} \frac{\partial \overline{\mathcal{L}}}{\partial u} \\ \frac{\partial \overline{\mathcal{L}}}{\partial v} \\ \frac{\partial \overline{\mathcal{L}}}{\partial w} \end{bmatrix} = \begin{bmatrix} \frac{\partial \overline{\mathcal{L}}}{\partial u_c} \\ \frac{\partial \overline{\mathcal{L}}}{\partial v_c} \\ \frac{\partial \overline{\mathcal{L}}}{\partial w_c} \end{bmatrix} ; \begin{bmatrix} \frac{\partial \overline{\mathcal{L}}}{\partial x_p} \\ \frac{\partial \overline{\mathcal{L}}}{\partial y_p} \\ \frac{\partial \overline{\mathcal{L}}}{\partial z_p} \end{bmatrix} = \begin{bmatrix} \frac{\partial \overline{\mathcal{L}}}{\partial x_c} \\ \frac{\partial \overline{\mathcal{L}}}{\partial x_c} \\ \frac{\partial \overline{\mathcal{L}}}{\partial x_c} \end{bmatrix} \tag{4.165}$$

and thus when the center of mass and the origin of the rigid body do not coincide, the transformed d'Alembert–Lagrange dynamic equations at the center of mass become:

$$\frac{d}{dt}\begin{bmatrix} \frac{\partial \overline{\mathcal{L}}}{\partial u_c} \\ \frac{\partial \overline{\mathcal{L}}}{\partial v_c} \\ \frac{\partial \overline{\mathcal{L}}}{\partial w_c} \end{bmatrix} + \begin{bmatrix} 0 & -r_b & q_b \\ r_b & 0 & -p_b \\ -q_b & p_b & 0 \end{bmatrix} \begin{bmatrix} \frac{\partial \overline{\mathcal{L}}}{\partial u_c} \\ \frac{\partial \overline{\mathcal{L}}}{\partial v_c} \\ \frac{\partial \overline{\mathcal{L}}}{\partial w_c} \end{bmatrix} - \begin{bmatrix} \frac{\partial \overline{\mathcal{L}}}{\partial x_c} \\ \frac{\partial \overline{\mathcal{L}}}{\partial y_c} \\ \frac{\partial \overline{\mathcal{L}}}{\partial z_c} \end{bmatrix} = F_B \tag{4.166}$$

Transformation for Rotational Motion

The transformed rotational d'Alembert–Lagrange dynamic equations, when the origin is at the center of mass, were previously shown to be (see Eq. 4.139):

$$\frac{d}{dt}\left\{\frac{\partial \overline{\mathcal{L}}}{\partial \omega}\right\} - L^T\left[\frac{\partial \overline{\mathcal{L}}}{\partial \Phi}\right] + \omega \times \left\{\frac{\partial \overline{\mathcal{L}}}{\partial \omega}\right\} = L^T Q_\tau$$

$$= \frac{d}{dt}\begin{bmatrix} \frac{\partial \overline{\mathcal{L}}}{\partial p_b} \\ \frac{\partial \overline{\mathcal{L}}}{\partial q_b} \\ \frac{\partial \overline{\mathcal{L}}}{\partial r_b} \end{bmatrix} - \begin{bmatrix} 1 & 0 & 0 \\ \sin\phi\tan\theta & \cos\phi & \frac{\sin\phi}{\cos\theta} \\ \cos\phi\tan\theta & -\sin\phi & \frac{\cos\phi}{\cos\theta} \end{bmatrix} \begin{bmatrix} \frac{\partial \overline{\mathcal{L}}}{\partial \phi} \\ \frac{\partial \overline{\mathcal{L}}}{\partial \theta} \\ \frac{\partial \overline{\mathcal{L}}}{\partial \psi} \end{bmatrix} + \begin{bmatrix} 0 & -r_b & q_b \\ r_b & 0 - p_b & \\ -q_b & p_b & 0 \end{bmatrix} \begin{bmatrix} \frac{\partial \overline{\mathcal{L}}}{\partial p_b} \\ \frac{\partial \overline{\mathcal{L}}}{\partial q_b} \\ \frac{\partial \overline{\mathcal{L}}}{\partial r_b} \end{bmatrix}$$

$$= \begin{bmatrix} 1 & 0 & 0 \\ \sin\phi\tan\theta & \cos\phi & \frac{\sin\phi}{\cos\theta} \\ \cos\phi\tan\theta & -\sin\phi & \frac{\cos\phi}{\cos\theta} \end{bmatrix} \begin{bmatrix} Q_{\tau\phi} \\ Q_{\tau\theta} \\ Q_{\tau\psi} \end{bmatrix} \tag{4.167}$$

Hence, we would expect that when the origin is not situated at the center of mass, the transformed rotational d'Alembert–Lagrange dynamic equations will have additional terms to account for the offset of the c.g. point from the origin. In fact the equations are:

$$\frac{d}{dt}\left(\frac{\partial \overline{\mathcal{L}}}{\partial \omega}\right) - L^T\frac{\partial \overline{\mathcal{L}}}{\partial \Phi} + \omega \times \left(\frac{\partial \overline{\mathcal{L}}}{\partial \omega}\right) + \left\{V_c \times \left\{\frac{\partial \overline{\mathcal{L}}}{\partial V_c}\right\}\right\} = L^T Q_\tau + R_c \times F_B$$

$$= \frac{d}{dt}\begin{bmatrix} \frac{\partial \overline{\mathcal{L}}}{\partial p_b} \\ \frac{\partial \overline{\mathcal{L}}}{\partial q_b} \\ \frac{\partial \overline{\mathcal{L}}}{\partial r_b} \end{bmatrix} - \begin{bmatrix} 1 & 0 & 0 \\ \sin\phi\tan\theta & \cos\phi & \frac{\sin\phi}{\cos\theta} \\ \cos\phi\tan\theta & -\sin\phi & \frac{\cos\phi}{\cos\theta} \end{bmatrix} \begin{bmatrix} \frac{\partial \overline{\mathcal{L}}}{\partial \phi} \\ \frac{\partial \overline{\mathcal{L}}}{\partial \theta} \\ \frac{\partial \overline{\mathcal{L}}}{\partial \psi} \end{bmatrix} + \begin{bmatrix} 0 & -r_b & q_b \\ r_b & 0 & -p_b \\ -q_b & p_b & 0 \end{bmatrix} \begin{bmatrix} \frac{\partial \overline{\mathcal{L}}}{\partial p_b} \\ \frac{\partial \overline{\mathcal{L}}}{\partial q_b} \\ \frac{\partial \overline{\mathcal{L}}}{\partial r_b} \end{bmatrix}$$

$$+\begin{bmatrix} 0 & -w_c & v_c \\ w_c & 0 & -u_c \\ -v_c & u_c & 0 \end{bmatrix}\begin{bmatrix} \frac{\partial \overline{\mathcal{L}}}{\partial u_c} \\ \frac{\partial \overline{\mathcal{L}}}{\partial v_c} \\ \frac{\partial \overline{\mathcal{L}}}{\partial w_c} \end{bmatrix}$$

$$= \begin{bmatrix} 1 & 0 & 0 \\ \sin\phi\tan\theta & \cos\phi & \frac{\sin\phi}{\cos\theta} \\ \cos\phi\tan\theta & -\sin\phi & \frac{\cos\phi}{\cos\theta} \end{bmatrix}\begin{bmatrix} Q_{\tau\phi} \\ Q_{\tau\theta} \\ Q_{\tau\psi} \end{bmatrix} + \begin{bmatrix} 0 & -z_c & y_c \\ z_c & 0 & -x_c \\ -y_c & x_c & 0 \end{bmatrix}\begin{bmatrix} F_x \\ F_y \\ F_z \end{bmatrix} \tag{4.168}$$

The term $R_c \times F_B$, where F_B is the vector of forces acting on the body at the origin of the body-centered coordinate system, is due to the moment induced by the force vector F_B and the distance to the c.g. point R_c. The term $\left\{\frac{\partial \overline{\mathcal{L}}}{\partial V_c}\right\}$ was shown above to be equal to:

$$\begin{bmatrix} \frac{\partial \overline{\mathcal{L}}}{\partial u_c} \\ \frac{\partial \overline{\mathcal{L}}}{\partial v_c} \\ \frac{\partial \overline{\mathcal{L}}}{\partial w_c} \end{bmatrix} = \begin{bmatrix} u_c \\ v_c \\ w_c \end{bmatrix} = V_c \tag{4.169}$$

and this implies that:

$$V_c \times \begin{bmatrix} \frac{\partial \overline{\mathcal{L}}}{\partial u_c} \\ \frac{\partial \overline{\mathcal{L}}}{\partial v_c} \\ \frac{\partial \overline{\mathcal{L}}}{\partial w_c} \end{bmatrix} = V_c \times V_c = 0 \tag{4.170}$$

Hence the transformed rotational d'Alembert–Lagrange dynamic equations, when the c.g. point is offset from the origin, are:

$$\frac{d}{dt}\left(\frac{\partial \overline{\mathcal{L}}}{\partial \omega}\right) - L^T \frac{\partial \overline{\mathcal{L}}}{\partial \Phi} + \omega \times \left(\frac{\partial \overline{\mathcal{L}}}{\partial \omega}\right) + \overbrace{\left\{V_c \times \left\{\frac{\partial \overline{\mathcal{L}}}{\partial V_c}\right\}\right\}}^{=0} = L^T Q_\tau + R_c \times F_B$$

$$= \frac{d}{dt}\begin{bmatrix} \frac{\partial \overline{\mathcal{L}}}{\partial p_b} \\ \frac{\partial \overline{\mathcal{L}}}{\partial q_b} \\ \frac{\partial \overline{\mathcal{L}}}{\partial r_b} \end{bmatrix} - \begin{bmatrix} 1 & 0 & 0 \\ \sin\phi\tan\theta & \cos\phi & \frac{\sin\phi}{\cos\theta} \\ \cos\phi\tan\theta & -\sin\phi & \frac{\cos\phi}{\cos\theta} \end{bmatrix} \begin{bmatrix} \frac{\partial \overline{\mathcal{L}}}{\partial \phi} \\ \frac{\partial \overline{\mathcal{L}}}{\partial \theta} \\ \frac{\partial \overline{\mathcal{L}}}{\partial \psi} \end{bmatrix} + \begin{bmatrix} 0 & -r_b & q_b \\ r_b & 0 & -p_b \\ -q_b & p_b & 0 \end{bmatrix} \begin{bmatrix} \frac{\partial \overline{\mathcal{L}}}{\partial p_b} \\ \frac{\partial \overline{\mathcal{L}}}{\partial q_b} \\ \frac{\partial \overline{\mathcal{L}}}{\partial r_b} \end{bmatrix}$$

$$+ \begin{bmatrix} 0 & -w_c & v_c \\ w_c & 0 & -u_c \\ -v_c & u_c & 0 \end{bmatrix} \begin{bmatrix} \frac{\partial \overline{\mathcal{L}}}{\partial u_c} \\ \frac{\partial \overline{\mathcal{L}}}{\partial v_c} \\ \frac{\partial \overline{\mathcal{L}}}{\partial w_c} \end{bmatrix}$$

$$= \begin{bmatrix} 1 & 0 & 0 \\ \sin\phi\tan\theta & \cos\phi & \frac{\sin\phi}{\cos\theta} \\ \cos\phi\tan\theta & -\sin\phi & \frac{\cos\phi}{\cos\theta} \end{bmatrix} \begin{bmatrix} Q_{\tau\phi} \\ Q_{\tau\theta} \\ Q_{\tau\psi} \end{bmatrix} + \begin{bmatrix} 0 & -z_c & y_c \\ z_c & 0 & -x_c \\ -y_c & x_c & 0 \end{bmatrix} \begin{bmatrix} F_x \\ F_y \\ F_z \end{bmatrix} \quad (4.171)$$

Example 5: Rigid Aircraft Dynamics [50, pp. 75–79]

We have shown previously that the velocity of any point P, within a rigid body with Cartesian coordinates $R_p = [x_p, y_p, z_p]$ and with respect to an inertial coordinate frame, is: $V_p = V_o + \omega \times R_p$, where $V_p = \begin{bmatrix} u & v & w \end{bmatrix}^T$, and V_o, the velocity at the origin of the coordinate frame within the rigid body in inertial coordinates, is: $V_o = \begin{bmatrix} \dot{x}_c & \dot{y}_c & \dot{z}_c \end{bmatrix}^T$. The equations for u, v, w express the x, y, z components of the

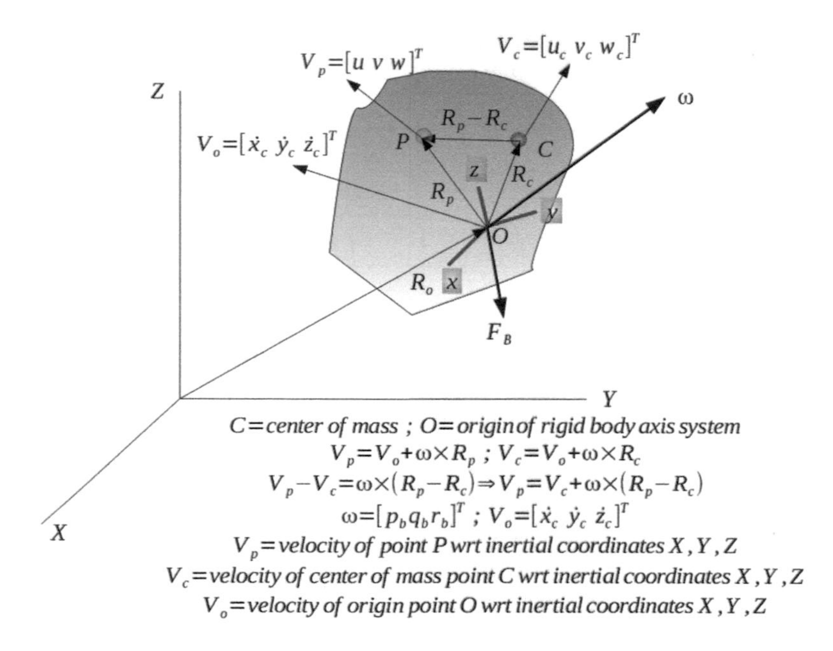

Fig. 4.4 Velocity of point P with respect to center of mass point C

inertial space velocity of point P in a rigid body (see Fig. 4.4). We can rewrite this velocity equation in the following way:

$$V_o = V_p - \omega \times R_p = V_p + R_p \times \omega \Rightarrow V_o = V_p + R_p \times \omega \tag{4.172}$$

or in matrix form:

$$\begin{bmatrix} \dot{x}_c \\ \dot{y}_c \\ \dot{z}_c \end{bmatrix} = \begin{bmatrix} u \\ v \\ w \end{bmatrix} - \begin{bmatrix} 0 & -r_b & q_b \\ r_b & 0 & -p_b \\ -q_b & p_b & 0 \end{bmatrix} \begin{bmatrix} x_p \\ y_p \\ z_p \end{bmatrix} = \begin{bmatrix} u + r_b y_p - q_b z_p \\ v + p_b z_p - r_b x_p \\ w + q_b x_p - p_b y_p \end{bmatrix} \tag{4.173}$$

Hence a general expression for the kinetic energy T is obtained by inserting these relations into $T = 1/2 \int \int_{v} \int (\dot{x}_c^2 + \dot{y}_c^2 + \dot{z}_c^2) dm$, where the triple integral signifies integration over the complete volume of the rigid body. On collecting terms, we have:

$$
\begin{aligned}
T = {} & \frac{1}{2}m(u^2+v^2+w^2) + \frac{p_b^2}{2}\iiint\limits_v (y_p^2+z_p^2)dm + \frac{q_b^2}{2}\iiint\limits_v (x_p^2+z_p^2)dm \\
& + \frac{r_b^2}{2}\iiint\limits_v (x_p^2+y_p^2)dm - p_b q_b \iiint\limits_v x_p y_p dm - p_b r_b \iiint\limits_v x_p z_p dm \\
& - q_b r_b \iiint\limits_v y_p z_p dm - u\left(\iiint\limits_v q_b z_p dm - \iiint\limits_v r_b y_p dm\right) \\
& - v\left(\iiint\limits_v r_b x_p dm - \iiint\limits_v p_b z_p dm\right) \\
& - w\left(\iiint\limits_v p_b y_p dm - \iiint\limits_v q_b x_p dm\right)
\end{aligned}
\tag{4.174}
$$

Notice that by definition: $\iiint\limits_v x_p dm = m x_c$; $\iiint\limits_v y_p dm = m y_c$; $\iiint\limits_v z_p dm = m z_c$. The kinetic energy T may be shown to be equivalent to:

$$
\begin{aligned}
T = {} & \frac{1}{2}m(u^2+v^2+w^2) \\
& + \frac{1}{2}\left[p_b^2 I_x + q_b^2 I_y + r_b^2 I_z - 2p_b q_b I_{xy} - 2p_b r_b I_{xz} - 2q_b r_b I_{yz}\right] \\
& + m\left[u(r_b y_c - q_b z_c) + v(p_b z_c - r_b x_c) + w(q_b x_c - p_b y_c)\right] \\
= {} & \frac{1}{2}m\begin{bmatrix} u & v & w \end{bmatrix}\begin{bmatrix} u \\ v \\ w \end{bmatrix} \\
& + \frac{1}{2}\left[p_b^2 I_x + q_b^2 I_y + r_b^2 I_z - 2p_b q_b I_{xy} - 2p_b r_b I_{xz} - 2q_b r_b I_{yz}\right] \\
& + m\begin{bmatrix} x_c & y_c & z_c \end{bmatrix}\begin{bmatrix} 0 & -r_b & q_b \\ r_b & 0 & -p_b \\ -q_b & p_b & 0 \end{bmatrix}\begin{bmatrix} u \\ v \\ w \end{bmatrix}
\end{aligned}
\tag{4.175}
$$

where m is the total mass of the body, and where x_c, y_c, z_c, are the coordinates the center of mass point C of the rigid body with respect to the origin O. The moments of inertia I_x, I_y, I_z are, respectively: $I_x = \int\int_v\int (y_p^2 + z_p^2)dm$, $I_y = \int\int_v\int (x_p^2 + z_p^2)dm$, $I_z = \int\int_v\int (x_p^2 + y_p^2)dm$ and the cross products of inertia I_{xy}, I_{xz}, I_{yz} are: $I_{xy} = \int\int_v\int x_p y_p dm$, $I_{xz} = \int\int_v\int x_p z_p dm$, $I_{yz} = \int\int_v\int y_p z_p dm$. It may be shown that:

$$\left[p_b^2 I_x + q_b^2 I_y + r_b^2 I_z - 2p_b q_b I_{xy} - 2p_b r_b I_{xz} - 2q_b r_b I_{yz}\right] = \omega^T I \omega$$

$$\omega = \begin{bmatrix} p_b \\ q_b \\ r_b \end{bmatrix}; \; I = \begin{bmatrix} I_x & -I_{xy} & -I_{xz} \\ -I_{xy} & I_{yy} & -I_{yz} \\ -I_{xz} & -I_{yz} & -I_z \end{bmatrix}$$

The foregoing results will now be used to derive the equations of motion of a rigid body aircraft. For use in the sequel, we recall the definitions of the Euler transformation matrices $T_{B/I}$ and $T_{I/B}$ which transform from an inertial to a body axis system and vice versa. The matrices are:

$$T_{B/I} = \begin{bmatrix} \cos\theta\cos\psi & \cos\theta\sin\psi & -\sin\theta \\ \sin\phi\sin\theta\cos\psi - \cos\phi\sin\psi & \sin\phi\sin\theta\sin\psi + \cos\phi\cos\psi & \sin\phi\cos\theta \\ \cos\phi\sin\theta\cos\psi + \sin\phi\sin\psi & \cos\phi\sin\theta\sin\psi - \sin\phi\cos\psi & \cos\phi\cos\theta \end{bmatrix}$$

$$T_{I/B} = \begin{bmatrix} \cos\psi\cos\theta & \cos\psi\sin\phi\sin\theta - \cos\phi\sin\psi & \sin\phi\sin\psi + \cos\phi\cos\psi\sin\theta \\ \cos\theta\sin\psi & \cos\phi\cos\psi + \sin\phi\sin\psi\sin\theta & \cos\phi\sin\psi\sin\theta - \cos\psi\sin\phi \\ -\sin\theta & \cos\theta\sin\phi & \cos\phi\cos\theta \end{bmatrix} \tag{4.176}$$

The translational equations were previously derived (see Eq. 4.166) and are of the form:

$$\frac{d}{dt}\left\{\frac{\partial \mathcal{L}}{\partial V_p}\right\} + \omega \times \left\{\frac{\partial \mathcal{L}}{\partial V_p}\right\} - \left\{\frac{\partial \mathcal{L}}{\partial R_p}\right\} = T_{B/I} Q_I = F_B$$

$$\Rightarrow \frac{d}{dt}\begin{bmatrix} \frac{\partial \mathcal{L}}{\partial u} \\ \frac{\partial \mathcal{L}}{\partial v} \\ \frac{\partial \mathcal{L}}{\partial w} \end{bmatrix} + \begin{bmatrix} 0 & -r_b & q_b \\ r_b & 0 & -p_b \\ -q_b & p_b & 0 \end{bmatrix}\begin{bmatrix} \frac{\partial \mathcal{L}}{\partial u} \\ \frac{\partial \mathcal{L}}{\partial v} \\ \frac{\partial \mathcal{L}}{\partial w} \end{bmatrix} - \begin{bmatrix} \frac{\partial \mathcal{L}}{\partial x_p} \\ \frac{\partial \mathcal{L}}{\partial y_p} \\ \frac{\partial \mathcal{L}}{\partial z_p} \end{bmatrix} = F_B \tag{4.177}$$

However the potential energy V is:

$$V = -mgZ_I = -mg\begin{bmatrix}0\;0\;1\end{bmatrix} T_{I/B}\begin{bmatrix} x_p \\ y_p \\ z_p \end{bmatrix}$$

$$= -mg\begin{bmatrix}-\sin\theta & \cos\theta\sin\phi & \cos\phi\cos\theta\end{bmatrix}\begin{bmatrix} x_p \\ y_p \\ z_p \end{bmatrix} \tag{4.178}$$

Hence,

$$\frac{\partial\overline{\mathcal{L}}}{\partial R_p} = -\frac{\partial V}{\partial R_p} = -\left(\frac{\partial V}{\partial Z_I}\right)\begin{bmatrix} \frac{\partial Z_I}{\partial x_p} \\ \frac{\partial Z_I}{\partial y_p} \\ \frac{\partial Z_I}{\partial z_p} \end{bmatrix} = mg\begin{bmatrix} -\sin\theta \\ \cos\theta\sin\phi \\ \cos\phi\cos\theta \end{bmatrix}; \quad \frac{\partial Z_I}{\partial R_p} = \begin{bmatrix} \frac{\partial Z_I}{\partial x_p} \\ \frac{\partial Z_I}{\partial y_p} \\ \frac{\partial Z_I}{\partial z_p} \end{bmatrix} \tag{4.179}$$

From Eq. 4.175, the kinetic energy T was shown to be:

$$\begin{aligned} T &= \frac{1}{2}m\underbrace{(u^2+v^2+w^2)}_{V_p^2} \\ &\quad + \frac{1}{2}\left[p_b^2 I_x + q_b^2 I_y + r_b^2 I_z - 2p_b q_b I_{xy} - 2p_b r_b I_{xz} - 2q_b r_b I_{yz}\right] \\ &\quad + m\left[u(r_b y_c - q_b z_c) + v(p_b z_c - r_b x_c) + w(q_b x_c - p_b y_c)\right] \end{aligned} \tag{4.180}$$

The partial derivatives of $\overline{\mathcal{L}}$ with respect to the vector of linear velocities u, v, w are deduced to be:

$$\left\{\frac{\partial\overline{\mathcal{L}}}{\partial V_p}\right\} = \left\{\frac{\partial T}{\partial V_p}\right\} = \begin{bmatrix} \frac{\partial T}{\partial u} \\ \frac{\partial T}{\partial v} \\ \frac{\partial T}{\partial w} \end{bmatrix} = m\begin{bmatrix} u + (y_c r_b - z_c q_b) \\ v + (z_c p_b - x_c r_b) \\ w + (x_c q_b - y_c p_b) \end{bmatrix}$$

$$= m\begin{bmatrix} u \\ v \\ w \end{bmatrix} + m\begin{bmatrix} y_c r_b - z_c q_b \\ z_c p_b - x_c r_b \\ x_c q_b - y_c p_b \end{bmatrix} = m\begin{bmatrix} u \\ v \\ w \end{bmatrix} + m\begin{bmatrix} 0 & -z_c & y_c \\ z_c & 0 & -x_c \\ -y_c & x_c & 0 \end{bmatrix}\begin{bmatrix} p_b \\ q_b \\ r_b \end{bmatrix} \tag{4.181}$$

Calculating the time derivative of $\left\{\frac{\partial T}{\partial V_p}\right\}$, we have:

$$\frac{d}{dt}\left\{\frac{\partial T}{\partial V_p}\right\} = m\begin{bmatrix} \dot{u} \\ \dot{v} \\ \dot{w} \end{bmatrix} + m\begin{bmatrix} 0 & -z_c & y_c \\ z_c & 0 & -x_c \\ -y_c & x_c & 0 \end{bmatrix}\begin{bmatrix} \dot{p}_b \\ \dot{q}_b \\ \dot{r}_b \end{bmatrix} \tag{4.182}$$

The term $\omega \times \left\{\frac{\partial T}{\partial V_p}\right\}$ is:

$$\omega \times \left\{\frac{\partial T}{\partial V_p}\right\} = \begin{bmatrix} 0 & -r_b & q_b \\ r_b & 0 & -p_b \\ -q_b & p_b & 0 \end{bmatrix}\begin{bmatrix} \frac{\partial T}{\partial u} \\ \frac{\partial T}{\partial v} \\ \frac{\partial T}{\partial w} \end{bmatrix} = m\begin{bmatrix} 0 & -r_b & q_b \\ r_b & 0 & -p_b \\ -q_b & p_b & 0 \end{bmatrix}\begin{bmatrix} u \\ v \\ w \end{bmatrix}$$

$$+m\begin{bmatrix} 0 & -r_b & q_b \\ r_b & 0 & -p_b \\ -q_b & p_b & 0 \end{bmatrix}\begin{bmatrix} 0 & -z_c & y_c \\ z_c & 0 & -x_c \\ -y_c & x_c & 0 \end{bmatrix}\begin{bmatrix} p_b \\ q_b \\ r_b \end{bmatrix}$$

$$= m\begin{bmatrix} 0 & -r_b & q_b \\ r_b & 0 & -p_b \\ -q_b & p_b & 0 \end{bmatrix}\begin{bmatrix} u \\ v \\ w \end{bmatrix} - m\begin{bmatrix} 0 & -r_b & q_b \\ r_b & 0 & -p_b \\ -q_b & p_b & 0 \end{bmatrix}\begin{bmatrix} z_c q_b - y_c r_b \\ x_c r_b - z_c p_b \\ y_c p_b - x_c q_b \end{bmatrix} \tag{4.183}$$

Combining the terms in Eqs. 4.182 and 4.183, the dynamic equations are:

$$\frac{d}{dt}\left\{\frac{\partial \overline{\mathcal{L}}}{\partial V_p}\right\} + \omega \times \left\{\frac{\partial \overline{\mathcal{L}}}{\partial V_p}\right\} - \left\{\frac{\partial \overline{\mathcal{L}}}{\partial R_p}\right\} = F_B$$

$$\Rightarrow m\begin{bmatrix} \dot{u} \\ \dot{v} \\ \dot{w} \end{bmatrix} + m\begin{bmatrix} 0 & -z_c & y_c \\ z_c & 0 & -x_c \\ -y_c & x_c & 0 \end{bmatrix}\begin{bmatrix} \dot{p}_b \\ \dot{q}_b \\ \dot{r}_b \end{bmatrix} + m\begin{bmatrix} 0 & -r_b & q_b \\ r_b & 0 & -p_b \\ -q_b & p_b & 0 \end{bmatrix}\begin{bmatrix} u \\ v \\ w \end{bmatrix}$$

$$-m\begin{bmatrix} 0 & -r_b & q_b \\ r_b & 0 & -p_b \\ -q_b & p_b & 0 \end{bmatrix}\begin{bmatrix} z_c q_b - y_c r_b \\ x_c r_b - z_c p_b \\ y_c p_b - x_c q_b \end{bmatrix} - mg\begin{bmatrix} -\sin\theta \\ \cos\theta\sin\phi \\ \cos\theta\cos\phi \end{bmatrix} = F_B \tag{4.184}$$

The rotational d'Alembert–Lagrange equations, for the situation where the center of mass and the origin of the body-fixed coordinate system do not coincide, were previously shown to be:

$$\frac{d}{dt}\left(\frac{\partial \overline{\mathcal{L}}}{\partial \omega}\right) - L^T\frac{\partial \overline{\mathcal{L}}}{\partial \Phi} + \omega \times \left(\frac{\partial \overline{\mathcal{L}}}{\partial \omega}\right) + \left\{V_c \times \left\{\frac{\partial \overline{\mathcal{L}}}{\partial V_c}\right\}\right\} = L^T Q_\tau + R_c \times F_B$$

$$\Rightarrow \frac{d}{dt}\begin{bmatrix} \frac{\partial \overline{\mathcal{L}}}{\partial p_b} \\ \frac{\partial \overline{\mathcal{L}}}{\partial q_b} \\ \frac{\partial \overline{\mathcal{L}}}{\partial r_b} \end{bmatrix} - \begin{bmatrix} 1 & 0 & 0 \\ \sin\phi\tan\theta & \cos\phi & \frac{\sin\phi}{\cos\theta} \\ \cos\phi\tan\theta & -\sin\phi & \frac{\cos\phi}{\cos\theta} \end{bmatrix}\begin{bmatrix} \frac{\partial \overline{\mathcal{L}}}{\partial \phi} \\ \frac{\partial \overline{\mathcal{L}}}{\partial \theta} \\ \frac{\partial \overline{\mathcal{L}}}{\partial \psi} \end{bmatrix} + \begin{bmatrix} 0 & -r_b & q_b \\ r_b & 0- & p_b \\ -q_b & p_b & 0 \end{bmatrix}\begin{bmatrix} \frac{\partial \overline{\mathcal{L}}}{\partial p_b} \\ \frac{\partial \overline{\mathcal{L}}}{\partial q_b} \\ \frac{\partial \overline{\mathcal{L}}}{\partial r_b} \end{bmatrix}$$

$$+\begin{bmatrix} 0 & -w_c & v_c \\ w_c & 0 & -u_c \\ -v_c & u_c & 0 \end{bmatrix}\begin{bmatrix} \frac{\partial \overline{\mathcal{L}}}{\partial u_c} \\ \frac{\partial \overline{\mathcal{L}}}{\partial v_c} \\ \frac{\partial \overline{\mathcal{L}}}{\partial w_c} \end{bmatrix}$$

$$= \begin{bmatrix} 1 & 0 & 0 \\ \sin\phi\tan\theta & \cos\phi & \frac{\sin\phi}{\cos\theta} \\ \cos\phi\tan\theta & -\sin\phi & \frac{\cos\phi}{\cos\theta} \end{bmatrix} \begin{bmatrix} Q_{\tau\phi} \\ Q_{\tau\theta} \\ Q_{\tau\psi} \end{bmatrix} + \begin{bmatrix} 0 & -z_C & y_C \\ z_C & 0 & -x_C \\ -y_C & x_C & 0 \end{bmatrix} \begin{bmatrix} F_x \\ F_y \\ F_z \end{bmatrix} \tag{4.185}$$

The kinetic energy $\bar{\boldsymbol{T}}$ may be shown to be:

$$\begin{aligned}
\bar{\boldsymbol{T}} &= \frac{1}{2}m(u^2+v^2+w^2) \\
&\quad + \frac{1}{2}\left[p_b^2 I_x + q_b^2 I_y + r_b^2 I_z - 2p_b q_b I_{xy} - 2p_b r_b I_{xz} - 2q_b r_b I_{yz}\right] \\
&\quad + m\left[u(r_b y_c - q_b z_c) + v(p_b z_c - r_b x_c) + w(q_b x_c - p_b y_c)\right] \\
&= \frac{1}{2}m\begin{bmatrix} u & v & w \end{bmatrix}\begin{bmatrix} u \\ v \\ w \end{bmatrix} \\
&\quad + \frac{1}{2}\left[p_b^2 I_x + q_b^2 I_y + r_b^2 I_z - 2p_b q_b I_{xy} - 2p_b r_b I_{xz} - 2q_b r_b I_{yz}\right] \\
&\quad + m\begin{bmatrix} x_c & y_c & z_c \end{bmatrix}\begin{bmatrix} 0 & -r_b & q_b \\ r_b & 0 & -p_b \\ -q_b & p_b & 0 \end{bmatrix}\begin{bmatrix} u \\ v \\ w \end{bmatrix}
\end{aligned} \tag{4.186}$$

where m is the total mass of the body, and where x_c, y_c, z_c, are the coordinates the center of mass point C of the rigid body with respect to the body-fixed origin O. The potential energy V is: $V = -mg\begin{bmatrix} 0 & 0 & 1 \end{bmatrix} T_{I/B}\begin{bmatrix} x_p \\ y_p \\ z_p \end{bmatrix} = -mgz_I$, and so the modified Lagrangian $\overline{\mathcal{L}} = \bar{\boldsymbol{T}} - V$ becomes:

$$\begin{aligned}
\overline{\mathcal{L}} &= \frac{1}{2}m(u^2+v^2+w^2) \\
&\quad + \frac{1}{2}\left[p_b^2 I_x + q_b^2 I_y + r_b^2 I_z - 2p_b q_b I_{xy} - 2p_b r_b I_{xz} - 2q_b r_b I_{yz}\right] \\
&\quad + m\left[u(r_b y_c - q_b z_c) + v(p_b z_c - r_b x_c) + w(q_b x_c - p_b y_c)\right] \\
&= \frac{1}{2}m\begin{bmatrix} u & v & w \end{bmatrix}\begin{bmatrix} u \\ v \\ w \end{bmatrix}
\end{aligned}$$

$$
\begin{aligned}
&+\frac{1}{2}\left[p_b^2 I_x + q_b^2 I_y + r_b^2 I_z - 2p_b q_b I_{xy} - 2p_b r_b I_{xz} - 2q_b r_b I_{yz}\right] \\
&+m\left[x_c\ y_c\ z_c\right]\begin{bmatrix} 0 & -r_b & q_b \\ r_b & 0 & -p_b \\ -q_b & p_b & 0 \end{bmatrix}\begin{bmatrix} u \\ v \\ w \end{bmatrix} \\
&+\underbrace{mg\left[-\sin\theta\ \cos\theta\sin\phi\ \cos\theta\cos\phi\right]\begin{bmatrix} x_p \\ y_p \\ z_p \end{bmatrix}}_{z_I}
\end{aligned}
\tag{4.187}
$$

The vector $\left\{\frac{\partial \overline{\mathcal{L}}}{\partial \omega}\right\}$ turns out to be:

$$
\begin{aligned}
\frac{\partial \overline{\mathcal{L}}}{\partial p_b} &= I_x p_b - I_{xy} q_b - I_{xz} r_b + m\left(v z_c - w y_c\right) \\
\frac{\partial \overline{\mathcal{L}}}{\partial q_b} &= I_y q_b - I_{xy} p_b - I_{yz} r_b + m\left(w x_c - u z_c\right) \\
\frac{\partial \overline{\mathcal{L}}}{\partial r_b} &= I_z r_b - I_{xz} p_b - I_{yz} q_b + m\left(u y_c - v x_c\right)
\end{aligned}
$$

$$
\Rightarrow \begin{bmatrix} \frac{\partial \overline{\mathcal{L}}}{\partial p_b} \\ \frac{\partial \overline{\mathcal{L}}}{\partial q_b} \\ \frac{\partial \overline{\mathcal{L}}}{\partial r_b} \end{bmatrix} = \begin{bmatrix} I_x & -I_{xy} & -I_{xz} \\ -I_{xy} & I_y & -I_{yz} \\ -I_{xz} & -I_{yz} & I_z \end{bmatrix}\begin{bmatrix} p_b \\ q_b \\ r_b \end{bmatrix} + m\begin{bmatrix} 0 & z_c & -y_c \\ -z_c & 0 & x_c \\ y_c & -x_c & 0 \end{bmatrix}\begin{bmatrix} u \\ v \\ w \end{bmatrix}
\tag{4.188}
$$

Differentiating the above result with respect to time leads to:

$$
\frac{d}{dt}\begin{bmatrix} \frac{\partial \overline{\mathcal{L}}}{\partial p_b} \\ \frac{\partial \overline{\mathcal{L}}}{\partial q_b} \\ \frac{\partial \overline{\mathcal{L}}}{\partial r_b} \end{bmatrix} = \begin{bmatrix} I_x & -I_{xy} & -I_{xz} \\ -I_{xy} & I_y & -I_{yz} \\ -I_{xz} & -I_{yz} & I_z \end{bmatrix}\begin{bmatrix} \dot{p}_b \\ \dot{q}_b \\ \dot{r}_b \end{bmatrix} + m\begin{bmatrix} 0 & z_c & -y_c \\ -z_c & 0 & x_c \\ y_c & -x_c & 0 \end{bmatrix}\begin{bmatrix} \dot{u} \\ \dot{v} \\ \dot{w} \end{bmatrix}
\tag{4.189}
$$

In order to calculate $\left[L^T\right]\left\{\frac{\partial \overline{\mathcal{L}}}{\partial \Phi}\right\}$, the following relationships are required:

$$\left[0\;0\;1\right] T_{I/B} = \left[-\sin\theta\;\cos\theta\sin\phi\;\cos\theta\cos\phi\right]$$

$$\Rightarrow \begin{bmatrix} \frac{\partial}{\partial\phi} \\ \frac{\partial}{\partial\theta} \\ \frac{\partial}{\partial\psi} \end{bmatrix} \left[0\;0\;1\right] T_{I/B} = \begin{bmatrix} 0 & \cos\theta\cos\phi & -\cos\theta\sin\phi \\ -\cos\theta & -\sin\theta\sin\phi & -\sin\theta\cos\phi \\ 0 & 0 & 0 \end{bmatrix}$$

$$\Rightarrow \begin{bmatrix} 1 & 0 & 0 \\ \sin\phi\tan\theta & \cos\phi & \frac{\sin\phi}{\cos\theta} \\ \cos\phi\tan\theta & -\sin\phi & \frac{\cos\phi}{\cos\theta} \end{bmatrix} \begin{bmatrix} \frac{\partial}{\partial\phi} \\ \frac{\partial}{\partial\theta} \\ \frac{\partial}{\partial\psi} \end{bmatrix} \left[0\;0\;1\right] T_{I/B}$$

$$= \begin{bmatrix} 0 & \cos\theta\cos\phi & -\cos\theta\sin\phi \\ -\cos\theta\cos\phi & 0 & -\sin\theta \\ \sin\phi\cos\theta & \sin\theta & 0 \end{bmatrix} = -\begin{bmatrix} 0 & -k_z & k_y \\ k_z & 0 & -k_x \\ -k_y & k_x & 0 \end{bmatrix} \tag{4.190}$$

where $\left[k_x\;k_y\;k_z\right]^T = \left[-\sin\theta\;\cos\theta\sin\phi\;\cos\theta\cos\phi\right]^T$.

However

$$\begin{bmatrix} 1 & 0 & 0 \\ \sin\phi\tan\theta & \cos\phi & \frac{\sin\phi}{\cos\theta} \\ \cos\phi\tan\theta & -\sin\phi & \frac{\cos\phi}{\cos\theta} \end{bmatrix} \begin{bmatrix} \frac{\partial}{\partial\phi} \\ \frac{\partial}{\partial\theta} \\ \frac{\partial}{\partial\psi} \end{bmatrix} \left[0\;0\;1\right] T_{I/B} \begin{bmatrix} x_p \\ y_p \\ z_p \end{bmatrix}$$

$$= -\begin{bmatrix} 0 & -k_z & k_y \\ k_z & 0 & -k_x \\ -k_y & k_x & 0 \end{bmatrix}\begin{bmatrix} x_p \\ y_p \\ z_p \end{bmatrix} = \begin{bmatrix} 1 & 0 & 0 \\ \sin\phi\tan\theta & \cos\phi & \frac{\sin\phi}{\cos\theta} \\ \cos\phi\tan\theta & -\sin\phi & \frac{\cos\phi}{\cos\theta} \end{bmatrix}\begin{bmatrix} \frac{\partial}{\partial\phi} \\ \frac{\partial}{\partial\theta} \\ \frac{\partial}{\partial\psi} \end{bmatrix}[0\ 0\ 1]\begin{bmatrix} x_I \\ y_I \\ z_I \end{bmatrix}$$

$$= \begin{bmatrix} 1 & 0 & 0 \\ \sin\phi\tan\theta & \cos\phi & \frac{\sin\phi}{\cos\theta} \\ \cos\phi\tan\theta & -\sin\phi & \frac{\cos\phi}{\cos\theta} \end{bmatrix}\begin{bmatrix} \frac{\partial z_I}{\partial\phi} \\ \frac{\partial z_I}{\partial\theta} \\ \frac{\partial z_I}{\partial\psi} \end{bmatrix} \tag{4.191}$$

Since

$$V = V(\phi, \theta) = -mgz_I \Rightarrow \frac{\partial\overline{\mathcal{L}}}{\partial\phi} = -\frac{\partial V}{\partial\phi} = mg\frac{\partial z_I}{\partial\phi}$$

$$\frac{\partial\overline{\mathcal{L}}}{\partial\theta} = -\frac{\partial V}{\partial\theta} = mg\frac{\partial z_I}{\partial\theta};\quad \frac{\partial\overline{\mathcal{L}}}{\partial\psi} = 0$$

the term $\left[L^T\right]\left\{\frac{\partial\overline{\mathcal{L}}}{\partial\Phi}\right\}$, from Eq. 4.191, is therefore equivalent to:

$$\begin{bmatrix} 1 & 0 & 0 \\ \sin\phi\tan\theta & \cos\phi & \frac{\sin\phi}{\cos\theta} \\ \cos\phi\tan\theta & -\sin\phi & \frac{\cos\phi}{\cos\theta} \end{bmatrix}\begin{bmatrix} \frac{\partial\overline{\mathcal{L}}}{\partial\phi} \\ \frac{\partial\overline{\mathcal{L}}}{\partial\theta} \\ \frac{\partial\overline{\mathcal{L}}}{\partial\psi} \end{bmatrix} = mg\begin{bmatrix} 1 & 0 & 0 \\ \sin\phi\tan\theta & \cos\phi & \frac{\sin\phi}{\cos\theta} \\ \cos\phi\tan\theta & -\sin\phi & \frac{\cos\phi}{\cos\theta} \end{bmatrix}\begin{bmatrix} \frac{\partial z_I}{\partial\phi} \\ \frac{\partial z_I}{\partial\theta} \\ \frac{\partial z_I}{\partial\psi} \end{bmatrix}$$

$$= -mg \begin{bmatrix} 0 & -k_z & k_y \\ k_z & 0 & -k_x \\ -k_y & k_x & 0 \end{bmatrix} \begin{bmatrix} x_p \\ y_p \\ z_p \end{bmatrix} \tag{4.192}$$

Hence the rotational d'Alembert–Lagrange equations are:

$$\frac{d}{dt}\left(\frac{\partial \overline{\mathcal{L}}}{\partial \omega}\right) - L^T \frac{\partial \overline{\mathcal{L}}}{\partial \Phi} + \omega \times \left(\frac{\partial \overline{\mathcal{L}}}{\partial \omega}\right) + \left\{V_c \times \left\{\frac{\partial \overline{\mathcal{L}}}{\partial V_c}\right\}\right\} = L^T Q_\tau + R_c \times F_B$$

$$\begin{aligned}
&\frac{d}{dt}\begin{bmatrix} \frac{\partial \overline{\mathcal{L}}}{\partial p_b} \\ \frac{\partial \overline{\mathcal{L}}}{\partial q_b} \\ \frac{\partial \overline{\mathcal{L}}}{\partial r_b} \end{bmatrix} - \begin{bmatrix} 1 & 0 & 0 \\ \sin\phi \tan\theta & \cos\phi & \frac{\sin\phi}{\cos\theta} \\ \cos\phi \tan\theta & -\sin\phi & \frac{\cos\phi}{\cos\theta} \end{bmatrix} \begin{bmatrix} \frac{\partial \overline{\mathcal{L}}}{\partial \phi} \\ \frac{\partial \overline{\mathcal{L}}}{\partial \theta} \\ \frac{\partial \overline{\mathcal{L}}}{\partial \psi} \end{bmatrix} + \begin{bmatrix} 0 & -r_b & q_b \\ r_b & 0 - p_b & \\ -q_b & p_b & 0 \end{bmatrix} \begin{bmatrix} \frac{\partial \overline{\mathcal{L}}}{\partial p_b} \\ \frac{\partial \overline{\mathcal{L}}}{\partial q_b} \\ \frac{\partial \overline{\mathcal{L}}}{\partial r_b} \end{bmatrix} \\
&\quad + \begin{bmatrix} 0 & -w_c & v_c \\ w_c & 0 & -u_c \\ -v_c & u_c & 0 \end{bmatrix} \begin{bmatrix} \frac{\partial \overline{\mathcal{L}}}{\partial u_c} \\ \frac{\partial \overline{\mathcal{L}}}{\partial v_c} \\ \frac{\partial \overline{\mathcal{L}}}{\partial w_c} \end{bmatrix} = \begin{bmatrix} 1 & 0 & 0 \\ \sin\phi \tan\theta & \cos\phi & \frac{\sin\phi}{\cos\theta} \\ \cos\phi \tan\theta & -\sin\phi & \frac{\cos\phi}{\cos\theta} \end{bmatrix} \begin{bmatrix} Q_{\tau\phi} \\ Q_{\tau\theta} \\ Q_{\tau\psi} \end{bmatrix} \\
&\quad + \begin{bmatrix} 0 & -z_c & y_c \\ z_c & 0 & -x_c \\ -y_c & x_c & 0 \end{bmatrix} \begin{bmatrix} F_x \\ F_y \\ F_z \end{bmatrix}
\end{aligned} \tag{4.193}$$

However from Eq. 4.189, $\frac{d}{dt}\left\{\frac{\partial\overline{\mathcal{L}}}{\partial\omega}\right\}$ is:

$$\frac{d}{dt}\begin{bmatrix}\frac{\partial\overline{\mathcal{L}}}{\partial p_b}\\ \frac{\partial\overline{\mathcal{L}}}{\partial q_b}\\ \frac{\partial\overline{\mathcal{L}}}{\partial r_b}\end{bmatrix}=\begin{bmatrix}I_x & -I_{xy} & -I_{xz}\\ -I_{xy} & I_y & -I_{yz}\\ -I_{xz} & -I_{yz} & I_z\end{bmatrix}\begin{bmatrix}\dot{p}_b\\ \dot{q}_b\\ \dot{r}_b\end{bmatrix}+m\begin{bmatrix}0 & z_c & -y_c\\ -z_c & 0 & x_c\\ y_c & -x_c & 0\end{bmatrix}\begin{bmatrix}\dot{u}\\ \dot{v}\\ \dot{w}\end{bmatrix} \tag{4.194}$$

Furthermore, from Eq. 4.188, $\omega\times\left\{\frac{\partial\overline{\mathcal{L}}}{\partial\omega}\right\}$ turns out to be:

$$\begin{bmatrix}0 & -r_b & q_b\\ r_b & 0 & -p_b\\ -q_b & p_b & 0\end{bmatrix}\begin{bmatrix}\frac{\partial\overline{\mathcal{L}}}{\partial p_b}\\ \frac{\partial\overline{\mathcal{L}}}{\partial q_b}\\ \frac{\partial\overline{\mathcal{L}}}{\partial r_b}\end{bmatrix}=\begin{bmatrix}0 & -r_b & q_b\\ r_b & 0 & -p_b\\ -q_b & p_b & 0\end{bmatrix}\begin{bmatrix}I_x & -I_{xy} & -I_{xz}\\ -I_{xy} & I_y & -I_{yz}\\ -I_{xz} & -I_{yz} & I_z\end{bmatrix}\begin{bmatrix}p_b\\ q_b\\ r_b\end{bmatrix}$$
$$+m\begin{bmatrix}0 & -r_b & q_b\\ r_b & 0 & -p_b\\ -q_b & p_b & 0\end{bmatrix}\begin{bmatrix}0 & z_c & -y_c\\ -z_c & 0 & x_c\\ y_c & -x_c & 0\end{bmatrix}\begin{bmatrix}u\\ v\\ w\end{bmatrix} \tag{4.195}$$

and $-\left[L^T\right]\left\{\frac{\partial\overline{\mathcal{L}}}{\partial\Phi}\right\}$ is:

$$mg\begin{bmatrix}0 & -k_z & k_y\\ k_z & 0 & -k_x\\ -k_y & k_x & 0\end{bmatrix}\begin{bmatrix}x_p\\ y_p\\ z_p\end{bmatrix}=-mg\begin{bmatrix}0 & -z_p & y_p\\ z_p & 0 & -x_p\\ -y_p & x_p & 0\end{bmatrix}\begin{bmatrix}k_x\\ k_y\\ k_z\end{bmatrix} \tag{4.196}$$

Combining all of the above results:

$$
\begin{aligned}
&\frac{d}{dt}\left(\frac{\partial \overline{\mathcal{L}}}{\partial \omega}\right) - L^T \frac{\partial \overline{\mathcal{L}}}{\partial \Phi} + \omega \times \left(\frac{\partial \overline{\mathcal{L}}}{\partial \omega}\right) + \left\{V_c \times \left\{\frac{\partial \overline{\mathcal{L}}}{\partial V_c}\right\}\right\} = L^T Q_\tau + R_c \times F_B \\
&\Rightarrow \begin{bmatrix} I_x & -I_{xy} & -I_{xz} \\ -I_{xy} & I_y & -I_{yz} \\ -I_{xz} & -I_{yz} & I_z \end{bmatrix} \begin{bmatrix} \dot{p}_b \\ \dot{q}_b \\ \dot{r}_b \end{bmatrix} + m \begin{bmatrix} 0 & z_c & -y_c \\ -z_c & 0 & x_c \\ y_c & -x_c & 0 \end{bmatrix} \begin{bmatrix} \dot{u} \\ \dot{v} \\ \dot{w} \end{bmatrix} \\
&+ \begin{bmatrix} 0 & -r_b & q_b \\ r_b & 0 & -p_b \\ -q_b & p_b & 0 \end{bmatrix} \begin{bmatrix} I_x & -I_{xy} & -I_{xz} \\ -I_{xy} & I_y & -I_{yz} \\ -I_{xz} & -I_{yz} & I_z \end{bmatrix} \begin{bmatrix} p_b \\ q_b \\ r_b \end{bmatrix} \\
&+ m \begin{bmatrix} 0 & -r_b & q_b \\ r_b & 0 & -p_b \\ -q_b & p_b & 0 \end{bmatrix} \begin{bmatrix} 0 & z_c & -y_c \\ -z_c & 0 & x_c \\ y_c & -x_c & 0 \end{bmatrix} \begin{bmatrix} u \\ v \\ w \end{bmatrix} \\
&= \begin{bmatrix} 1 & 0 & 0 \\ \sin\phi \tan\theta & \cos\phi & \frac{\sin\phi}{\cos\theta} \\ \cos\phi \tan\theta & -\sin\phi & \frac{\cos\phi}{\cos\theta} \end{bmatrix} \begin{bmatrix} Q_{\tau\phi} \\ Q_{\tau\theta} \\ Q_{\tau\psi} \end{bmatrix} \\
&+ mg \begin{bmatrix} 0 & -z_p & y_p \\ z_p & 0 & -x_p \\ -y_p & x_p & 0 \end{bmatrix} \begin{bmatrix} k_x \\ k_y \\ k_z \end{bmatrix} + \begin{bmatrix} 0 & z_c & -y_c \\ -z_c & 0 & x_c \\ y_c & -x_c & 0 \end{bmatrix} \begin{bmatrix} F_x \\ F_y \\ F_z \end{bmatrix} \qquad (4.197)
\end{aligned}
$$

Example 6: Quadcopter Equations of Motion [17, pp. 1–18]

The quadrotor model has been presented previously as an example (see Chap. 3, page 92). In the present approach, the modeling procedure will be based upon the quasi-velocities instead of the generalized coordinates. It is assumed that *the center of gravity coincides with the origin of the body-fixed coordinate system* . The quasi-velocities are non-holonomic in the sense that they are not the derivatives of any generalized coordinates (generalized Euler angles or position in inertial coordinates). The solution procedure will be outlined in the sequel. The quasi-velocities are defined to be:

1. $\{\dot{\Gamma}_1\}^T = \left[u \; v \; w\right]^T$
2. $\{\dot{\Gamma}_2\}^T = \left[p_b \; q_b \; r_b\right]^T$

and are all nonlinear functions of the generalized coordinates $q_1 = \left[X \; Y \; Z\right]^T$, $q_2 = \left[\phi \; \theta \; \psi\right]^T$, respectively, and their time derivatives, as listed below:

$$\{\dot{\Gamma}_1\} = \begin{bmatrix} u \\ v \\ w \end{bmatrix} = T_{B/I} \begin{bmatrix} \dot{X} \\ \dot{Y} \\ \dot{Z} \end{bmatrix}$$

$$= \begin{bmatrix} \cos\theta\cos\psi & \cos\theta\sin\psi & -\sin\theta \\ \sin\phi\sin\theta\cos\psi - \cos\phi\sin\psi & \sin\phi\sin\theta\sin\psi + \cos\phi\cos\psi & \sin\phi\cos\theta \\ \cos\phi\sin\theta\cos\psi + \sin\phi\sin\psi & \cos\phi\sin\theta\sin\psi - \sin\phi\cos\psi & \cos\phi\cos\theta \end{bmatrix} \begin{bmatrix} \dot{X} \\ \dot{Y} \\ \dot{Z} \end{bmatrix}$$

$$\{\dot{\Gamma}_2\} = \begin{bmatrix} p_b \\ q_b \\ r_b \end{bmatrix} = \begin{bmatrix} \dot{\phi} - \dot{\psi}\sin\theta \\ \dot{\theta}\cos\phi + \dot{\psi}\cos\theta\sin\phi \\ -\dot{\theta}\sin\phi + \dot{\psi}\cos\theta\cos\phi \end{bmatrix} = \begin{bmatrix} 1 & 0 & -\sin\theta \\ 0 & \cos\phi & \cos\theta\sin\phi \\ 0 & -\sin\phi & \cos\theta\cos\phi \end{bmatrix} \begin{bmatrix} \dot{\phi} \\ \dot{\theta} \\ \dot{\psi} \end{bmatrix} \tag{4.198}$$

Once again, note that the quasi-velocities cannot be integrated to obtain the generalized coordinates. The kinetic energy $\bar{\boldsymbol{T}}$ is of the form:

$$\bar{\boldsymbol{T}} = \frac{1}{2}\left[p_b \; q_b \; r_b\right] \begin{bmatrix} I_x & 0 & 0 \\ 0 & I_y & 0 \\ 0 & 0 & I_z \end{bmatrix} \begin{bmatrix} p_b \\ q_b \\ r_b \end{bmatrix} + \frac{1}{2} m \left[u \; v \; w\right] \begin{bmatrix} u \\ v \\ w \end{bmatrix}$$

$$= \frac{1}{2}\left(I_x p_b^2 + I_y q_b^2 + I_z r_b^2\right) + \frac{m}{2}\left(u^2 + v^2 + w^2\right) \tag{4.199}$$

The potential energy is: $V = -mgZ$, and so the Lagrangian $\overline{\mathcal{L}}$ is:

$$\overline{\mathcal{L}} = \bar{T} - V = \frac{1}{2}\left(I_x p_b^2 + I_y q_b^2 + I_z r_b^2\right) + \frac{m}{2}\left(u^2 + v^2 + w^2\right) + mgZ$$

$$Z = \begin{bmatrix} -\sin\theta & \cos\theta\sin\phi & \cos\theta\cos\phi \end{bmatrix} \begin{bmatrix} x_c \\ y_c \\ z_c \end{bmatrix} \tag{4.200}$$

The d'Alembert–Lagrange equations, when the center of mass and the origin of the body-centered coordinate system coincide, were shown to be of the form:

$$\frac{d}{dt}\left(\frac{\partial\overline{\mathcal{L}}}{\partial V_B}\right) + \omega \times \frac{\partial\overline{\mathcal{L}}}{\partial V_B} - \frac{\partial\overline{\mathcal{L}}}{\partial X_B} = \underbrace{T_{B/I} Q_I}_{F_B}$$

$$\frac{d}{dt}\left\{\frac{\partial\overline{\mathcal{L}}}{\partial\omega}\right\} - L^T\left\{\frac{\partial\overline{\mathcal{L}}}{\partial\Phi}\right\} + \omega \times \left\{\frac{\partial\overline{\mathcal{L}}}{\partial\omega}\right\} = L^T Q_\tau$$

$$V_B = \begin{bmatrix} u \\ v \\ w \end{bmatrix}; \quad X_B = \begin{bmatrix} x_c \\ y_c \\ z_c \end{bmatrix}; \quad Q_\tau = \begin{bmatrix} Q_{\tau\phi} \\ Q_{\tau\theta} \\ Q_{\tau\psi} \end{bmatrix}$$

$$\Phi = \begin{bmatrix} \phi \\ \theta \\ \psi \end{bmatrix}; \quad \omega = \begin{bmatrix} p_b \\ q_b \\ r_b \end{bmatrix}; \quad L^T = \begin{bmatrix} 1 & 0 & 0 \\ \sin\phi\tan\theta & \cos\phi & \frac{\sin\phi}{\cos\theta} \\ \cos\phi\tan\theta & -\sin\phi & \frac{\cos\phi}{\cos\theta} \end{bmatrix} \tag{4.201}$$

Hence, the force terms in the d'Alembert–Lagrange equations are as follows:

$$\frac{d}{dt}\left(\frac{\partial\overline{\mathcal{L}}}{\partial V_B}\right) = m\begin{bmatrix} \dot{u} \\ \dot{v} \\ \dot{w} \end{bmatrix}$$

$$\omega \times \frac{\partial \overline{\mathcal{L}}}{\partial V_B} = m \begin{bmatrix} 0 & -r_b & q_b \\ r_b & 0 & -p_b \\ -q_b & p_b & 0 \end{bmatrix} \begin{bmatrix} u \\ v \\ w \end{bmatrix} = m \begin{bmatrix} wq_b - vr_b \\ ur_b - wp_b \\ vp_b - uq_b \end{bmatrix}$$

$$\frac{\partial \overline{\mathcal{L}}}{\partial X_B} = mg \begin{bmatrix} \frac{\partial \overline{\mathcal{L}}}{\partial x_c} \\ \frac{\partial \overline{\mathcal{L}}}{\partial y_c} \\ \frac{\partial \overline{\mathcal{L}}}{\partial z_c} \end{bmatrix} = mg \begin{bmatrix} -\sin\theta \\ \cos\theta\sin\phi \\ \cos\theta\cos\phi \end{bmatrix}; \; F_B = \begin{bmatrix} 0 \\ 0 \\ f_1 + f_2 + f_3 + f_4 \end{bmatrix} \tag{4.202}$$

Similarly the moment terms in the d'Alembert–Lagrange equations (see Eq. 4.202) are:

$$\frac{d}{dt}\left\{\frac{\partial \overline{\mathcal{L}}}{\partial \omega}\right\} = \begin{bmatrix} I_x \dot{p}_b \\ I_y \dot{q}_b \\ I_z \dot{r}_b \end{bmatrix}$$

$$-L^T \left\{\frac{\partial \overline{\mathcal{L}}}{\partial \Phi}\right\} = -\begin{bmatrix} 1 & 0 & 0 \\ \sin\phi\tan\theta & \cos\phi & \frac{\sin\phi}{\cos\theta} \\ \cos\phi\tan\theta & -\sin\phi & \frac{\cos\phi}{\cos\theta} \end{bmatrix} \begin{bmatrix} \frac{\partial \overline{\mathcal{L}}}{\partial \phi} \\ \frac{\partial \overline{\mathcal{L}}}{\partial \theta} \\ \frac{\partial \overline{\mathcal{L}}}{\partial \psi} \end{bmatrix} \tag{4.203}$$

The remaining terms are:

$$\begin{bmatrix} x_x \\ y_c \\ z_z \end{bmatrix} = \begin{bmatrix} x & x & -\sin\theta \\ x & x & \sin\phi\cos\theta \\ x & x & \cos\phi\cos\theta \end{bmatrix} \begin{bmatrix} 0 \\ 0 \\ Z \end{bmatrix}$$

$$\Rightarrow x_c = -Z\sin\theta; \; y_c = Z\cos\theta\sin\phi; \; z_c = Z\cos\theta\cos\phi \tag{4.204}$$

where x signifies a don't care condition, hence:

$$
\begin{aligned}
\frac{\partial \overline{\mathcal{L}}}{\partial \phi} &= mg(y_c \cos\theta \cos\phi - z_c \cos\theta \sin\phi) \\
&= mgZ(\cos^2\theta \cos\phi \sin\phi - \cos^2\theta \cos\phi \sin\phi) = 0 \\
\frac{\partial \overline{\mathcal{L}}}{\partial \theta} &= -mg(x_c \cos\theta + z_c \cos\phi \sin\theta + y_c \sin\phi \sin\theta) \\
&= -mgZ(-\sin\theta \cos\theta + \cos^2\phi \sin\theta \cos\theta + \sin^2\phi \cos\theta \sin\theta) = 0 \\
\frac{\partial \overline{\mathcal{L}}}{\partial \psi} &= 0 \Rightarrow -L^T \left\{ \frac{\partial \overline{\mathcal{L}}}{\partial \Phi} \right\} = - \begin{bmatrix} 1 & 0 & 0 \\ \sin\phi \tan\theta & \cos\phi & \frac{\sin\phi}{\cos\theta} \\ \cos\phi \tan\theta & -\sin\phi & \frac{\cos\phi}{\cos\theta} \end{bmatrix} \begin{bmatrix} \frac{\partial \overline{\mathcal{L}}}{\partial \phi} \\ \frac{\partial \overline{\mathcal{L}}}{\partial \theta} \\ \frac{\partial \overline{\mathcal{L}}}{\partial \psi} \end{bmatrix} \\
&= - \begin{bmatrix} 1 & 0 & 0 \\ \sin\phi \tan\theta & \cos\phi & \frac{\sin\phi}{\cos\theta} \\ \cos\phi \tan\theta & -\sin\phi & \frac{\cos\phi}{\cos\theta} \end{bmatrix} \begin{bmatrix} 0 \\ 0 \\ 0 \end{bmatrix} = \begin{bmatrix} 0 \\ 0 \\ 0 \end{bmatrix}
\end{aligned}
\tag{4.205}
$$

and furthermore:

$$
\begin{aligned}
\omega \times \left\{ \frac{\partial \overline{\mathcal{L}}}{\partial \omega} \right\} &= \begin{bmatrix} 0 & -r_b & q_b \\ r_b & 0 & -p_b \\ -q_b & p_b & 0 \end{bmatrix} \begin{bmatrix} I_x p_b \\ I_y q_b \\ I_z r_b \end{bmatrix} = \begin{bmatrix} (I_z - I_y) q_b r_b \\ (I_x - I_z) p_b r_b \\ (I_y - I_x) p_b q_b \end{bmatrix} \\
&= \begin{bmatrix} (I_z - I_y) q_b r_b \\ (I_x - I_z) p_b r_b \\ 0 \end{bmatrix} \\
I_x &= I_y
\end{aligned}
\tag{4.206}
$$

Combining all of the above (Eqs. 4.203, 4.205, and 4.206), the moment equations become:

$$\begin{bmatrix} I_x \dot{p}_b \\ I_y \dot{q}_b \\ I_z \dot{r}_b \end{bmatrix} + \begin{bmatrix} 0 \\ 0 \\ 0 \end{bmatrix} + \begin{bmatrix} (I_z - I_y) q_b r_b \\ (I_x - I_z) p_b r_b \\ 0 \end{bmatrix} = L^T Q_\tau = \begin{bmatrix} \tau_\phi \\ \tau_\theta \\ \tau_\psi \end{bmatrix} \tag{4.207}$$

Close examination of Fig. 3.10 reveals that the roll torque is a function of $f_2 - f_4$, the pitch torque depends upon $f_1 - f_3$, and the yaw torque is the sum of all of the individual torques produced by the individual motors, that is, $\tau_{M_1} - \tau_{M_2} + \tau_{M_3} - \tau_{M_4}$. The motor torques $\tau_{m_i}, i = 1, 2, \ldots, 4$ are actually the reaction torques due to shaft acceleration and aerodynamic drag of the body of the quadcopter, including the rotor blades. The aerodynamic drag opposes the motor torque, thus resulting in the following torque equation for each motor/rotor combination:

$$I_{rotor} \dot{\omega} = \tau_{m_i} - \tau_{drag_i} \quad i = 1, 2, 3, 4 \tag{4.208}$$

where I_{rotor} is the moment of inertia of the motor and rotor combination in the direction of the motor's rotation (motor's z axis), ω is the rotation rate of the ith motor, and τ_{m_i} is the torque produced by the ith motor.

From basic fluid dynamics, the drag force equation, describing the frictional force which opposes motion in the direction of the velocity vector of the rotor, is:

$$D = \frac{1}{2} \rho C_D A_b v^2$$

where C_D is the drag coefficient of the blade,A_b is the cross sectional area of the rotor blade, and $v = \omega \times R$ is the linear velocity of the blade at its tip (R is the radius of the rotor blade). The torque caused by the rotor blade's profile drag is therefore:

$$\tau_{drag_i} = D \times R = \frac{1}{2} \rho C_D A_b v^2 R = \frac{1}{2} \rho C_D A_b (\omega R)^2 R = b\omega^2$$

For quasi-stationary maneuvers, ω is constant, which implies that $I_{rotor} \dot{\omega} = 0$ and so the motor's reaction torque equals the drag induced torque or $\tau_{m_i} = \tau_{drag_i}$. In order to maneuver, the quadrotor must adjust the speed of its motors as follows:

1. Forward pitch motion is obtained by increasing the speed of motor m_3 while reducing the speed of motor m_1.
2. Similarly, roll motion is obtained by using motors m_2 and m_4.
3. Yaw motion is obtained by increasing the torque τ_{m_1} and τ_{m_3} of motors m_1 and m_3, respectively, while decreasing the torques τ_{m_2} and τ_{m_4} of motors m_2 and m_4, respectively.

For maneuvers where ω is constant, the aerodynamic drag moment equals the motor torque, since:

$$I_{rot} \underbrace{\dot{\omega}}_{=0} = \tau_{M_i} - \tau_{drag} \Rightarrow \tau_{M_i} = \tau_{drag} \tag{4.209}$$

In order to obtain forward pitch motion, motor m_3 must rotate more quickly than motor m_1. For yawing motion, the torques of motors m_1 and m_3 are increased, while the torques of motors m_2 and m_4 are decreased. Roll motion is achieved by increasing the speed of motor m_2 while decreasing the speed of motor m_4 and vice versa.

Torques

There are two distinct sources for torques which act on the quadcopter's body (we have neglected aerodynamic torques, except for the blade profile drag torques). The first source is due to the torques exerted by the motors on the body, or the motors' reaction torques τ_{m_i}, which were shown to be related directly to the profile drag of the rotor blades and are proportional to each individual rotor's rotation rate $b\omega^2$. The sum of all of the four motor reaction torques is the torque which will tend to turn the vehicle about the body's $\hat{k}$ axis (see Fig. 3.10). Mathematically, this may be written as: $\sum_{i=1}^{4}(-1)^{i+1}\tau_{m_i} = \tau_{yaw}$. The torque which gives rise to angular roll is directly related to the imbalance between the forces f_2 and f_2, that is, $(f_2 - f_4)l = \tau_{roll}$. A similar force imbalance situation arises for the angular pitch; however, forces f_1 and f_3 are involved, leading to: $(f_1 - f_3)l = \tau_{pitch}$. The second source of externally applied torques is due to the "gyro effect" of each individual motor and rotor blade combination (see Beer et al. [4, pp. 1185]). The rotating rotor may be looked upon as a gyro, and when it is subject to an angular rotation rate perpendicular to its axis of rotation or perpendicular to the angular momentum vector which defines its rotation, a torque results which is perpendicular to both the angular momentum vector and the applied angular rate. In mathematical terms, we have:

$$\tau_{gyro} = \Omega \times H$$

$$H\hat{k} = \sum_{i=1}^{4} H_i\hat{k} = \sum_{i=1}^{4} I_{m_i}\omega_i\hat{k};\quad \Omega = \begin{bmatrix} p \\ q \\ r \end{bmatrix}$$

$$\tau_{gyro_{roll}} = -q\hat{j} \times H\hat{k} = -q\sum_{i=1}^{4} I_{m_i}\omega_i\hat{i};$$

$$\tau_{gyro_{pitch}} = -p\hat{i} \times H\hat{k} = -p\sum_{i=1}^{4} I_{m_i}\omega_i\hat{j}$$

The generalized torques may be written as:

$$
\tau = \begin{bmatrix} \tau_\psi \\ \tau_\theta \\ \tau_\phi \end{bmatrix} = \begin{bmatrix} \tau_{yaw} \\ \tau_{pitch} + \tau_{gyro_{pitch}} \\ \tau_{roll} + \tau_{gyro_{roll}} \end{bmatrix}
$$

$$
= \begin{bmatrix} \tau_{m1} - \tau_{m2} + \tau_{m3} - \tau_{m4} \\ (f_2 - f_4)l \\ (f_3 - f_1)l \end{bmatrix} - \begin{bmatrix} 0 \\ p\hat{i} \times H\hat{k} \\ q\hat{j} \times H\hat{k} \end{bmatrix}
$$

$$
\Rightarrow \begin{bmatrix} \tau_\psi \\ \tau_\theta \\ \tau_\phi \end{bmatrix} = \begin{bmatrix} \tau_{m1} - \tau_{m2} + \tau_{m3} - \tau_{m4} \\ (f_2 - f_4)l \\ (f_3 - f_1)l \end{bmatrix} - \begin{bmatrix} 0 \\ p\sum_{i=1}^{4} I_{m_i}\omega_i \\ q\sum_{i=1}^{4} I_{m_i}\omega_i \end{bmatrix} \tag{4.210}
$$

where l is the distance from the center of any motor to the center of gravity of the system. The moment equations then become:

$$
\begin{bmatrix} I_x\dot{p}_b \\ I_y\dot{q}_b \\ I_z\dot{r}_b \end{bmatrix} + \begin{bmatrix} (I_z - I_y)q_b r_b \\ (I_x - I_z)p_b r_b \\ 0 \end{bmatrix} = L^T Q_\tau = \begin{bmatrix} \tau_\phi \\ \tau_\theta \\ \tau_\psi \end{bmatrix}
$$

$$
= \begin{bmatrix} (f_3 - f_1)l - q\sum_{i=1}^{4} I_{m_i}\omega_i \\ (f_2 - f_4)l - p\sum_{i=1}^{4} I_{m_i}\omega_i \\ \tau_{m_1} - \tau_{m_2} + \tau_{m_3} - \tau_{m_4} \end{bmatrix} \tag{4.211}
$$

Chapter 5
Conclusions

In this book an attempt was made to review the basics of Newtonian mechanics (see Chap. 2), and introduce some of the key concepts involved in formulating Lagrangian dynamics such as *virtual work, kinetic energy, the principle of d'Alembert for dynamical systems, the mathematics of conservative forces, generalized coordinates, generalized forces, constraints, both holonomic and non-holonomic, the extended Hamilton's principle, etc.* (Chaps. 2 and 3). The treatment of a particular class of non-holonomic constraints, where the quasi-velocities can be modeled as linear functions of the time derivatives of the generalized coordinates, is dealt with in Chap. 4 by methods introduced by Whittaker [52], Meirovitch [23], and Cameron and Book [8]. While this approach is sound, it is also somewhat cumbersome as was demonstrated in examples 1 and 2 of Sect. 4.3.The method of Prof. Ranjan Vepa in Sects. 4.4 and 4.5 is more suited for deriving equations of motion. His scheme is based upon transforming the Lagrangian, which is a function of the generalized coordinates and generalized velocities, that is, $\mathcal{L}(q, \dot{q})$, into the Lagrangian containing both the quasi-coordinates and quasi-velocities, that is, $\overline{\mathcal{L}}(\Gamma, \dot{\Gamma})$, where Γ is the vector of quasi-coordinates and $\dot{\Gamma}$ is the vector of quasi-velocities, respectively. The advantage of this approach is that the derivation of the equations of motion turns out to be far less cumbersome. Throughout the text, examples have been presented to illustrate the concepts involved. Although the presentation is mathematically sound, the approach taken in this text was intermediate and did not cover many important topics such as the calculus of variations as exemplified by Cornelius Lanczos' book "The Variational Principles of Mechanics" [19].There is a more comprehensive and more mathematically oriented (and perhaps more advanced) treatment of the subject of mechanics in the form of Arnold's book "Mathematical Methods of Classical Mechanics" [2], with topics ranging from Lagrangian mechanics, variational calculus, Lagrangian mechanics on manifolds, differential forms, Lie algebras of vector fields, and so on. As mentioned in Sect. 3.2, a more exhaustive approach to the subject of non-holonomic systems, their characterization, identification, and control based on the following topics found

A. W. Pila, *Introduction To Lagrangian Dynamics*,
https://doi.org/10.1007/978-3-030-22378-6_5

in differential geometry and related subject matter such as Lie groups, Lie algebras, etc. would include the following topics among others:

- Manifolds, Differentiable manifolds, manifolds and maps
- Tangent vectors, spaces, vector fields
- Fiber bundles
- Differential k-forms
- Exterior derivatives
- Jacobi–Lie brackets, Lie groups
- Vector fields and flows
- Lie brackets and Frobenius' theorem, the Lie algebra associated with a Lie group, actions of Lie groups, Canonical coordinates on a Lie group
- Tangent spaces and tangent maps
- Cotangent spaces and cotangent maps
- Differential forms
- The exponential map
- The geometry of the Euclidean group, metric properties of SE(3), volume forms on SE(3)
- Lie groups and robot kinematics

The interested reader is encouraged to pursue these topics in greater detail by referring to the works by Murray et al. "A Mathematical Introduction to Robotic Manipulation" [25], Bullo and Lewis "Geometric Control of Mechanical Systems Modeling, Analysis, and Design for Simple Mechanical Control Systems" [7], Bloch et al. "Nonholonomic Mechanics and Control" [5], Soltakhanov et al. "Mechanics of Non-Holonomic Systems—A New Class of Control Systems" [35], to name but a few. The area of robotics has also borrowed heavily from these advanced mathematical methods, for example, Siciliano et al. in the "Springer Handbook of Robotics—2nd Ed." [33] and Siciliano et al. in the text "Robotics—Modelling, Planning and Control" [34], both discuss in detail the subject of non-holonomic trajectory planning. In addition, the control of robotic systems is replete with variational approaches in the form of optimal control theory, such as appears, for instance, in the book by Bloch et al. "Nonholonomic Mechanics and Control" [5] and the work of Soltakhanov et al. "Mechanics of Non-Holonomic Systems—A New Class of Control Systems [35]."

Bibliography

1. Anon, *Nonholonomic System* (Wikipedia, 2017). https://en.wikipedia.org/wiki/Nonholonomic_system. Accessed 30 May 2017
2. V.I. Arnold, *Mathematical Methods of Classical Mechanics*, 2nd edn. (Springer, New York, 1989)
3. H. Baruh, *Analytical Dynamics*, International edn. (WCB/McGraw Hill, Singapore, 1999)
4. F.P. Beer, E.R. Johnston Jr., D.F. Mazurek, P.J. Cornwell, E.R. Eisenberg, *Vector Mechanics for Engineers-Statics and Dynamics*, 9th edn. (McGraw-Hill, New York, 2010)
5. A.M. Bloch, J. Baillieul, P.E. Crouch, J.E. Marsden, D. Zenkov, *Nonholonomic Mechanics and Control*, 2nd edn. (Springer, New York, 2015)
6. K.R. Britting, *Inertial Navigation Systems Analysis* (Wiley, New York, 1971)
7. F. Bullo, A.D. Lewis, *Geometric Control of Mechanical Systems: Modeling, Analysis, and Design for Simple Mechanical Control Systems* (Springer Science + Business Media Inc., New York, 2004)
8. J.M. Cameron, W.J. Book, Modeling mechanisms with nonholonomic joints using the boltzmann-hamel equations. Int. J. Robot. Res. **16**(1), 47–59 (1997)
9. J. Deyst, J.P. How, *MIT OpenCourseWare Lecture Notes: 16.61 Aerospace Dynamics-Lecture 7, Lagrange's Equations* (MIT, Cambridge, 2003). https://ocw.mit.edu
10. I. Fantoni, R. Lozano, *Non-linear Control for Underactuated Mechanical Systems* (Springer, London, 2002)
11. R. Fitzpatrick, *Newtonian Dynamics* (Lulu Press, Morrisville, 2011)
12. A. Gibiansky, *Quadcopter Dynamics and Simulation* (2012). http://andrew.gibiansky.com/blog/physics/quadcopter-dynamics/. Accessed 27 February 2018
13. J. Ginsberg, *Engineering Dynamics* (Cambridge University Press, London, 2008)
14. D.T. Greenwood, *Advanced Dynamics* (Cambridge University Press, Cambridge, 2003)
15. W. Greiner, *Classical Mechanics- Systems of Particles and Hamiltonian Dynamics*, 2nd edn. (Springer, Berlin, 2010)
16. E. Have, R.O. Nielsen, B.T. Nielsen, The double pendulum, in *The Niels Bohr Institute Publication* (2013), p. 4
17. J. Kim, M.S. Kang, S. Park, Accurate modeling and robust hovering control for a quad-rotor vtol aircraft. J. Intell. Robot. Syst. **57**(9), 1–18 (2010)
18. V. Kumar, J.L. Bassani, *Holonomic and Nonholonomic Constraints - course MEAM, Lecture 10* (University of Pennsylvania, Pennsylvania, 2000). https://alliance.seas.upenn.edu/~meam535/cgi-bin/pmwiki/uploads/Main/Constraints10.pdf
19. C. Lanczos, *The Variational Principles of Mechanics* (University of Toronto Press, Toronto, 1949)

A. W. Pila, *Introduction To Lagrangian Dynamics*,
https://doi.org/10.1007/978-3-030-22378-6

20. D.B. Marghitu, *Mechanisms and Robots Analysis with MATLAB* (Springer, London, 2009)
21. M.T. Mason, *Mechanics of Robotic Manipulation* (MIT Press, Cambridge, 2001)
22. M.T. Mason, *Mechanics of Manipulation-course 16-741, Lecture 5* (Carnegie Mellon University, Pittsburgh, 2016). http://www.cs.cmu.edu/afs/cs/academic/class/16741-s07/www/
23. L. Meirovitch, *Methods of Analytical Dynamics* (McGraw-Hill, New York, 1970)
24. L. Meirovitch, *Fundamentals of Vibrations* (McGraw-Hill, New York, 2001)
25. R.M. Murray, Z. Li, S.S. Sastry, *A Mathematical Introduction to Robotic Manipulation* (CRC Press, Boca Raton, 1994)
26. A. Noureldin, T.B. Karamat, J. Georgy, *Fundamentals of Inertial Navigation, Satellite-Based Positioning and Their Integration* (Springer, Berlin, 2013)
27. T. Peacock, N. Hadjiconstantinou, *MIT OpenCourseWare Lecture Notes: 2.003J/1.053J - Dynamics and Control I - Lecture 10* (MIT, Cambridge, 2007). https://ocw.mit.edu
28. C. Poole H. Goldstein, J. Safko, *Classical Mechanics*, 3rd edn. (Addison Wesley, Boston, 2002)
29. L. Rodolfo, G. Carrillo, A.E.D. López, R. Lozano, C.Pégard, *Quad Rotorcraft Control Vision-Based Hovering and Navigation* (Springer, London, 2013)
30. W.W. Rouse Ball, *A Short Account of the History of Mathematics*, 4th edn. (Dover Press, New York, 1908)
31. R. Seddon, S. Newman, *Basic Helicopter Aerodynamics*, 3rd edn. (Wiley, Chichester, 2011)
32. J.M. Selig, *Geometrical Methods in Robotics* (Springer, Berlin, 1996)
33. B. Siciliano, O. Khatib (eds.), *Springer Handbook of Robotics*, 2nd edn. (Springer, Berlin, 2016)
34. B. Siciliano, L. Sciavicco, L. Villani, G. Oriolo, *Robotics-Modelling, Planning and Control*. Advanced Textbooks in Control and Signal Processing, 2nd edn. (Springer, Berlin, 2008)
35. Sh.Kh. Soltakhanov, M.P. Yushkov, S.A. Zegzhda, *Mechanics of non-holonomic systems-A New Class of control systems* (Springer, Berlin, 2009)
36. B. van Brunt, *The Calculus of Variations* (Springer, New York, 2004)
37. J. Vandiver, D. Gossard, *MIT OpenCourseWare: 2.003SC Engineering Dynamics. - Problem Set 6 - Solutions* (MIT, Cambridge, 2011). https://ocw.mit.edu
38. J. Vandiver, D. Gossard, *MIT OpenCourseWare: 2.003SC Engineering Dynamics. - Recitation 8 Notes: Cart and Pendulum (Lagrange)* (MIT, Cambridge, 2011). https://ocw.mit.edu
39. J. Vandiver, D. Gossard, *MIT OpenCourseWare: 2.003SC Engineering Dynamics. - Recitation 9 Notes: Generalized Forces with Double Pendulum Example* (MIT, Cambridge, 2011). https://ocw.mit.edu
40. J. Vandiver, D. Gossard, *MIT OpenCourseWare: 2.003SC Engineering Dynamics. - Video of Lecture 10: Equations of Motion, Torque, Angular Momentum of Rigid Bodies* (MIT, Cambridge, 2011). https://ocw.mit.edu
41. J. Vandiver, D. Gossard, *MIT OpenCourseWare: 2.003SC Engineering Dynamics. - Video of Lecture 12: Problem Solving Methods for Rotating Rigid Bodies*. (MIT, Cambridge, 2011). https://ocw.mit.edu
42. J. Vandiver, D. Gossard, *MIT OpenCourseWare: 2.003SC Engineering Dynamics. - Video of Lecture 15: Introduction to Lagrange With Examples* (MIT, Cambridge, 2011). https://ocw.mit.edu
43. J. Vandiver, D. Gossard, *MIT OpenCourseWare: 2.003SC Engineering Dynamics. - Video of Lecture 16: Kinematic Approach to Finding Generalized Forces* (MIT, Cambridge, 2011). https://ocw.mit.edu
44. J. Vandiver, D. Gossard, *MIT OpenCourseWare: 2.003SC Engineering Dynamics. - Video of Lecture 17: Practice Finding EOM Using Lagrange Equations* (MIT, Cambridge, 2011). https://ocw.mit.edu
45. J. Vandiver, D. Gossard, *MIT OpenCourseWare: 2.003SC Engineering Dynamics. - Video of Lecture 5: Angular Momentum* (MIT, Cambridge, 2011). https://ocw.mit.edu
46. J. Vandiver, D. Gossard, *MIT OpenCourseWare: 2.003SC Engineering Dynamics. - Video of Recitation 8: Cart and Pendulum Lagrange Method* (MIT, Cambridge, 2011). https://ocw.mit.edu

47. J. Vandiver, D. Gossard, *MIT OpenCourseWare: 2.003SC Engineering Dynamics. - Video of Recitation 9: Generalized Forces* (MIT, Cambridge, 2011). https://ocw.mit.edu
48. J. Vandiver, D. Gossard, *MIT OpenCourseWare: 2.003SC Engineering Dynamics. - Video on Notation* (MIT, Cambridge, 2011). https://ocw.mit.edu. Accessed 17 September 2016
49. J. Vandiver, D. Gossard, *MIT OpenCourseWare Lecture Notes: 2.003SC Engineering Dynamics. -An Introduction To Lagrange Equations* (MIT, Cambridge, 2011). https://ocw.mit.edu
50. R. Vepa, *Nonlinear Control of Robots and Unmanned Aerial Vehicles* (Taylor & Francis-CRC Press, London, 2017)
51. D. Wells, *SCHAUM'S Outline of Lagrangian Dynamics* (McGraw-Hill, New York, 1967)
52. E.T. Whittaker, *A Treatise on the Analytical Dynamics of Particles and Rigid Bodies*, 2nd edn (Cambridge University Press, Cambridge, 1917)
53. S. Widnall, J. Deyst, E. Greitzer, *MIT OpenCourseWare Lecture Notes: 16.07 Dynamics. - Lecture L20 - Energy Methods* (MIT, Cambridge, 2009). https://ocw.mit.edu

Index

A
Aerodynamic drag, 243
Angular momentum, 40, 139
  of a rigid body in three dimensions, 56
  of a system of particles about its mass center, 48
Angular motion of a rigid body constrained to rotate about a point, 181
Atwood's machine, 147

B
Back-emf, 94
Basic quadcopter maneuvers, 95
Blade profile drag, 93
Body angular rates, 168

C
Carnival ride problem, 37
Cart and pendulum, 113
Central motion, 161
Centripetal acceleration, 37, 39
Centripetal term, 110
Centroidal mass moments of inertia, 58
Centroidal mass products of inertia, 58
Coefficients of Θ_{ij}, 172
Completeness of generalized coordinates, 69
Configuration constraint, 66
Conservation of energy, 43
Conservative forces, 78, 81, 105
Consistency check, 111
Constraint equation, 65
Constraint force, 65
Constraint relations, 65
Contact force, 65
Coriolis, 112
  acceleration, 37, 39
  force, 110
Coriolis-centripetal vector, 100
Couple, 52
Curved path, 42

D
D'Alembert-Lagrange's equations, 115, 164, 180, 181, 188, 190, 198
D'Alembert-Lagrangian dynamics, 96
Damper, 116
Dashpot, 116
Degrees of freedom, 65, 69
Dependent virtual displacements, 165
Differential form, 72
Differential geometry, 74
Direction cosines, 3
Direction cosines l, m, n, 19
Double pendulum, 27, 89
Drag coefficient, 95

E
Effective force, 78
Elemental mass m', 21
Energy, 51
Equality constraints, 65
Equations of motion, 25
Equilibrium, 75
Equivalent forces, 121
Equivalent forces and torques, 144
Euler angles, 3

A. W. Pila, *Introduction To Lagrangian Dynamics*,
https://doi.org/10.1007/978-3-030-22378-6

Euler angular rates, 168
Eulerian acceleration, 39
Eulerian terms, 112
Euler's theorem for homogeneous functions, 159
Euler transformation, 98, 164
Extended Hamilton's principle, 80, 118

F
Falling stick problem, 149
Forces which do no work, 53
Free body diagram, 146
Friction, 149

G
Generalized constraint force, 120
Generalized coordinates, 64, 76, 114, 123, 167
Generalized forces, 77, 123, 124, 141, 145
Generalized non-conservative forces, 83
Generalized torques, 99
Generalized velocities, 168
Gyroscopic effect, 96

H
Hamiltonian, 151
  canonical equations, 155
  function, 153
Height differences, 108
Hessian, 152
Hockey puck problem, 139, 145
Holonomic, 118
  constraint, 67, 164
  systems, 71

I
Independence of generalized coordinates, 27, 65, 69, 81
Independent virtual displacement, 76
Inequalities, 71
Inequality constraint, 69
Inertial force, 78
Infinitesimal changes in position, 75
Infinitesimal mass dm, 22
Input torques τ_1, τ_2 and τ_3, 187
Integration by parts, 84
Internal force, 45
Involution, 152

J
Jacobian, 152
Jacobian matrix, 209

K
Kinetic energy, 21, 24, 28, 42, 83
  modified, 181
  of a rigid body
    in planar motion-non-centroidal rotation, 55
    in three dimensions, 59

L
Lagrange multipliers, 118, 119, 165
Lagrangian, 24, 103, 115, 118
Lagrangian $\overline{\mathcal{L}}$, 189
Laws of statics, 111
Legendre transformation, 151
Linear mass distribution ρ, 29
Linear momentum L, 39

M
Mass-spring one degree of freedom system, 86
Mass-spring two degrees of freedom system, 87
Mass spring dashpot (damper) single degree of freedom system, 105
Modified kinetic energy, 169
Modified Lagrangian, 232
Moment of inertia of the motor and rotor combination, 95
Moment of momentum, 40
Moments of inertia I_x, I_y, I_z, 22
Momentum theory, 94
Motion of the center of mass of a system of particles, 47
Motor's electric torque, 94
Motor torques, 243

N
Newton's second law, 45
Newton's third law, 45
Non-conservative forces, 81, 105, 116, 123
Non-holonomic constraints, 67, 121, 164, 187
Non-holonomic joints, 181
Non-holonomic systems, 71
Non-integrable constraints, 72
Notation system II, 32

O
$\omega_{/O}$-rotation of rigid body wrt O, 33
ω_O-rotation of rigid body wrt O, 33

P
Parallel axis theorem, 24, 146
Particle angular momentum $h_{B/O}$, 34
Particle linear momentum $P_{B/O}$, 33
Pendulum, 23
    with a mass and spring, 107
    with a plane of symmetry, 146
Perfect differential, 80
Pfaffian form, 67, 121
Pitch torque, 93
Potential energy, 83
The principle of d'Alembert, 77
Principle of virtual work, 76, 164
Products of inertia I_{xy}, I_{xz}, I_{yz}, 22

Q
Quad copter, 92, 239
    power, 94
    torque, 244
Quadrotor's controls, 93
Quasi-coordinates, 168
Quasi-velocities, 168, 169
Quasi-velocity Lagrangian for origin $= c.g.$, 199
Quasi-velocity Lagrangian for origin $\neq c.g.$, 215

R
Rate variables, 71
Reaction force, 65
Reaction torques, 93
Resultant force, 75
Rheonomic constraint, 67
Rigid aircraft dynamics, 225
Rigid body moments of inertia, 226, 228
Robotics, 73
Roll-pitch-yaw quasi-velocities w_x, w_y, w_z, 181
Roll torque, 93
Rotational equations of motion, 99
Rotational kinetic energy, 110

S
Scleronomic constraint, 67
Simple pendulum, 88
Six degrees of freedom, 208
Spherical pendulum, 65
Spring, 41
System's kinetic energy, 114
System's potential energy, 114

T
Tangential force component, 42
Tangential velocity, 42
Three-wheeled mobile robot, 187
Time derivative of Θ_{ij}, 172
Time derivative of a rotation matrix, 200
Torques, 96, 139
    rigid body, 37
    $\tau_{B/A}$, 36
    wrt c.g. $\tau_{/G}$, 36
Total mechanical energy, 44
Total system kinetic energy, 110
Transformed Lagrangian for rotations, 208
Transformed Lagrangian for translations, 204
Translational equations of motion, 99
Tricycle, 187
    forward velocity v, 188
    front wheel lateral velocity v_R, 188
    rear wheel lateral velocity v_{Yi}, 188

V
Variation, 80, 118
Variation in kinetic energy, 83
Vector inner product, 175
Vector of quasi-velocities, 189
Virtual displacements, 75, 116, 121
Virtual work, 75–78, 80, 81, 83, 111, 124, 143, 145, 177

W
Weight W, 41
Wheel rolling without slipping, 72
Work, 41, 51
    of a weight W, 41
    total work over a path, 41